C0-BGO-516

# MATRICES FOR SCIENTISTS AND ENGINEERS

# MATRICES FOR SCIENTISTS AND ENGINEERS

W.W. BELL M.Sc., Ph.D

*Senior Lecturer in Engineering*
*University of Aberdeen*

VAN NOSTRAND REINHOLD COMPANY

New York - Cincinnati - Toronto - London - Melbourne

ISBN 0 442 30081 6 Cloth
ISBN 0 442 30082 4 Paperback

Library of Congress Catalog Card No. 75-14491



Published by Van Nostrand Reinhold Company Ltd.,
Molly Millars Lane, Wokingham, Berkshire, England.

Published in 1975 by Van Nostrand Reinhold Company.
A Division of Litton Educational Publishing, Inc.
450 West 33rd Street, New York, NY 10001, U.S.A.

Van Nostrand Reinhold Limited
1410 Birchmount Road, Scarborough, Ontario M1P 2E7, Canada.

Van Nostrand Reinhold Australia Pty. Limited,
17 Queen Street, Mitcham, Victoria 3132, Australia.

Library of Congress Cataloging In Publication Data

Bell, William Wallace.
Matrices for scientists and engineers.

Includes index.
1. Matrices. I. Title.
QA188.B44 1975 512.9'43 75-14491
ISBN 0-442-30081-6
ISBN 0-442-30082-4 pbk.

Printed in Great Britain by
William Clowes and Sons Limited,
London, Colchester and Beccles.

# Contents

# Preface

The aim of this work is to present in one volume all the basic material in the theory of matrices and determinants needed by an undergraduate student of science or engineering. In addition the book is suitable for use as a reference text or self-instruction manual for post-graduate workers who may wish to refresh their knowledge of matrix algebra or who have reason to learn the subject for the first time.

The aim throughout has been to combine a simplicity of approach with mathematical respectability. Thus in addition to the class of reader mentioned above the book could well also appeal to students of pure and applied mathematics. It is hoped that the book will prove attractive to those who are looking for a treatment of matrix algebra intermediate between the rather advanced approach designed for readers with a fair degree of mathematical sophistication and the other extreme of the inadequate and mathematically imprecise elementary introduction to the subject with a view to immediate application.

The emphasis is on mathematics which has applications in many branches of science and engineering, but not on dealing explicitly with these applications themselves. It is believed to be advantageous for a student to obtain a good grounding in mathematical techniques before applying them, rather than being introduced to a new piece of mathematics in an ad hoc fashion for the solution of one particular problem. The belief is that in this manner the student is better able to apply his mathematical knowledge to a variety of situations; there is the added advantage that the reader does not get involved with applications which are of no interest to him.

The work is organized in four chapters. The first is concerned with determinants; after dealing with determinants of orders 2 and 3 and their properties the results derived are extended to determinants of arbitrary order. The second chapter is devoted to basic matrix algebra; after definitions are given of matrices and elementary operations with them the book proceeds to treat special matrices, partitioning, inversion, elementary transformations, quadratic forms and coordinate transformations. Chapter three is concerned with the important topic of systems of simultaneous linear equations; the existence and uniqueness of solutions is discussed and the more important methods of solution are described. The final chapter is on eigenvalue theory. All the important properties of eigenvalues and eigenvectors are proved and discussed and a brief mention is made of numerical methods for their determination.

Throughout the text we use the symbol □ to indicate 'end of definition' and the symbol ■ to indicate 'end of theorem'. There are a number of worked examples to illustrate the principles and techniques presented. The exercises for the reader to attempt are considered to be an integral part of the text; an adequate number is provided to ensure that the reader gains practice and understanding.

Aberdeen, 1975. W. W. Bell.

# 1. Determinants

## 1.1 Determinants of Order 2 and 3 and their Properties

In many scientific and engineering contexts we are faced with the problem of solving $n$ linear equations in $n$ unknowns. These might be written in the form

$$\begin{aligned}
a_{11}x_1 + a_{12}x_2 + a_{13}x_3 + \ldots + a_{1n}x_n &= b_1,\\
a_{21}x_1 + a_{22}x_2 + \qquad\quad \ldots + a_{2n}x_n &= b_2,\\
&\ \vdots\\
a_{n1}x_1 + a_{n2}x_2 + \qquad\quad \ldots + a_{nn}x_n &= b_n.
\end{aligned} \tag{1.1.1}$$

We wish to obtain, if possible, an expression for each $x_i$ in terms of the $a$'s and the $b$'s. Observe that the notation we use is such as to give the coefficients of the unknowns two suffices; the first refers to the equation in which it appears and the second to the unknown which it multiplies. Thus $a_{ij}$ is the coefficient of $x_j$ in the $i$th equation. Similarly $b_i$ is the constant on the right-hand side of the $i$th equation.

We proceed by looking first at the simplest situations and then generalizing from these. Thus first of all let us consider the case of two equations in two unknowns:

$$\begin{aligned}
a_{11}x_1 + a_{12}x_2 &= b_1,\\
a_{21}x_1 + a_{22}x_2 &= b_2.
\end{aligned} \tag{1.1.2}$$

Eliminating $x_1$ by multiplying the first equation by $a_{21}$, the second by $a_{11}$ and subtracting gives

$$(a_{12}a_{21} - a_{11}a_{22})x_2 = (b_1a_{21} - a_{11}b_2).$$

Similarly, eliminating $x_2$ will give

$$(a_{11}a_{22} - a_{12}a_{21})x_1 = (b_1a_{22} - a_{12}b_2).$$

Thus, provided $a_{11}a_{22} - a_{12}a_{21} \neq 0$ we obtain

$$x_1 = \frac{b_1a_{22} - a_{12}b_2}{a_{11}a_{22} - a_{12}a_{21}}, \qquad x_2 = \frac{a_{11}b_2 - b_1a_{21}}{a_{11}a_{22} - a_{12}a_{21}}.$$

We introduce a special notation for the 'cross-products' appearing in numerator and denominator. We define a determinant of order two by

$$\begin{vmatrix} \alpha & \beta \\ \gamma & \delta \end{vmatrix} = \alpha\delta - \beta\gamma, \tag{1.1.3}$$

and then we have

$$x_1 = \frac{\begin{vmatrix} b_1 & a_{12} \\ b_2 & a_{22} \end{vmatrix}}{\begin{vmatrix} a_{11} & a_{12} \\ a_{21} & a_{22} \end{vmatrix}}, \qquad x_2 = \frac{\begin{vmatrix} a_{11} & b_1 \\ a_{21} & b_2 \end{vmatrix}}{\begin{vmatrix} a_{11} & a_{12} \\ a_{21} & a_{22} \end{vmatrix}}. \tag{1.1.4}$$

These results can be written in the form

$$\frac{x_1}{\begin{vmatrix} b_1 & a_{12} \\ b_2 & a_{22} \end{vmatrix}} = \frac{x_2}{\begin{vmatrix} a_{11} & b_1 \\ a_{21} & b_2 \end{vmatrix}} = \frac{1}{\begin{vmatrix} a_{11} & a_{12} \\ a_{21} & a_{22} \end{vmatrix}} \tag{1.1.5}$$

where if either of the determinants in the first two denominators is zero we take it to imply that the corresponding numerator is zero (by equations (1.1.4)). Note that there is a useful mnemonic for the construction of the determinants appearing in equation (1.1.5). The last determinant (corresponding to numerator of value 1) is just obtained directly from the coefficients of the $x$'s in equation (1.1.2). The determinant whose numerator is $x_1$ is obtained from the coefficients by omitting the column pertaining to $x_1$ and replacing it by the column of constants $b_i$; similarly the determinant whose numerator is $x_2$ is obtained by omitting the column of coefficients of $x_2$ and replacing it by the column of constants $b_i$.

Equations (1.1.5) thus represent a neat way of writing down the solution to the system of equations (1.1.2), the solution being guaranteed to exist provided

$$\begin{vmatrix} a_{11} & a_{12} \\ a_{21} & a_{22} \end{vmatrix} \neq 0.$$

We now attempt to generalize this result somewhat by proceeding to the case of three equations in three unknowns:

$$\begin{aligned} a_{11}x_1 + a_{12}x_2 + a_{13}x_3 &= b_1, \\ a_{21}x_1 + a_{22}x_2 + a_{23}x_3 &= b_2, \\ a_{31}x_1 + a_{32}x_2 + a_{33}x_3 &= b_3. \end{aligned} \tag{1.1.6}$$

Eliminating $x_3$ in turn from the second and third, from the first and third and from the first and second of these equations gives

$$\begin{aligned} (-a_{21}a_{33} + a_{23}a_{31})x_1 + (-a_{22}a_{33} + a_{23}a_{32})x_2 &= -b_2a_{33} + b_3a_{23}, \\ (a_{11}a_{33} - a_{13}a_{31})x_1 + (a_{12}a_{33} - a_{13}a_{32})x_2 &= b_1a_{33} - b_3a_{13}, \\ (-a_{11}a_{23} + a_{13}a_{21})x_1 + (-a_{12}a_{23} + a_{13}a_{22})x_2 &= -b_1a_{23} + b_2a_{13}. \end{aligned}$$

If we now multiply the first of these by $a_{12}$, the second by $a_{22}$, the third by $a_{32}$ and add we obtain

$$\begin{aligned}&(-a_{12}a_{21}a_{33}+a_{12}a_{23}a_{31}+a_{11}a_{22}a_{33}-a_{13}a_{22}a_{31}-a_{11}a_{23}a_{32}+a_{13}a_{21}a_{32})x_1\\&+(-a_{12}a_{22}a_{33}+a_{12}a_{23}a_{32}+a_{12}a_{22}a_{33}-a_{13}a_{22}a_{32}-a_{12}a_{23}a_{32}+a_{13}a_{22}a_{32})x_2\\&\quad=-b_2a_{12}a_{33}+b_3a_{12}a_{23}+b_1a_{22}a_{33}-b_3a_{22}a_{13}-b_1a_{23}a_{32}+b_2a_{13}a_{32}.\end{aligned}\tag{1.1.7}$$

We note that we have eliminated $x_2$ since its coefficient reduces to zero, and we may thus write the resulting equation in the form

$$\frac{x_1}{\Delta_1}=\frac{1}{\Delta},$$

where

$$\Delta=-a_{12}a_{21}a_{33}+a_{12}a_{23}a_{31}+a_{11}a_{22}a_{33}-a_{13}a_{22}a_{31}-a_{11}a_{23}a_{32}+a_{13}a_{21}a_{32},$$

and

$$\Delta_1=-a_{12}b_2a_{33}+a_{12}a_{23}b_3+b_1a_{22}a_{33}-a_{13}a_{22}b_3-b_1a_{23}a_{32}+a_{13}b_2a_{32}.$$

Again this is valid provided $\Delta \neq 0$; if $\Delta_1 = 0$ it is taken to mean that $x_1 = 0$.

We see that the expression for $\Delta_1$ is obtained from that for $\Delta$ by replacing $a_{11}, a_{21}$ and $a_{31}$ respectively by $b_1, b_2$ and $b_3$ (i.e. replacing the coefficients of $x_1$ by the constants on the right-hand side of the given equations (1.1.6)).

We now introduce the definition and notation for a determinant of order 3:

$$\begin{aligned}\begin{vmatrix}a_{11}&a_{12}&a_{13}\\a_{21}&a_{22}&a_{23}\\a_{31}&a_{32}&a_{33}\end{vmatrix}&=a_{11}a_{22}a_{33}-a_{11}a_{23}a_{32}-a_{12}a_{21}a_{33}+a_{12}a_{23}a_{31}\\&\quad+a_{13}a_{21}a_{32}-a_{13}a_{22}a_{31}\\&=a_{11}(a_{22}a_{33}-a_{23}a_{32})-a_{12}(a_{21}a_{33}-a_{23}a_{31})\\&\quad+a_{13}(a_{21}a_{32}-a_{23}a_{31})\\&=a_{11}\begin{vmatrix}a_{22}&a_{23}\\a_{32}&a_{33}\end{vmatrix}-a_{12}\begin{vmatrix}a_{21}&a_{23}\\a_{31}&a_{33}\end{vmatrix}+a_{13}\begin{vmatrix}a_{21}&a_{22}\\a_{31}&a_{32}\end{vmatrix}\end{aligned}\tag{1.1.8}$$

Making use of this we see that

$$\Delta=\begin{vmatrix}a_{11}&a_{12}&a_{13}\\a_{21}&a_{22}&a_{23}\\a_{31}&a_{32}&a_{33}\end{vmatrix}\quad\text{and}\quad\Delta_1=\begin{vmatrix}b_1&a_{12}&a_{13}\\b_2&a_{22}&a_{23}\\b_3&a_{32}&a_{33}\end{vmatrix}.$$

Thus we have

$$\frac{x_1}{\begin{vmatrix}b_1&a_{12}&a_{13}\\b_2&a_{22}&a_{23}\\b_3&a_{32}&a_{33}\end{vmatrix}}=\frac{1}{\begin{vmatrix}a_{11}&a_{12}&a_{13}\\a_{21}&a_{22}&a_{23}\\a_{31}&a_{32}&a_{33}\end{vmatrix}},$$

and in a similar manner we should obtain

$$\frac{x_1}{\begin{vmatrix} b_1 & a_{12} & a_{13} \\ b_2 & a_{22} & a_{23} \\ b_3 & a_{32} & a_{33} \end{vmatrix}} = \frac{x_2}{\begin{vmatrix} a_{11} & b_1 & a_{13} \\ a_{21} & b_2 & a_{23} \\ a_{31} & b_3 & a_{33} \end{vmatrix}} = \frac{x_3}{\begin{vmatrix} a_{11} & a_{12} & b_1 \\ a_{21} & a_{22} & b_2 \\ a_{31} & a_{32} & b_3 \end{vmatrix}} = \frac{1}{\begin{vmatrix} a_{11} & a_{12} & a_{13} \\ a_{21} & a_{22} & a_{23} \\ a_{31} & a_{32} & a_{33} \end{vmatrix}}. \tag{1.1.9}$$

This expression for the solution of equations (1.1.6) is known as Cramer's rule. It must be understood to be valid only if

$$\begin{vmatrix} a_{11} & a_{12} & a_{13} \\ a_{21} & a_{22} & a_{23} \\ a_{31} & a_{32} & a_{33} \end{vmatrix} \neq 0;$$

if any of the other determinants is zero, this is taken to mean that the variable appearing in the corresponding numerator must be zero.

At the moment, then, we have a determinant of order 2 defined by equation (1.1.3) and a determinant of order 3 defined by equation (1.1.8). Before generalizing to determinants of order $n$ which will lead to Cramer's rule for a system of $n$ linear equations in $n$ unknowns, we first investigate some properties of determinants of order 3, all of which we shall later generalize.

At first sight the definition of equation (1.1.8) appears rather unsystematic. We can, however, show that it is completely systematic if we introduce the idea of odd and even permutations. A re-arrangement of the ordered triple (1, 2, 3) is said to be an even permutation if it is accomplished by an even number of interchanges and an odd permutation if accomplished by an odd number of interchanges. Thus, for example, (2, 3, 1) is even because the number of interchanges required from (1, 2, 3) is 2:

(1, 2, 3)
× × interchanging items marked

(2, 1, 3)
× × interchanging items marked.

(2, 3, 1)

And (3, 2, 1) is odd because the number of interchanges is 1:

(1, 2, 3)
× × interchanging items marked.

(3, 2, 1)

We may thus rewrite equation (1.1.8) as the following formal definition.

DEFINITION 1.1

The determinant of order 3 is defined by

$$\begin{vmatrix} a_{11} & a_{12} & a_{13} \\ a_{21} & a_{22} & a_{23} \\ a_{31} & a_{32} & a_{33} \end{vmatrix} = \sum_P \pm a_{1i} a_{2j} a_{3k}$$

where the summation is over all permutations $(i, j, k)$ of $(1, 2, 3)$; the + sign is taken if the permutation is even and the − sign if it is odd. □

That this definition is equivalent to equation (1.1.8) is easily seen by writing down all permutations of (1, 2, 3). Since there are 3 items there are 3! = 6 permutations altogether. These, together with the number of interchanges necessary to achieve them, and their consequent character (odd or even) are:

| | | |
|---|---|---|
| (1, 2, 3) | 0 | Even |
| (1, 3, 2) | 1 | Odd |
| (3, 2, 1) | 1 | Odd |
| (2, 1, 3) | 1 | Odd |
| (3, 1, 2) | 2 | Even |
| (2, 3, 1) | 2 | Even |

Thus, by definition 1.1 we have

$$\begin{vmatrix} a_{11} & a_{12} & a_{13} \\ a_{21} & a_{22} & a_{23} \\ a_{31} & a_{32} & a_{33} \end{vmatrix} = a_{11}a_{22}a_{33} - a_{11}a_{23}a_{32} - a_{13}a_{22}a_{31} - a_{12}a_{21}a_{33} + a_{13}a_{21}a_{32} + a_{12}a_{23}a_{31}$$

which is exactly the same as equation (1.1.8).

We shall now see how it is possible to express (in many ways) a determinant of order 3 in terms of determinants of order 2. We first require two more definitions.

DEFINITION 1.2

The minor of the element $a_{ij}$ in the determinant

$$\begin{vmatrix} a_{11} & a_{12} & a_{13} \\ a_{21} & a_{22} & a_{23} \\ a_{31} & a_{32} & a_{33} \end{vmatrix}$$

is the determinant of order 2 obtained by omitting the row and column containing $a_{ij}$ (i.e. the $i$th row and the $j$th column). □

For example, the minor of $a_{11}$ is

$$\begin{vmatrix} a_{22} & a_{23} \\ a_{32} & a_{33} \end{vmatrix} \quad \text{obtained from} \quad \begin{vmatrix} \not{a}_{11} & \not{a}_{12} & \not{a}_{13} \\ \not{a}_{21} & a_{22} & a_{23} \\ \not{a}_{31} & a_{32} & a_{33} \end{vmatrix},$$

and the minor of $a_{32}$ is

$$\begin{vmatrix} a_{11} & a_{13} \\ a_{21} & a_{23} \end{vmatrix} \quad \text{obtained from} \quad \begin{vmatrix} a_{11} & \not{a}_{12} & a_{13} \\ a_{21} & \not{a}_{22} & a_{23} \\ \not{a}_{31} & \not{a}_{32} & \not{a}_{33} \end{vmatrix}.$$

DEFINITION 1.3

The cofactor of the element $a_{ij}$ in the determinant

$$\begin{vmatrix} a_{11} & a_{12} & a_{13} \\ a_{21} & a_{22} & a_{23} \\ a_{31} & a_{32} & a_{33} \end{vmatrix}$$

is its minor multiplied by $(-1)^{i+j}$. We denote the cofactor by $\alpha_{ij}$. □

Since $(-1)^{1+1} = 1$, $(-1)^{1+2} = -1$, $(-1)^{1+3} = 1$ etc the cofactor of any element is obtained from its minor by multiplying by either +1 or −1 according to the following scheme:

$$\begin{matrix} + & - & + \\ - & + & - \\ + & - & + \end{matrix}$$

*Examples*

1.1.1 Obtain the minors and cofactors of all the elements in the second row of the determinant

$$\begin{vmatrix} 1 & 7 & 8 \\ -1 & 2 & 4 \\ 3 & -1 & 7 \end{vmatrix}$$

(i) Element in second row and first column (−1)

$$\text{Minor} = \begin{vmatrix} 7 & 8 \\ -1 & 7 \end{vmatrix} \qquad \text{obtained from} \qquad \begin{vmatrix} 1 & 7 & 8 \\ -1 & 2 & 4 \\ 3 & -1 & 7 \end{vmatrix}$$

$= 49 + 8$ (by equation (1.1.3))

$= 57$

Cofactor $= -57$.

(ii) Element in second row and second column (2)

$$\text{Minor} = \begin{vmatrix} 1 & 8 \\ 3 & 7 \end{vmatrix} \qquad \text{obtained from} \qquad \begin{vmatrix} 1 & 7 & 8 \\ -1 & 2 & 4 \\ 3 & -1 & 7 \end{vmatrix}$$

$= 7 - 24$ (from equation (1.1.3))

$= -17$

Cofactor $= -17$.

(iii) Element in second row and third column (4)

$$\text{Minor} = \begin{vmatrix} 1 & 7 \\ 3 & -1 \end{vmatrix} \qquad \text{obtained from} \qquad \begin{vmatrix} 1 & 7 & 8 \\ 1 & 2 & 4 \\ 3 & -1 & 7 \end{vmatrix}$$

$$= -1 - 21 \qquad \text{(from equation (1.1.3))}$$
$$= -22$$

Cofactor = 22.

1.1.2 Obtain $\alpha_{32}$ for the determinant

$$\begin{vmatrix} a & h & g \\ h & b & f \\ g & f & c \end{vmatrix}.$$

$\alpha_{32}$ is the cofactor of the element in the third row and second column. Thus

$$\alpha_{32} = -\begin{vmatrix} a & g \\ h & f \end{vmatrix}$$
$$= -(af - hg)$$
$$= hg - af.$$

1.1.3 Show that Definition 1.1, suitably modified, applies also to determinants of order 2.

The definition, in a form applicable to determinants of order 2 would be:

$$\begin{vmatrix} a_{11} & a_{12} \\ a_{21} & a_{22} \end{vmatrix} = \sum_P \pm a_{1i}a_{2j}$$

where the sum is over all permutations $(i, j)$ of $(1, 2)$; the + sign is chosen if the permutation is even and the – sign if the permutation is odd.

Since there are two items in $(1, 2)$ there are $2! = 2$ permutations. These are:

| | | | |
|---|---|---|---|
| (1, 2) | 0 | interchange | Even |
| (2, 1) | 1 | interchange | Odd. |

Thus

$$\sum_P \pm a_{1i}a_{ij} = a_{11}a_{22} - a_{12}a_{21}$$

which gives exactly the same expression as equation (1.1.3).

We are now in a position to write the determinant of order 3 in terms of determinants of order 2 (in fact its cofactors). The appropriate result is in the following theorem.

THEOREM 1.1

For the determinant

$$\Delta = \begin{vmatrix} a_{11} & a_{12} & a_{13} \\ a_{21} & a_{22} & a_{23} \\ a_{31} & a_{32} & a_{33} \end{vmatrix}$$

we have

$$\begin{aligned}
\Delta &= a_{11}\alpha_{11} + a_{12}\alpha_{12} + a_{13}\alpha_{13} && \text{(expansion by 1st row)} \\
&= a_{21}\alpha_{21} + a_{22}\alpha_{22} + a_{23}\alpha_{23} && \text{(expansion by 2nd row)} \\
&= a_{31}\alpha_{31} + a_{32}\alpha_{32} + a_{33}\alpha_{33} && \text{(expansion by 3rd row)} \\
&= a_{11}\alpha_{11} + a_{21}\alpha_{21} + a_{31}\alpha_{31} && \text{(expansion by 1st column)} \\
&= a_{12}\alpha_{12} + a_{22}\alpha_{22} + a_{32}\alpha_{32} && \text{(expansion by 2nd column)} \\
&= a_{13}\alpha_{13} + a_{23}\alpha_{23} + a_{33}\alpha_{33} && \text{(expansion by 3rd column)}
\end{aligned}$$

(we see that each expression consists of the sum of all the elements in a particular row or column, each multiplied by its own cofactor).

On the other hand, taking the sum of all the elements in a particular row (or column) each multiplied by the corresponding cofactor of another row (or column) gives zero. For example

$$a_{11}\alpha_{21} + a_{12}\alpha_{22} + a_{13}\alpha_{23} = 0$$

(elements of 1st row taken with cofactors of 2nd row)

$$a_{13}\alpha_{11} + a_{23}\alpha_{21} + a_{33}\alpha_{31} = 0$$

(elements in 3rd column taken with cofactors of 1st column). This is the so-called *Rule of False Cofactors*. All the above results can be summed up by:

$$\begin{aligned}
\Delta &= \sum_{j=1}^{3} a_{ij}\alpha_{ij} && (i = 1, 2 \text{ or } 3) && \text{expansion by rows} \\
&= \sum_{i=1}^{3} a_{ij}\alpha_{ij} && (j = 1, 2 \text{ or } 3) && \text{expansion by columns} \\
\sum_{j=1}^{3} a_{ij}\alpha_{kj} &= 0 && (i \neq k) && \text{false cofactors for rows} \\
\sum_{i=1}^{3} a_{ij}\alpha_{ik} &= 0 && (j \neq k) && \text{false cofactors for columns.}
\end{aligned}$$

*Proof*

We prove only that $\Delta = a_{11}\alpha_{11} + a_{12}\alpha_{12} + a_{13}\alpha_{13}$; all the other results could be proved in an exactly similar way. We have

$$a_{11}\alpha_{11} + a_{12}\alpha_{12} + a_{13}\alpha_{13} = a_{11}\begin{vmatrix} a_{22} & a_{23} \\ a_{32} & a_{33} \end{vmatrix} - a_{12}\begin{vmatrix} a_{21} & a_{23} \\ a_{31} & a_{33} \end{vmatrix}$$

$$+ a_{13}\begin{vmatrix} a_{21} & a_{22} \\ a_{31} & a_{32} \end{vmatrix} \qquad \text{(by definitions 1.3 and 1.2)}$$

$$= a_{11}(a_{22}a_{33} - a_{23}a_{32}) - a_{12}(a_{21}a_{33} - a_{23}a_{31}) + a_{13}(a_{21}a_{32} - a_{22}a_{31})$$

$$= a_{11}a_{22}a_{33} - a_{11}a_{23}a_{32} - a_{12}a_{21}a_{33} + a_{12}a_{23}a_{31} + a_{13}a_{21}a_{32} - a_{13}a_{22}a_{31}$$

$$= \Delta, \qquad \text{(from equation (1.1.8)).}$$ ■

We may use the above theorem to evaluate determinants of order 3, taking advantage of any zeros which occur in the determinant so as to minimize the calculation.

*Examples*

1.1.4 Evaluate

$$\Delta = \begin{vmatrix} 2 & 0 & 3 \\ 2 & 4 & 1 \\ -8 & 9 & 6 \end{vmatrix}.$$

Expanding along the first row we have, by Theorem 1.1

$$\Delta = 2\begin{vmatrix} 4 & 1 \\ 9 & 6 \end{vmatrix} - 0\begin{vmatrix} 2 & 1 \\ -8 & 6 \end{vmatrix} + 3\begin{vmatrix} 2 & 4 \\ -8 & 9 \end{vmatrix}$$

$$= 2(24 - 9) - 0 + 3(18 + 32)$$

$$= 30 + 150$$

$$= 180.$$

1.1.5 Evaluate

$$\Delta = \begin{vmatrix} 7 & 8 & -6 \\ 4 & 2 & 0 \\ -5 & 1 & 7 \end{vmatrix}.$$

Expanding down the third column to take advantage of the 0 in it we obtain, again using theorem 1.1

$$\Delta = -6\begin{vmatrix} 4 & 2 \\ -5 & 1 \end{vmatrix} - 0\begin{vmatrix} 7 & 8 \\ -5 & 1 \end{vmatrix} + 7\begin{vmatrix} 7 & 8 \\ 4 & 2 \end{vmatrix}$$

$$= -6(4 + 10) + 7(14 - 32)$$

$$= -84 - 126$$

$$= -2$$

1.1.6 Evaluate

$$\Delta = \begin{vmatrix} 1 & 6 & 5 \\ 2 & 4 & -1 \\ 3 & 8 & 4 \end{vmatrix}.$$

This time there is no especially simple row or column to use, so we may as well expand along the first row

$$\Delta = 1\begin{vmatrix} 4 & -1 \\ 8 & 4 \end{vmatrix} - 6\begin{vmatrix} 2 & -1 \\ 3 & 4 \end{vmatrix} + 5\begin{vmatrix} 2 & 4 \\ 3 & 8 \end{vmatrix}$$

$$= (16 + 8) - 6(8 + 3) + 5(16 - 12)$$

$$= 24 - 66 + 20$$

$$= -22.$$

We shall now proceed to prove various properties of determinants of order 3. It is trivial to show that these properties hold also for determinants of order 2, and we shall see later that they all generalize to determinants of any order.

THEOREM 1.2

A determinant in which all the elements of a row (or column) are zero, itself has the value zero.

*Proof*

This result follows immediately from Theorem 1.1 by expanding along the row (or down the column) in question. ■

THEOREM 1.3

If two rows (or columns) of a determinant are interchanged the value of the determinant is multiplied by $-1$.

*Proof*

Let us consider

$$\Delta = \begin{vmatrix} a_1 & a_2 & a_3 \\ b_1 & b_2 & b_3 \\ c_1 & c_2 & c_3 \end{vmatrix}.$$

We shall prove the result for the interchanging of rows 2 and 3. The proof for any other rows (or columns) would be exactly similar.

Carrying out the interchange gives the determinant $\Delta'$ where

$$\Delta' = \begin{vmatrix} a_1 & a_2 & a_3 \\ c_1 & c_2 & c_3 \\ b_1 & b_2 & b_3 \end{vmatrix}$$

$$= a_1(c_2b_3 - c_3b_2) - a_2(c_1b_3 - c_3b_1) + a_3(c_1b_2 - c_2b_1)$$

(expanding along the first row)

$$= a_1b_3c_2 - a_1b_2c_3 - a_2b_3c_1 + a_2b_1c_3 + a_3b_2c_1 - a_3b_1c_2.$$

Also

$$\Delta = \begin{vmatrix} a_1 & a_2 & a_3 \\ b_1 & b_2 & b_3 \\ c_1 & c_2 & c_3 \end{vmatrix}$$

$$= a_1(b_2c_3 - b_3c_2) - a_2(b_1c_3 - b_3c_1) + a_3(b_1c_2 - b_2c_1)$$

$$= a_1b_2c_3 - a_1b_3c_2 - a_2b_1c_3 + a_2b_3c_1 + a_3b_1c_2 - a_3b_2c_1.$$

Examining the two expressions for $\Delta'$ and $\Delta$ we see that $\Delta' = -\Delta$, as required. ■

THEOREM 1.4

A determinant in which all corresponding elements in any two rows (or columns) are equal, has the value zero.

*Proof*

We make use of Theorem 1.3. Interchanging the two rows (or columns) in question changes the sign of the determinant. It also leaves it unaltered, since the two rows (or columns) are identical. Thus, if $\Delta$ is the value of the determinant we have $\Delta = -\Delta$ and hence $\Delta = 0$, as required. ■

THEOREM 1.5

If the elements of a row (or column) of a determinant are each multiplied by a factor $\lambda$, then the value of the determinant is multiplied by $\lambda$.

*Proof*

We consider the first row multiplied by $\lambda$; the proof for other rows and columns is exactly similar. Consider

$$\Delta = \begin{vmatrix} a_{11} & a_{12} & a_{13} \\ a_{21} & a_{22} & a_{23} \\ a_{31} & a_{32} & a_{33} \end{vmatrix}.$$

Then the determinant obtained from $\Delta$ by multiplying all the elements in the first row by $\lambda$ is given by

$$\Delta' = \begin{vmatrix} \lambda a_{11} & \lambda a_{12} & \lambda a_{13} \\ a_{21} & a_{22} & a_{23} \\ a_{31} & a_{32} & a_{33} \end{vmatrix}$$

$$= \lambda a_{11} \begin{vmatrix} a_{22} & a_{23} \\ a_{32} & a_{33} \end{vmatrix} - \lambda a_{12} \begin{vmatrix} a_{21} & a_{23} \\ a_{31} & a_{33} \end{vmatrix} + \lambda a_{13} \begin{vmatrix} a_{21} & a_{22} \\ a_{31} & a_{32} \end{vmatrix}$$

(expanding along the first row)

$$= \lambda \left( a_{11} \begin{vmatrix} a_{22} & a_{23} \\ a_{32} & a_{33} \end{vmatrix} - a_{12} \begin{vmatrix} a_{21} & a_{23} \\ a_{31} & a_{33} \end{vmatrix} + a_{13} \begin{vmatrix} a_{21} & a_{22} \\ a_{31} & a_{32} \end{vmatrix} \right)$$

$$= \lambda \begin{vmatrix} a_{11} & a_{12} & a_{13} \\ a_{21} & a_{22} & a_{23} \\ a_{31} & a_{32} & a_{33} \end{vmatrix}$$

$= \lambda\Delta$, as required. ■

THEOREM 1.6

If the elements in any row (or column) of a determinant are expressed as sums of two quantities, then the determinant can be expressed as the sum of two determinants. For example,

$$\begin{vmatrix} x_1 + y_1 & x_2 + y_2 & x_3 + y_3 \\ b_1 & b_2 & b_3 \\ c_1 & c_2 & c_3 \end{vmatrix} = \begin{vmatrix} x_1 & x_2 & x_3 \\ b_1 & b_2 & b_3 \\ c_1 & c_2 & c_3 \end{vmatrix} + \begin{vmatrix} y_1 & y_2 & y_3 \\ b_1 & b_2 & b_3 \\ c_1 & c_2 & c_3 \end{vmatrix}.$$

*Proof*

We give here the proof for the illustration above; a similar proof can obviously be given for any other row or column. We have

$$\begin{vmatrix} x_1 + y_1 & x_2 + y_2 & x_3 + y_3 \\ b_1 & b_2 & b_3 \\ c_1 & c_2 & c_3 \end{vmatrix} = (x_1 + y_1)\begin{vmatrix} b_2 & b_3 \\ c_2 & c_3 \end{vmatrix} - (x_2 + y_2)\begin{vmatrix} b_1 & b_3 \\ c_1 & c_3 \end{vmatrix}$$

$$+ (x_3 + y_3)\begin{vmatrix} b_1 & b_2 \\ c_1 & c_2 \end{vmatrix}$$

$$= x_1\begin{vmatrix} b_2 & b_3 \\ c_2 & c_3 \end{vmatrix} - x_2\begin{vmatrix} b_1 & b_3 \\ c_1 & c_3 \end{vmatrix} + x_3\begin{vmatrix} b_1 & b_2 \\ c_1 & c_2 \end{vmatrix}$$

$$+ y_1\begin{vmatrix} b_2 & b_3 \\ c_2 & c_3 \end{vmatrix} - y_2\begin{vmatrix} b_1 & b_3 \\ c_1 & c_3 \end{vmatrix} + y_3\begin{vmatrix} b_1 & b_2 \\ c_1 & c_2 \end{vmatrix}$$

$$= \begin{vmatrix} x_1 & x_2 & x_3 \\ b_1 & b_2 & b_3 \\ c_1 & c_2 & c_3 \end{vmatrix} + \begin{vmatrix} y_1 & y_2 & y_3 \\ b_1 & b_2 & b_3 \\ c_1 & c_2 & c_3 \end{vmatrix} \quad \text{as required.}$$

■

THEOREM 1.7

The value of a determinant is unaltered by adding to the elements of any row (or column) a constant multiple of the corresponding elements of any other row (or column).

*Proof*

We show that adding the elements of row 2 multiplied by $\lambda$ to the elements of row 1 leaves the determinant unaltered (for other rows and columns the proof is exactly similar). Thus if

$$\Delta = \begin{vmatrix} a_1 & a_2 & a_3 \\ b_1 & b_2 & b_3 \\ c_1 & c_2 & c_3 \end{vmatrix}$$

we wish to show that

$$\Delta = \begin{vmatrix} a_1 + \lambda b_1 & a_2 + \lambda b_2 & a_3 + \lambda b_3 \\ b_1 & b_2 & b_3 \\ c_1 & c_2 & c_3 \end{vmatrix}.$$

Now, we have

$$\begin{vmatrix} a_1 + \lambda b_1 & a_2 + \lambda b_2 & a_3 + \lambda b_3 \\ b_1 & b_2 & b_3 \\ c_1 & c_2 & c_3 \end{vmatrix}$$

$$= \begin{vmatrix} a_1 & a_2 & a_3 \\ b_1 & b_2 & b_3 \\ c_1 & c_2 & c_3 \end{vmatrix} + \begin{vmatrix} \lambda b_1 & \lambda b_2 & \lambda b_3 \\ b_1 & b_2 & b_3 \\ c_1 & c_2 & c_3 \end{vmatrix} \quad \text{by Theorem 1.6}$$

$$= \begin{vmatrix} a_1 & a_2 & a_3 \\ b_1 & b_2 & b_3 \\ c_1 & c_2 & c_3 \end{vmatrix} + \lambda \begin{vmatrix} b_1 & b_2 & b_3 \\ b_1 & b_2 & b_3 \\ c_1 & c_2 & c_3 \end{vmatrix} \quad \text{by Theorem 1.5}$$

$$= \begin{vmatrix} a_1 & a_2 & a_3 \\ b_1 & b_2 & b_3 \\ c_1 & c_2 & c_3 \end{vmatrix} \quad \text{by Theorem 1.4}$$

$$= \Delta \quad \text{as required.}$$

■

THEOREM 1.8

The value of a determinant is unaltered if rows and columns are interchanged. That is,

$$\begin{vmatrix} a_1 & a_2 & a_3 \\ b_1 & b_2 & b_3 \\ c_1 & c_2 & c_3 \end{vmatrix} = \begin{vmatrix} a_1 & b_1 & c_1 \\ a_2 & b_2 & c_2 \\ a_3 & b_3 & c_3 \end{vmatrix}.$$

*Proof*

We have, expanding across the first row

$$\begin{vmatrix} a_1 & a_2 & a_3 \\ b_1 & b_2 & b_3 \\ c_1 & c_2 & c_3 \end{vmatrix} = a_1(b_2c_3 - b_3c_2) - a_2(b_1c_3 - b_3c_1) + a_3(b_1c_2 - b_2c_1).$$

Also, expanding down the first column of the second determinant gives

$$\begin{vmatrix} a_1 & b_1 & c_1 \\ a_2 & b_2 & c_2 \\ a_3 & b_3 & c_3 \end{vmatrix} = a_1(b_2c_3 - b_3c_2) - a_2(b_1c_3 - b_3c_1) + a_3(b_1c_2 - b_2c_1).$$

Hence the two determinants are equal as required. ■

We may take advantage of the above theorems (especially Theorem 1.7) to help in the evaluation of determinants; the technique is to try to introduce zeros in some row or column in order to simplify the expansion along that row or down that column.

*Examples*

1.1.7 Evaluate the following determinants:

(i) $\begin{vmatrix} 2 & 1 & 3 \\ 4 & 2 & 8 \\ 7 & -1 & 2 \end{vmatrix}$ (ii) $\begin{vmatrix} 4 & 1 & 3 \\ 3 & -6 & 4 \\ 7 & 25 & -9 \end{vmatrix}$

(iii) $\begin{vmatrix} a & b & c \\ b & c & a \\ a+b & b+c & c+a \end{vmatrix}$ (iv) $\begin{vmatrix} 1 & 1 & 1 \\ 2x & 2y & 2z \\ x^2 & y^2 & z^2 \end{vmatrix}$.

(i) We make use of Theorem 1.7 to introduce zeros in the second column as follows:

$$\begin{vmatrix} 2 & 1 & 3 \\ 4 & 2 & 8 \\ 7 & -1 & 2 \end{vmatrix} = \begin{vmatrix} 2+7 & 1+(-1) & 3+2 \\ 4 & 2 & 8 \\ 7 & -1 & 2 \end{vmatrix} \quad \text{(Row 1 + Row 3)}$$

$$= \begin{vmatrix} 9 & 0 & 5 \\ 4 & 2 & 8 \\ 7 & -1 & 2 \end{vmatrix}$$

$$= \begin{vmatrix} 9 & 0 & 5 \\ 4+2\times 7 & 2+2\times(-1) & 8+2\times 2 \\ 7 & -1 & 2 \end{vmatrix} \quad \text{(Row 2 + 2 x Row 3)}$$

$$= \begin{vmatrix} 9 & 0 & 5 \\ 18 & 0 & 12 \\ 7 & -1 & 2 \end{vmatrix}$$

$$= -(-1)\begin{vmatrix} 9 & 5 \\ 18 & 12 \end{vmatrix} \quad \text{(expanding down the second column)}$$

$$= 9\begin{vmatrix} 1 & 5 \\ 2 & 12 \end{vmatrix}$$

$$= 9(12 - 10)$$

$$= 18.$$

(ii) Here we introduce zeros into the first row:

$$\begin{vmatrix} 4 & 1 & 3 \\ 3 & -6 & 4 \\ 7 & 25 & -9 \end{vmatrix} = \begin{vmatrix} 0 & 1 & 3 \\ 27 & -6 & 4 \\ -93 & 25 & -9 \end{vmatrix} \qquad \text{(Column 1} - 4 \times \text{column 2)}$$

$$= \begin{vmatrix} 0 & 1 & 0 \\ 27 & -6 & 22 \\ -93 & 25 & -84 \end{vmatrix} \qquad \text{(Column 3} - 3 \times \text{column 2)}$$

$$= -\begin{vmatrix} 27 & 22 \\ -93 & -84 \end{vmatrix} \qquad \text{(expanding across the first row)}$$

$$= -6\begin{vmatrix} 9 & 11 \\ -31 & -42 \end{vmatrix}$$

$$= -6(-9 \times 42 + 11 \times 31)$$

$$= (-6) \times (-37)$$

$$= 222.$$

(iii) This time we see that simplification is achieved by subtracting row 2 from row 3:

$$\begin{vmatrix} a & b & c \\ b & c & a \\ a+b & b+c & c+a \end{vmatrix} = \begin{vmatrix} a & b & c \\ b & c & a \\ a & b & c \end{vmatrix} \qquad \text{(Row 3} - \text{Row 2)}$$

$$= 0 \qquad \text{(by Theorem 1.4).}$$

(iv) We first remove the factor of 2 from the second row and obtain

$$\begin{vmatrix} 1 & 1 & 1 \\ 2x & 2y & 2z \\ x^2 & y^2 & z^2 \end{vmatrix} = 2\begin{vmatrix} 1 & 1 & 1 \\ x & y & z \\ x^2 & y^2 & z^2 \end{vmatrix} \qquad \text{(by Theorem 1.5)}$$

$$= 2\begin{vmatrix} 0 & 0 & 1 \\ x-z & y-z & z \\ x^2-z^2 & y^2-z^2 & z^2 \end{vmatrix} \qquad \begin{pmatrix} \text{Column 1} - \text{column 3} \\ \text{Column 3} - \text{column 3} \end{pmatrix}$$

$$= 2(x-z)(y-z)\begin{vmatrix} 0 & 0 & 1 \\ 1 & 1 & z \\ x+z & y+z & z^2 \end{vmatrix} \qquad \text{(again by Theorem 1.5)}$$

$$= 2(x-z)(y-z)\{(y+z)-(x+z)\} \qquad \text{(expanding along the first row)}$$

$$= 2(x-z)(y-z)(y-x)$$

$$= 2(y-z)(z-x)(x-y).$$

1.1.8 Use Cramer's rule to solve the system of equations

$$x + y + z = 5$$
$$2x - y + 3z = 13$$
$$-x + 3y + z = 1.$$

By Cramer's rule (equation (1.1.9)) we have

$$\frac{x}{\begin{vmatrix} 5 & 1 & 1 \\ 13 & -1 & 3 \\ 1 & 3 & 1 \end{vmatrix}} = \frac{y}{\begin{vmatrix} 1 & 5 & 1 \\ 2 & 13 & 3 \\ -1 & 1 & 1 \end{vmatrix}} = \frac{z}{\begin{vmatrix} 1 & 1 & 5 \\ 2 & -1 & 13 \\ -1 & 3 & 1 \end{vmatrix}} = \frac{1}{\begin{vmatrix} 1 & 1 & 1 \\ 2 & -1 & 3 \\ -1 & 3 & 1 \end{vmatrix}}.$$

Thus we must evaluate the four determinants appearing in the denominator.
Now,

$$\begin{vmatrix} 5 & 1 & 1 \\ 13 & -1 & 3 \\ 1 & 3 & 1 \end{vmatrix} = \begin{vmatrix} 0 & 1 & 0 \\ 18 & -1 & 4 \\ -14 & 3 & -2 \end{vmatrix}$$

$$= -\begin{vmatrix} 18 & 4 \\ -14 & -2 \end{vmatrix}$$

$$= -20,$$

$$\begin{vmatrix} 1 & 5 & 1 \\ 2 & 13 & 3 \\ -1 & 1 & 1 \end{vmatrix} = \begin{vmatrix} 1 & 6 & 2 \\ 2 & 15 & 5 \\ -1 & 0 & 0 \end{vmatrix}$$

$$= 3\begin{vmatrix} 1 & 2 & 2 \\ 2 & 5 & 5 \\ -1 & 0 & 0 \end{vmatrix}$$

$= 0$ (by Theorem 1.4 since second and third columns are equal),

$$\begin{vmatrix} 1 & 1 & 5 \\ 2 & -1 & 13 \\ -1 & 3 & 1 \end{vmatrix} = \begin{vmatrix} 1 & 0 & 0 \\ 2 & -3 & 3 \\ -1 & 4 & 6 \end{vmatrix}$$

$$= \begin{vmatrix} -3 & 3 \\ 4 & 6 \end{vmatrix}$$

$$= -30$$

and

$$\begin{vmatrix} 1 & 1 & 1 \\ 2 & -1 & 3 \\ -1 & 3 & 1 \end{vmatrix} = \begin{vmatrix} 1 & 0 & 0 \\ 2 & -3 & 1 \\ -1 & 4 & 2 \end{vmatrix}$$

$$= \begin{vmatrix} -3 & 1 \\ 4 & 2 \end{vmatrix}$$

$$= -10.$$

Thus we have

$$\frac{x}{-20} = \frac{y}{0} = \frac{z}{-30} = \frac{1}{-10}$$

which gives

$$x = 2, \qquad y = 0, \qquad z = 3.$$

*Exercises*

1.1.1 Evaluate the following determinants:

$$\text{(i)} \begin{vmatrix} 3 & 6 & 8 \\ -1 & 7 & 2 \\ 5 & -2 & 0 \end{vmatrix} \qquad \text{(ii)} \begin{vmatrix} 99 & 100 & 101 \\ 100 & 101 & 102 \\ 101 & 102 & 103 \end{vmatrix} \qquad \text{(iii)} \begin{vmatrix} a & b & c \\ b & c & a \\ c & a & b \end{vmatrix}.$$

1.1.2 Use Cramer's rule to solve the following systems of equations:

(i) $x + y + z = 5$

$2x + 3y - z = 0$

$x + 2y + z = 11$

(ii) $2x + 3y - z = 6$

$x + 2y + z = 3$

$x + y + 2z = 3.$

## 1.2 Determinants of General Order and their Properties

We now wish to generalize the results of Section 1.1 to determinants of any order. This is most readily accomplished by generalizing Definition 1.1 as follows.

DEFINITION 1.4

A determinant of order $n$ is defined by

$$\begin{vmatrix} a_{11} & a_{12} & a_{13} & \cdots & a_{1n} \\ a_{21} & a_{22} & & \cdots & a_{2n} \\ \cdot & & & & \\ \cdot & & & & \\ \cdot & & & & \\ a_{n1} & \cdots & & \cdots & a_{nn} \end{vmatrix} = \sum_P \pm a_{1s_1} a_{2s_2} a_{3s_3}, \ldots, a_{ns_n}$$

where the summation is over all permutations $(s_1, s_2, s_3, \ldots, s_n)$ of $(1, 2, 3, \ldots, n)$; the + sign is taken if the permutation is even and the − sign if it is odd. (We recall that a permutation is even if accomplished by an even number of interchanges and odd if accomplished by an odd number.) □

We note that since there are $n!$ permutations of $n$ objects there will be $n!$ terms in the sum in the above definition.

*Example*

1.2.1 Determinant of order 5:

$$\Delta = \begin{vmatrix} a_{11} & a_{12} & a_{13} & a_{14} & a_{15} \\ a_{21} & a_{22} & a_{23} & a_{24} & a_{25} \\ a_{31} & a_{32} & a_{33} & a_{34} & a_{35} \\ a_{41} & a_{42} & a_{43} & a_{44} & a_{45} \\ a_{51} & a_{52} & a_{53} & a_{54} & a_{55} \end{vmatrix}.$$

Here there are $5! = 120$ terms in the sum; we do not wish to write these out in full but typical ones would be

$$\begin{array}{cccccc} + & a_{11} & a_{22} & a_{34} & a_{45} & a_{53} \\ - & a_{12} & a_{23} & a_{31} & a_{45} & a_{54} \\ + & a_{14} & a_{22} & a_{35} & a_{41} & a_{53}. \end{array}$$

The signs are obtained by noting the number of interchanges necessary to bring the sequence of second suffices back to natural order (1, 2, 3, 4, 5).
Thus:

(1, 2, 4, 5, 3)
(1, 2, 3, 5, 4)
(1, 2, 3, 4, 5) 2 interchanges (even) and hence + sign

(2, 3, 1, 5, 4)
(1, 3, 2, 5, 4)
(1, 2, 3, 5, 4)
(1, 2, 3, 4, 5) 3 interchanges (odd) and hence − sign

(4, 2, 5, 1, 3)

(1, 2, 5, 4, 3)

(1, 2, 3, 4, 5) 2 interchanges (even) and hence + sign.

From the definition we see that one of the $n!$ terms in the sum will always be the product of the elements along the diagonal, $a_{11}a_{22}a_{33} \ldots a_{nn}$, and this occurs with a + sign. This is because here $(s_1, s_2, s_3, \ldots s_n) = (1, 2, 3, \ldots n)$ and this is a permutation of $(1, 2, 3, \ldots n)$ which requires 0 (even) interchanges to give the natural order.

We note immediately that because there are $n!$ terms it is completely impracticable to write out the expansion in full for even moderate values of $n$. This is readily seen when we realize that

$$5! = 120$$

$$10! = 3\,628\,800$$

$$15! \simeq 1{\cdot}3 \times 10^{10}$$

$$20! \simeq 2{\cdot}4 \times 10^{18}.$$

We also note that this implies the complete impracticability of basing a computer program for the evaluation of a determinant on this expansion. For if we assume that multiplications take $10^{-6}$ seconds and take no account of time for additions and any other overheads of the method, we should be faced with times as follows (remembering that the $n$th order determinant has $n!$ terms each with $(n-1)$ products):

| Order of determinant | Time |
|---|---|
| 5 | $4 \times 5! \times 10^{-6}$ seconds $\simeq 5 \times 10^{-4}$ seconds |
| 10 | $9 \times 10! \times 10^{-6}$ seconds $\simeq 32$ seconds |
| 15 | $14 \times 15! \times 10^{-6}$ seconds $\simeq 21$ days |
| 20 | $19 \times 20! \times 10^{-6}$ seconds $\simeq 10^7$ years |

We are now in a position to generalize the results of Section 1.1. We first give general definitions of minor and cofactor.

DEFINITION 1.5

The *minor* of the element $a_{ij}$ in the determinant of order $n$ given by

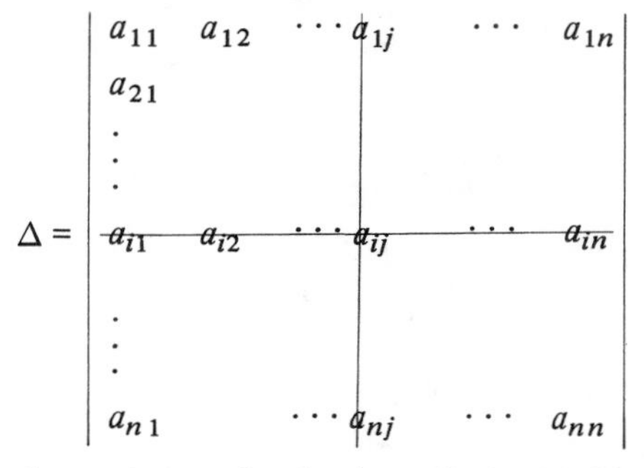


is the determinant of order $(n-1)$ obtained by omitting the row and column passing through $a_{ij}$ (i.e. the $i$th row and $j$th column) as indicated above.

Thus, from Definition 1.4, we have

$$\text{Minor of } a_{ij} = \sum_P \pm a_{1s_1} a_{2s_2} \ldots a_{i-1\,s_{i-1}} a_{i+1\,s_{i+1}} \ldots a_{ns_n}$$

where the summation is over all permutations $P$ of $(1, 2, \ldots j-1, j+1 \ldots n)$ □

DEFINITION 1.6

The *cofactor* of the element $a_{ij}$ in the general determinant of order $n$ is defined by

$$\text{Cofactor of } a_{ij} = (-1)^{i+j} \times (\text{minor of } a_{ij}).$$ □

We denote the cofactor by $\alpha_{ij}$ and see that it is obtained by multiplying the minor by the appropriate sign from the 'chequer board' pattern

$$\begin{vmatrix} + & - & + & - & + & - & \cdots \\ - & + & - & + & - & + & \\ + & - & + & & & & \\ - & & & & & & \\ + & & & & & & \\ \cdot & & & & & & \\ \cdot & & & & & & \\ \cdot & & & & & & \end{vmatrix}.$$

We are now in a position to generalize Theorem 1.1 concerning the expansion of determinants of order 3 in terms of those of order 2. The result we shall now prove concerns the expansion of a determinant of order $n$ in terms of determinants of order $(n-1)$.

THEOREM 1.9

If

$$\Delta = \begin{vmatrix} a_{11} & a_{12} & a_{13} & \cdots & a_{1n} \\ a_{21} & & & & \\ a_{31} & & & & \\ \cdot & & & & \\ \cdot & & & & \\ \cdot & & & & \\ a_{n1} & \cdots & \cdots & \cdots & a_{nn} \end{vmatrix}$$

and if $\alpha_{ij}$ is the cofactor of the element $a_{ij}$ then

(i) $\Delta = \sum_{j=1}^{n} a_{ij}\alpha_{ij}$ $\quad (1 \leqslant i \leqslant n)$—expansion along the $i$th row.

(ii) $\Delta = \sum_{i=1}^{n} a_{ij}\alpha_{ij}$ $\quad (1 \leqslant j \leqslant n)$—expansion down the $j$th column.

(iii) $0 = \sum_{j=1}^{n} a_{ij}\alpha_{kj}$ $\quad (i \neq k)$—rule of false cofactors.

(iv) $0 = \sum_{i=1}^{n} a_{ij}\alpha_{ik}$ $\quad (j \neq k)$—rule of false cofactors.

*Proof*

(i) We have, from Definitions 1.5 and 1.6,

$$\alpha_{ij} = (-1)^{i+j} \sum_P \pm a_{1s_1} a_{2s_2} \ldots a_{i-1s_{i-1}} a_{i+1s_{i+1}} \ldots a_{ns_n}$$

where the sum $P$ is over all permutations $(s_1, s_2, \ldots s_{i-1}, s_{i+1}, \ldots s_n)$ of $(1, 2, 3, \ldots j-1, j+1, \ldots n)$.

Hence

$$a_{ij}\alpha_{ij} = (-1)^{i+j} \sum_P \pm a_{1s_1} a_{2s_2} \ldots a_{i-1s_{i-1}} a_{ij} a_{i+1s_{i+1}} \ldots a_{ns_n}$$

and so

$$\sum_{j=1}^{n} a_{ij}\alpha_{ij} = \sum_{j=1}^{n} (-1)^{i+j} \sum_P \pm a_{1s_1} a_{2s_2} \ldots a_{i-1s_{i-1}} a_{ij} a_{i+1s_{i+1}} \ldots a_{ns_n}.$$

To complete the proof we have to show that this is the same as

$$\sum_{P'} \pm a_{1s_1} a_{2s_2} \ldots a_{i-1s_{i-1}} a_{is_i} a_{i+1s_{i+1}} \ldots a_{ns_n}$$

where the sum $P'$ is over all permutations $(s_1, s_2, \ldots s_n)$ of $(1, 2, \ldots n)$.

We first note that summing over $j$ is just what is necessary to give all permutations of the quantities $(1, 2, \ldots n)$ from the $n-1$ quantities $(1, 2, \ldots j-1, j+1, \ldots n)$.

Next we note that if

$$(s_1, s_2, \ldots s_{i-1}, s_{i+1}, \ldots s_n)$$

is an even (or odd) permutation of the $(n-1)$ quantities

$$(1, 2, \ldots j-1, j+1, \ldots n)$$

then

$$(s_1, s_2, \ldots s_{i-1}, j, s_{i+1} \ldots s_n)$$

is a permutation of the $n$ quantities $(1, 2, \ldots n)$ accomplished by a further $(i+j-2)$ interchanges; for $j$ may be put into its correct position by $(i-1)$ interchanges moving $j$ to the left by one each time and then $(j-1)$ interchanges moving $j$ to the right each time. Hence the parity (+1 if even, −1 if odd) of the permutation

$$(s_1, s_2, \ldots s_{i-1}, j, s_{i+1}, \ldots s_n)$$

is obtained from the parity of

$$(s_1, s_2, \ldots s_{i-1}, s_{i+1}, \ldots s_n)$$

by multiplying by the factor $(-1)^{i+j-2} = (-1)^{i+j}$ and this is just the factor which enables us to say that

$$\sum_{j=1}^{n} (-1)^{i+j} \sum_P \pm a_{1s_1} a_{2s_2} \ldots a_{i-1s_{i-1}} a_{ij} a_{i+1s_{i+1}} \ldots a_{ns_n}$$

$$= \sum_{P'} \pm a_{1s_1} a_{2s_2} \ldots a_{ns_n}$$

and hence the proof is completed.

(ii) This proof is similar to (i) above and there is no need to give the details here.

(iii) We shall prove later (Theorem 1.12) that Theorem 1.4 holds true for determinants of the $n$th order—a determinant is zero if it has the corresponding elements in two rows (or columns) equal. We can make use of this result to prove the rule of false cofactors.

For $\sum_{j=1}^{n} a_{ij}\alpha_{kj}$ is just the expansion along the $k$th row of the determinant obtained from

$$\begin{vmatrix} a_{11} & a_{12} & \cdots & a_{1n} \\ a_{21} & & & \\ \cdot & & & \\ \cdot & & & \\ \cdot & & & \\ a_{n1} & \cdots & \cdots & a_{nn} \end{vmatrix}$$

by replacing the elements $a_{kj}$ $(j = 1, 2, \ldots n)$ in the $k$th row by the corresponding elements $a_{ij}$ $(j = 1, 2, \ldots n)$ from the $i$th row. Thus $\sum_{j=1}^{n} a_{ij}\alpha_{kj}$ is just the expansion of a determinant which has corresponding elements in the $i$th and $k$th rows equal, and hence is zero.

(iv) The proof is similar to (iii) above, this time the argument proceeding in terms of the expansion of a determinant with two equal columns. ■

We are now in a position to generalize Theorems 1.2–1.8 to determinants of $n$th order. We provide the proofs in the following theorems.

THEOREM 1.10

A determinant in which all the elements of a row (or column) are zero, itself has the value zero.

*Proof*

We shall provide the proof for rows. Suppose, then, that all the elements in the $i$th row are zero, so that $a_{ij} = 0$ $(j = 1, 2, \ldots n)$. Then

$$\begin{aligned} \Delta &= \sum_{P} \pm a_{1s_1} a_{2s_2} \ldots a_{is_i} \ldots a_{ns_n} \qquad \text{(by Definition 1.4)} \\ &= 0 \end{aligned}$$

since every term in the sum contains a factor $a_{is_i}$ which is zero since $a_{ij} = 0$ for all $j$. ■

THEOREM 1.11

If two rows (or columns) of a determinant are interchanged, the value of the determinant is multiplied by $-1$.

*Proof*

Suppose we interchange the $p$th and $q$th rows so that we wish to consider

$$\Delta = \begin{vmatrix} a_{11} & a_{12} & \dots & a_{1n} \\ a_{21} & & & \\ \cdot & & & \\ \cdot & & & \\ \cdot & & & \\ a_{p1} & a_{p2} & \dots & a_{pn} \\ \cdot & & & \\ \cdot & & & \\ \cdot & & & \\ a_{q1} & a_{q2} & \dots & a_{qn} \\ \cdot & & & \cdot \\ \cdot & & & \cdot \\ \cdot & & & \cdot \\ a_{n1} & \dots & \dots & a_{nn} \end{vmatrix} \begin{matrix} \\ \\ \\ \\ \\ \leftarrow p\text{th row} \\ \\ \\ \\ \leftarrow q\text{th row} \\ \\ \\ \\ \\ \end{matrix}$$

and

$$\Delta' = \begin{vmatrix} a_{11} & a_{12} & \dots & a_{1n} \\ a_{21} & & & \\ \cdot & & & \\ \cdot & & & \\ \cdot & & & \\ a_{q1} & a_{q2} & \dots & a_{qn} \\ \cdot & & & \\ \cdot & & & \\ \cdot & & & \\ a_{p1} & a_{p2} & \dots & a_{pn} \\ \cdot & & & \\ \cdot & & & \\ \cdot & & & \\ a_{n1} & a_{n2} & \dots & a_{nn} \end{vmatrix} \begin{matrix} \\ \\ \\ \\ \\ \leftarrow p\text{th row} \\ \\ \\ \\ \leftarrow q\text{th row} \\ \\ \\ \\ \\ \end{matrix}$$

From Definition 1.4 we have

$$\Delta = \sum_P \pm a_{1s_1} a_{2s_2} \dots a_{ps_p} \dots a_{qs_q} \dots a_{ns_n}$$

and

$$\Delta' = \sum_P \pm a_{1s_1} a_{2s_2} \dots a_{qs_q} \dots a_{ps_p} \dots a_{ns_n}.$$

It thus follows that each term in $\Delta'$ is the same as a corresponding term of $\Delta$; but if

$$(s_1, s_2, \dots s_p, \dots s_q, \dots s_n)$$

is an even (odd) permutation then

$$(s_1, s_2, \dots s_q, \dots s_p, \dots s_n)$$

is odd (even) since it involves one more interchange. Thus there will be a change of sign associated with each term, and as a result we have $\Delta' = -\Delta$ as required. ■

THEOREM 1.12

A determinant in which all corresponding elements in any two rows (or columns) are equal has the value zero.

*Proof*

We make use of Theorem 1.11. Interchanging the two rows (or columns) in question changes the sign of the determinant. It also leaves it unaltered, since the two rows (or columns) are identical.

Thus, if $\Delta$ is the value of the determinant we have $\Delta = -\Delta$, and hence $\Delta = 0$, as required. ■

THEOREM 1.13

If the elements of a row (or column) of a determinant are each multiplied by a factor $\lambda$, then the value of the determinant is multiplied by $\lambda$.

*Proof*

Consider

$$\Delta = \begin{vmatrix} a_{11} & a_{12} & a_{13} & \cdots & a_{1n} \\ a_{21} & & & & \\ a_{31} & & & & \\ \cdot & & & & \\ \cdot & & & & \\ \cdot & & & & \\ a_{n1} & \cdots & \cdots & \cdots & a_{nn} \end{vmatrix}$$

and suppose we multiply the $i$th row by $\lambda$ so that we are interested in

$$\Delta' = \begin{vmatrix} a_{11} & a_{12} & \cdots & a_{1n} \\ a_{21} & & & \\ \cdot & & & \\ \cdot & & & \\ \cdot & & & \\ \lambda a_{i1} & \lambda a_{i2} & \cdots & \lambda a_{in} \\ \cdot & & & \\ \cdot & & & \\ \cdot & & & \\ a_{n1} & \cdots & \cdots & a_{nn} \end{vmatrix}.$$

But

$$\Delta' = \sum_P \pm a_{1s_1} a_{2s_2} \ldots a_{i-1s_{i-1}} (\lambda a_{is_i}) a_{i+1s_{i+1}} \ldots a_{ns_n} \qquad \text{(by Definition 1.4)}$$

$$= \lambda \sum_P \pm a_{is_1} a_{2s_2} \ldots a_{i-1\,s_{i-1}} a_{is_i} a_{i+1\,s_{i+1}} \ldots a_{ns_n}$$

$$= \lambda \Delta$$

as required. The proof for columns would be similar. ■

THEOREM 1.14

If the elements in any row or column of a determinant are expressed as sums of two quantities, then the determinant can be expressed as the sum of two determinants; for example if $a_{ij} = b_{ij} + c_{ij}$ $(j = 1, 2, \ldots n)$ then

$$\Delta = \begin{vmatrix} a_{11} & a_{12} & \cdots & a_{1n} \\ \cdot & & & \\ \cdot & & & \\ \cdot & & & \\ a_{i1} & a_{i2} & \cdots & a_{in} \\ \cdot & & & \\ \cdot & & & \\ \cdot & & & \\ a_{n1} & a_{n2} & \cdots & a_{nn} \end{vmatrix}$$

$$= \begin{vmatrix} a_{11} & a_{12} & \cdots & a_{1n} \\ \cdot & & & \\ \cdot & & & \\ \cdot & & & \\ b_{i1} + c_{i1} & b_{i2} + c_{i2} & \cdots & b_{in} + c_{in} \\ \cdot & & & \\ \cdot & & & \\ \cdot & & & \\ a_{n1} & a_{n2} & \cdots & a_{nn} \end{vmatrix}$$

$$= \begin{vmatrix} a_{11} & a_{12} & \cdots & a_{1n} \\ \cdot & & & \\ \cdot & & & \\ \cdot & & & \\ b_{i1} & b_{i2} & \cdots & b_{in} \\ \cdot & & & \\ \cdot & & & \\ \cdot & & & \\ a_{n1} & a_{n2} & \cdots & a_{nn} \end{vmatrix} + \begin{vmatrix} a_{11} & a_{12} & \cdots & a_{1n} \\ \cdot & & & \\ \cdot & & & \\ \cdot & & & \\ c_{i1} & c_{i2} & \cdots & c_{in} \\ \cdot & & & \\ \cdot & & & \\ \cdot & & & \\ a_{n1} & a_{n2} & \cdots & a_{nn} \end{vmatrix}$$

*Proof*

We have

$$\Delta = \sum_P \pm a_{1s_1} a_{2s_2} \cdots a_{is_i} \cdots a_{ns_n}$$

$$= \sum_P \pm a_{1s_1} a_{2s_2} \cdots (b_{is_i} + c_{is_i}) \cdots a_{ns_n}$$

$$= \sum_P \pm a_{1s_1} a_{2s_2} \cdots a_{i-1s_{i-1}} b_{is_i} a_{i+1s_{i+1}} \cdots a_{ns_n}$$

$$+ \sum_P \pm a_{1s_1} a_{2s_2} \cdots a_{i-1s_{i-1}} c_{is_i} a_{i+1s_{i+1}} \cdots a_{ns_n}$$

which proves the required result for rows. The proof for columns would be similar. ∎

THEOREM 1.15

The value of a determinant is unaltered by adding to the elements of any row (or column) a constant multiple of the corresponding elements of any other row (or column).

*Proof*

We show that adding a constant multiple of the $q$th row to the $p$th row leaves a determinant unaltered.

Thus we are interested in

$$\Delta = \begin{vmatrix} a_{11} & a_{12} & \cdots & a_{1n} \\ a_{21} & & & \\ \cdot & & & \\ \cdot & & & \\ \cdot & & & \\ a_{p1} & \cdots\cdots & & a_{pn} \\ \cdot & & & \\ \cdot & & & \\ \cdot & & & \\ a_{q1} & \cdots\cdots & & a_{qn} \\ \cdot & & & \\ \cdot & & & \\ \cdot & & & \\ a_{n1} & \cdots\cdots & & a_{nn} \end{vmatrix}$$

and

$$\Delta' = \begin{vmatrix} a_{11} & a_{12} & \cdots & a_{1n} \\ a_{21} & a_{22} & & \\ \cdot & & & \\ \cdot & & & \\ \cdot & & & \\ a_{p1} + \lambda a_{q1} & a_{p2} + \lambda a_{q2} & \cdots & a_{pn} + \lambda a_{qn} \\ \cdot & & & \\ \cdot & & & \\ \cdot & & & \\ a_{q1} & a_{q2} & \cdots & a_{qn} \\ \cdot & & & \\ \cdot & & & \\ \cdot & & & \\ a_{n1} & a_{n2} & \cdots & a_{nn} \end{vmatrix}.$$

But we have

$$\Delta' = \begin{vmatrix} a_{11} & a_{12} & \cdots & a_{1n} \\ a_{21} & & & \\ \vdots & & & \\ a_{p1} & a_{p2} & \cdots & a_{pn} \\ \vdots & & & \\ a_{q1} & a_{q2} & \cdots & a_{qn} \\ \vdots & & & \\ a_{n1} & \cdots & \cdots & a_{nn} \end{vmatrix} + \lambda \begin{vmatrix} a_{11} & a_{12} & \cdots & a_{1n} \\ a_{21} & & & \\ \vdots & & & \\ a_{q1} & a_{q2} & \cdots & a_{qn} \\ \vdots & & & \\ a_{q1} & a_{q2} & & a_{qn} \\ \vdots & & & \\ a_{n1} & \cdots & \cdots & a_{nn} \end{vmatrix}$$

by Theorems 1.14 and 1.13

$= \Delta + 0$ by Theorem 1.12

$= \Delta,$ as required.

The proof for columns is similar. ■

THEOREM 1.16

The value of a determinant is unaltered if rows and columns are interchanged.

*Proof*

We wish to consider

$$\Delta = \begin{vmatrix} a_{11} & a_{12} & \cdots & a_{1n} \\ a_{21} & a_{22} & \cdots & a_{2n} \\ a_{31} & & & \\ \vdots & & & \\ a_{n1} & & \cdots & a_{nn} \end{vmatrix}$$

$$\Delta' = \begin{vmatrix} a_{11} & a_{21} & a_{31} & \cdots & a_{n1} \\ a_{12} & a_{22} & & & \\ a_{13} & & & & \\ \vdots & & & & \\ a_{1n} & a_{2n} & \cdots & \cdots & a_{nn} \end{vmatrix} = \begin{vmatrix} b_{11} & b_{12} & b_{13} & \cdots & b_{1n} \\ b_{21} & b_{22} & & & \\ \vdots & & & & \\ b_{n1} & & & & b_{nn} \end{vmatrix}$$

say, where $b_{ij} = a_{ji}$. By Definition 1.4 we have

$$\Delta = \sum_{P} \pm a_{1s_1} a_{2s_2} \ldots a_{ns_n}$$

and

$$\Delta' = \sum_{P'} \pm b_{1s_1} b_{2s_2} \ldots b_{ns_n} = \sum_{P'} \pm a_{s_1 1} a_{s_2 2} \ldots a_{s_n n}$$

by definition of the $b_{ij}$.

This last sum contains the same $n!$ terms as the sum constituting $\Delta$; what we still require to show is that the signs are the same. For instance, in a fourth-order determinant the term $a_{12}a_{23}a_{34}a_{41}$ appears in $\Delta$ and has its sign calculated from the parity of the permutation (2, 3, 4, 1); the same term appears in $\Delta'$ as $a_{41}a_{12}a_{22}a_{34}$ (with second suffices in natural order) and has its sign calculated from the parity of the permutation (4, 1, 2, 3). What we need to do is show in general that these two permutations have the same parity.

Now we see that (2, 3, 4, 1) is reduced to natural order by the sequence of interchanges

(2, 3, 4, 1)

(1, 3, 4, 2) interchange 1st and 4th items

(1, 2, 4, 3) interchange 2nd and 3rd items

(1, 2, 3, 4) interchange 3rd and 4th items.

But (4, 1, 2, 3) is reduced to natural order by the sequence

(4, 1, 2, 3)

(1, 4, 2, 3) interchange 1 and 4

(1, 4, 3, 2) interchange 2 and 3

(1, 2, 3, 4) interchange 3 and 4.

So what we do to labelled items in the first permutation we do to the items themselves in the second; we hence reduce to natural order by the same number of interchanges and thus the two permutations have the same parity. It follows that the two terms have the same sign in both $\Delta$ and $\Delta'$.

The same argument can be applied to the general case, since if the term

$$a_{1s_1} a_{2s_2} \ldots a_{ns_n}$$

has its elements reordered so that the column suffices come into natural order yielding

$$a_{t_1 1} a_{t_2 2} a_{t_3 3} \ldots a_{t_n n}$$

then the permutations $(s_1, s_2, \ldots s_n)$ and $(t_1, t_2, \ldots t_n)$ are related in the same way as (2, 3, 4, 1) and (4, 1, 2, 3) in the example above.

This is because the reordering is accomplished by interchanging the index $s_i$ with place occupied.

Thus the two expansions of $\Delta$ and $\Delta'$ are the same, and the theorem is proved. ■

We now have enough results at our disposal to give the general form of Cramer's rule.

THEOREM 1.17 (Cramer's rule)

The solution of the system of equations

$$\begin{aligned} a_{11}x_1 + a_{12}x_2 + \ldots + a_{1n}x_n &= b_1 \\ a_{21}x_1 + a_{22}x_2 + \ldots + a_{2n}x_n &= b_2 \\ &\vdots \\ a_{n1}x_1 + a_{n2}x_2 + \ldots + a_{nn}x_n &= b_n \end{aligned}$$

is given by

$$\frac{x_1}{\Delta_1} = \frac{x_2}{\Delta_2} = \ldots = \frac{x_i}{\Delta_i} = \ldots = \frac{x_n}{\Delta_n} = \frac{1}{\Delta}$$

where

$$\Delta = \begin{vmatrix} a_{11} & a_{12} & \ldots & a_{1n} \\ a_{21} & & & \\ \vdots & & & \\ a_{n1} & a_{n2} & \ldots & a_{nn} \end{vmatrix}$$

and $\Delta_i$ is the determinant obtained from $\Delta$ by replacing the $i$th column by the column of $b$'s.

The solution is valid provided $\Delta \neq 0$, and $\Delta_i = 0$ is taken to mean that $x_i = 0$.

*Proof*

The result is proved if we can show that $x_i = \Delta_i/\Delta$ $(i = 1, 2, \ldots n)$.

If we multiply the $i$th equation by $\alpha_{ij}$, the cofactor of $a_{ij}$ in $\Delta$, we obtain

$$\alpha_{ij} \sum_{k=1}^{n} a_{ik}x_k = \alpha_{ij}b_i.$$

Summing all these equations for $i = 1, 2, \ldots n$ gives

$$\sum_{i=1}^{n} \alpha_{ij} \sum_{k=1}^{n} a_{ik}x_k = \sum_{i=1}^{n} \alpha_{ij}b_i$$

and reversing the order of summation yields

$$\sum_{k=1}^{n} \left( \sum_{i=1}^{n} a_{ik}\alpha_{ij} \right) x_k = \sum_{i=1}^{n} b_i\alpha_{ij}.$$

But by Theorem 1.9 we have

$$\sum_{i=1}^{n} a_{ik}\alpha_{ij} = \begin{cases} 0 & \text{if} \quad k \neq j \\ \Delta & \text{if} \quad k = j \end{cases}$$

and so the sum over $k$ on the left-hand side gives just one term, that for which $k = j$; this means that we have

$$\Delta x_j = \sum_{i=1}^{n} b_i\alpha_{ij}.$$

But the right-hand side of this equation is just the expansion down the $j$th column of the determinant obtained from $\Delta$ by replacing its $j$th column by $b_1, b_2, \ldots b_n$; in other words it is just $\Delta_j$ as defined above.

So we have

$$\Delta x_j = \Delta_j$$

or

$$x_j = \frac{\Delta_j}{\Delta}$$

as required, provided $\Delta \neq 0$. We see also that in this form $\Delta_j = 0$ leads immediately to the result $x_j = 0$. ■

Finally, we prove a special result concerning the expansion of determinants which will be of use to us in Chapter 2.

THEOREM 1.18 (Laplace's expansion)

Consider the determinant of order $n$

$$\Delta = \begin{vmatrix} a_{11} & a_{12} & \cdots & a_{1n} \\ a_{21} & a_{22} & & \\ \cdot & & & \\ \cdot & & & \\ \cdot & & & \\ a_{n1} & \cdots & \cdots & a_{nn} \end{vmatrix}.$$

Choose $m < n$ and consider the first $m$ rows of $\Delta$:

$$\Delta = \begin{vmatrix} a_{11} & a_{12} & \cdots & a_{1n} \\ a_{21} & & & \\ \cdot & & & \\ \cdot & & & \\ \cdot & & & \\ a_{m1} & \cdots & & a_{mn} \\ \hline a_{m+1\,1} & & \cdots & a_{m+1\,n} \\ \cdot & & & \\ \cdot & & & \\ \cdot & & & \\ a_{n1} & & \cdots & a_{nn} \end{vmatrix}$$

By choosing $m$ columns from the $n$ available we can form a determinant of order $m$; this can be done in ${}^nC_m = n!/m!(n-m)!$ different ways. Let us call a typical such determinant $\Delta_i^{(m)}$ and for each $\Delta_i^{(m)}$ let us define a minor as the determinant $\Delta_i^{(n-m)}$ of order $(n-m)$ obtained by omitting the $m$ columns of $\Delta_i^{(m)}$ from the last $(n-m)$ rows of $\Delta$.

Then we have

$$\Delta = \sum_i \pm \Delta_i^{(m)} \Delta_i^{(n-m)}$$

where the sum is over all choices of $\Delta_i^{(m)}$ and the + or – sign has to be chosen in some suitable manner.

*Proof*

We have from Definition 1.4

$$\Delta = \sum_P p(s_1, s_2, \ldots s_n) a_{1s_1} a_{2s_2} a_{3s_3} \ldots a_{ns_n}$$

where $p(s_1, s_2, \ldots s_n)$ is the parity of the permutation $(s_1, s_2, \ldots s_n)$. Thus

$$\Delta = \sum_{P'} p(s_1, s_2, \ldots s_n) \left( \sum_{Q_1} a_{1s_1} a_{2s_2} \ldots a_{ms_m} \right) \left( \sum_{Q_2} a_{m+1s_{m+1}} \ldots a_{ns_n} \right)$$

where the sum $Q_1$ is over all permutations $(s_1, s_2, \ldots s_m)$ of a fixed subset of $(1, 2, \ldots n)$, the sum $Q_2$ is over all permutations $(s_{m+1}, s_{m+2}, \ldots s_n)$ of $(1, 2, \ldots n)$ minus the subset chosen for $(s_1, s_2, \ldots s_m)$ and the sum $P'$ is over all possible choices for the fixed subset $(s_1, s_2, \ldots s_m)$.

But

$$p(s_1, s_2, \ldots s_n) = \pm p(s_1, s_2, \ldots s_m) p(s_{m+1} \ldots s_n)$$

the + or – sign being chosen according to the number of interchanges necessary to bring to natural order the permutation $(s_1, s_2, \ldots s_m)(s_{m+1}, \ldots s_m)$ when each subset is itself in natural order. Hence it follows that

$$\Delta = \sum_{P'} \pm \left( \sum_{Q_1} p(s_1, s_2, \ldots s_m) a_{1s_1} a_{2s_2} \ldots a_{ms_m} \right)$$
$$\times \left( \sum_{Q_2} p(s_{m+1}, \ldots s_n) a_{m+1s_{m+1}} \ldots a_{ns_n} \right)$$
$$= \sum_i \pm \Delta_i^{(m)} \Delta_i^{(n-m)}$$

where $\Delta_i^{(m)}$ and $\Delta_i^{(n-m)}$ are as defined above. ■

This result can be of use in evaluating determinants which contain large blocks of zeros. Its use in practice will be illustrated in the following examples.

*Examples*

1.2.2 Evaluate the determinants

$$\text{(i)}\ \Delta = \begin{vmatrix} 1 & 3 & -1 & 2 \\ 7 & 13 & 5 & 9 \\ 4 & 8 & -6 & 11 \\ 1 & 2 & -1 & 0 \end{vmatrix} \qquad \text{(ii)}\ \Delta = \begin{vmatrix} x & 1 & 1 & 1 \\ 1 & x & 1 & 1 \\ 1 & 1 & x & 1 \\ 1 & 1 & 1 & x \end{vmatrix}.$$

We proceed by removing any common factor from row or column (by Theorem 1.13), then reducing to zero as many elements as possible in a particular row or column (by Theorem 1.15) and then expanding along that row or down that column (by Theorem 1.9). Thus we have:

(i)

$$\Delta = \begin{vmatrix} 0 & 0 & -1 & 0 \\ 12 & 28 & 5 & 19 \\ -2 & -10 & -6 & -1 \\ 0 & -1 & -1 & -2 \end{vmatrix} \qquad \begin{matrix} \text{Col 1} = \text{col 1} + \text{col 3} \\ \text{Col 2} = \text{col 2} + 3 \text{ col 3} \\ \text{Col 4} = \text{col 4} + 2 \text{ col 3} \end{matrix}$$

$$= -1 \begin{vmatrix} 12 & 28 & 19 \\ -2 & -10 & -1 \\ 0 & -1 & -2 \end{vmatrix}$$

$$= - \begin{vmatrix} 12 & 28 & -37 \\ -2 & -10 & 19 \\ 0 & -1 & 0 \end{vmatrix} \qquad \text{Col 3} = \text{col 3} - 2 \text{ col 2}$$

$$= (-1) \begin{vmatrix} 12 & -37 \\ -2 & 19 \end{vmatrix}$$

$$= -2 \begin{vmatrix} 6 & -37 \\ -1 & 19 \end{vmatrix}$$

$$= -2(6 \times 19 - 1 \times 37)$$

$$= -154.$$

(ii)

$$\Delta = \begin{vmatrix} 3+x & 1 & 1 & 1 \\ 3+x & x & 1 & 1 \\ 3+x & 1 & x & 1 \\ 3+x & 1 & 1 & x \end{vmatrix} \qquad \text{Col 1} = \text{col 1} + \text{col 2} + \text{col 3} + \text{col 4}$$

$$= (3+x) \begin{vmatrix} 1 & 1 & 1 & 1 \\ 1 & x & 1 & 1 \\ 1 & 1 & x & 1 \\ 1 & 1 & 1 & x \end{vmatrix}$$

$$= (3+x) \begin{vmatrix} 1 & 1 & 1 & 1 \\ 0 & x-1 & 0 & 0 \\ 0 & 0 & x-1 & 0 \\ 0 & 0 & 0 & x-1 \end{vmatrix} \qquad \begin{matrix} \text{Row 2} = \text{row 2} - \text{row 1} \\ \text{Row 3} = \text{row 3} - \text{row 1} \\ \text{Row 4} = \text{row 4} - \text{row 1} \end{matrix}$$

$$= (3+x) \begin{vmatrix} x-1 & 0 & 0 \\ 0 & x-1 & 0 \\ 0 & 0 & x-1 \end{vmatrix}$$

$$= (3+x)(x-1)^3.$$

1.2.3 Use Laplace's expansion to evaluate

$$\Delta = \begin{vmatrix} 1 & 0 & 0 & 0 & -7 \\ 2 & 0 & 0 & 0 & 3 \\ 6 & -41 & 8 & 6 & 8 \\ 8 & 2 & 7 & 2 & 35 \\ 3 & 3 & 0 & 1 & 28 \end{vmatrix}.$$

In Theorem 1.18 we choose $m = 2$. Then the only non-zero $\Delta_i^{(m)}$ is formed from the 1st and 5th columns (all others contain a complete column of zeros) and we obtain

$$\Delta = \pm \begin{vmatrix} 1 & -7 \\ 2 & 3 \end{vmatrix} \begin{vmatrix} -41 & 8 & 6 \\ 2 & 7 & 2 \\ 3 & 0 & 1 \end{vmatrix}.$$

The sign has to be chosen carefully; we know from the remarks on page 19 that the $2 \times 2$ and $3 \times 3$ determinants will give rise to a term $1 \times 3 \times (-41) \times 7 \times (1)$. In the original determinant this corresponds to $a_{11}a_{25}a_{32}a_{43}a_{54}$ and hence must be signed according to the parity of (1, 5, 2, 3, 4). In fact this is odd:

(1, 5, 2, 3, 4)

(1, 2, 5, 3, 4)

(1, 2, 3, 5, 4)

(1, 2, 3, 4, 5) (3 interchanges).

Thus it must appear with a minus sign and so we have

$$\Delta = - \begin{vmatrix} 1 & -7 \\ 2 & 3 \end{vmatrix} \begin{vmatrix} -41 & 8 & 6 \\ 2 & 7 & 2 \\ 3 & 0 & 1 \end{vmatrix}.$$

But

$$\begin{vmatrix} 1 & -7 \\ 2 & 3 \end{vmatrix} = 3 + 14 = 17$$

and

$$\begin{vmatrix} -41 & 8 & 6 \\ 2 & 7 & 2 \\ 3 & 0 & 1 \end{vmatrix} = \begin{vmatrix} -59 & 8 & 6 \\ -4 & 7 & 2 \\ 0 & 0 & 1 \end{vmatrix} \qquad \text{Col 1} = \text{col 1} - 3 \text{ col 3}$$

$$= \begin{vmatrix} -59 & 8 \\ -4 & 7 \end{vmatrix}$$

$$= -59 \times 7 + 4 \times 8$$

$$= -381.$$

Thus, finally

$$\begin{aligned}\Delta &= -17 \times 381\\ &= -6477.\end{aligned}$$

*Exercises*

1.2.1 Evaluate the determinant

$$\begin{vmatrix} 2 & 2 & 3 & 4 \\ 4 & 5 & -1 & 2 \\ 6 & 3 & 9 & 0 \\ 5 & 2 & 6 & 3 \end{vmatrix}.$$

1.2.2 Find the value of $x$ such that

$$\begin{vmatrix} x-1 & 7 & 9 & 3 \\ 1 & 0 & 2 & 5 \\ 2x+2 & 6 & 8 & 3 \\ -2 & 1 & 1 & 0 \end{vmatrix} = 0.$$

1.2.3 Use Cramer's rule to solve the following system of equations for $x, y, z, u$:

$$\begin{aligned} x + y + z + u &= 2\\ x + y + 2z + 2u &= 4\\ 2x + 2y + z - u &= 2\\ x + 2y + z + 2u &= 1. \end{aligned}$$

1.2.4 Use Laplace's expansion to evaluate the determinant

$$\begin{vmatrix} 4 & 7 & 0 & 3 & 0 \\ 5 & 6 & 0 & 4 & 0 \\ 3 & 2 & 2 & 0 & 1 \\ 1 & 4 & 3 & 0 & 2 \\ 2 & 6 & 1 & 0 & 3 \end{vmatrix}.$$

# 2. Basic Matrix Algebra

## 2.1 Matrices and Basic Definitions

In this section we introduce the fundamental concepts underlying matrix algebra. We define what is meant by a matrix and explain some of the terminology used throughout matrix theory.

DEFINITION 2.1

A *matrix* is a rectangular array of items. □

The 'items' occurring in Definition 2.1 are generally numbers or function values (real or complex), and square brackets are used to denote the extent of the array. Thus

$$\begin{bmatrix} 1 & 2 & 5 & 7 \\ -3 & 0 & 8 & -6 \end{bmatrix}$$

is a single matrix with two rows and four columns, while

$$\begin{bmatrix} 1 & 2 \\ -3 & 0 \end{bmatrix} \begin{bmatrix} 5 & 7 \\ 8 & -6 \end{bmatrix}$$

consists of two matrices side by side, each with two rows and two columns.

DEFINITION 2.2

An individual item occurring in a matrix is called an *element* of the matrix. □

DEFINITION 2.3

If a matrix has $m$ rows and $n$ columns, it is said to be of *order* $m \times n$. If a matrix has both $n$ rows and $n$ columns it is said to be *square of order* $n$. □

DEFINITION 2.4

A matrix with only one column is called a *column vector*; similarly a matrix with only one row is called a *row vector*. □

*Examples*

2.1.1
$$\begin{bmatrix} 1 & 2 & 3 \\ 7 & 8 & -10 \end{bmatrix}$$

is a matrix of order 2 x 3.

2.1.2
$$\begin{bmatrix} a & h & g \\ h & b & f \\ g & f & c \end{bmatrix}$$

is a square matrix of order 3.

2.1.3
$$\begin{bmatrix} \cos\theta \\ \sin\theta \\ \tan\theta \end{bmatrix}$$

is a matrix of order 3 x 1. It is also a column vector.

In any branch of mathematics, symbols are introduced for the quantities of interest, and conventions arise as to the symbolism used. Matrix algebra is no exception, and matrices are usually denoted by bold capital Roman letters—**A**, **B**, **C** etc. Thus we can call the matrices in the examples (1), (2) and (3) above respectively **A**, **B** and **C**; then we should write

$$\mathbf{A} = \begin{bmatrix} 1 & 2 & 3 \\ 7 & 8 & -10 \end{bmatrix},$$

$$\mathbf{B} = \begin{bmatrix} a & h & g \\ h & b & f \\ g & f & c \end{bmatrix},$$

and

$$\mathbf{C} = \begin{bmatrix} \cos\theta \\ \sin\theta \\ \tan\theta \end{bmatrix}.$$

The element appearing in the $i$th row and $j$th column of a matrix **A** we shall denote by $A_{ij}$. We are thus putting suffices on the symbol representing the matrix to indicate the individual elements of the matrix; note that the *first* suffix refers to the *row* in which the element appears while the *second* suffix refers to the *column.*

An exception to the above convention often occurs in the case of row and column vectors. Here bold lower case letters are used for the vector, and individual elements are denoted by a single suffix. This means that if **u** is a column vector, $u_i$ is the element in the $i$th row whereas if $\mathbf{v}$ is a row vector, $v_i$ is the element in the $i$th column. Thus if $\mathbf{u}$ is a column vector with $n$ rows we have

$$\mathbf{u} = \begin{bmatrix} u_1 \\ u_2 \\ u_3 \\ \cdot \\ \cdot \\ \cdot \\ u_n \end{bmatrix}$$

and if $\mathbf{v}$ is a row vector with $n$ columns we have

$$\mathbf{v} = [v_1 \quad v_2 \quad v_3 \quad \dots \quad v_n].$$

If $\mathbf{A}$ is a square matrix, the elements $A_{ij}$ with $i = j$ are called diagonal elements (or elements on the diagonal). Thus if $\mathbf{A}$ is square of order 3 the diagonal elements are those shown underlined below:

$$\begin{bmatrix} \underline{A_{11}} & A_{12} & A_{13} \\ A_{21} & \underline{A_{22}} & A_{23} \\ A_{31} & A_{32} & \underline{A_{33}} \end{bmatrix}$$

*Examples*

2.1.4 If

$$\mathbf{A} = \begin{bmatrix} 1 & 2 & 3 \\ 7 & 8 & 10 \end{bmatrix}$$

use the double suffix notation to write down all the elements. We have

$$A_{11} = 1 \qquad A_{12} = 2 \qquad A_{13} = 3$$
$$A_{21} = 7 \qquad A_{21} = 8 \qquad A_{23} = 10.$$

2.1.5 If $\mathbf{X}$ is a square matrix of order 3 and if $X_{ij} = 2i - j$ obtain the matrix $\mathbf{X}$.

We have

$$\mathbf{X} = \begin{bmatrix} X_{11} & X_{12} & X_{13} \\ X_{21} & X_{22} & X_{23} \\ X_{31} & X_{32} & X_{33} \end{bmatrix}$$

since $\mathbf{X}$ is square of order 3.

Thus

$$\mathbf{X} = \begin{bmatrix} 2 \times 1 - 1 & 2 \times 1 - 2 & 2 \times 1 - 3 \\ 2 \times 2 - 1 & 2 \times 2 - 2 & 2 \times 2 - 3 \\ 2 \times 3 - 1 & 2 \times 3 - 2 & 2 \times 3 - 3 \end{bmatrix}$$

since $X_{ij} = 2i - j$

$$= \begin{bmatrix} 1 & 0 & -1 \\ 3 & 2 & 1 \\ 5 & 4 & 3 \end{bmatrix}.$$

2.1.6 If $\mathbf{u}$ is the column vector

$$\begin{bmatrix} 1 \\ \cos^2 \theta \\ \sin^2 \theta \end{bmatrix}$$

evaluate $u_1 + u_2 + u_3$.

Using the single suffix notation for vectors we have

$$u_1 = 1, \qquad u_2 = \cos^2\theta, \qquad u_3 = \sin^2\theta$$

and hence

$$\begin{aligned} u_1 + u_2 + u_3 &= 1 + \cos^2\theta + \sin^2\theta \\ &= 2. \end{aligned}$$

It is also common to use lower case letters to denote individual matrix elements; this means that the element in the $i$th row and $j$th column of the matrix $\mathbf{A}$ may also be denoted by $a_{ij}$ as well as $A_{ij}$. Both notations have their advantages, and we shall use both in this book. Thus if $\mathbf{A}$ is a matrix of order $m \times n$ we have

$$\mathbf{A} = \begin{bmatrix} a_{11} & a_{12} & \cdots & a_{1n} \\ a_{21} & a_{22} & \cdots & a_{2n} \\ \cdot & & & \\ \cdot & & & \\ \cdot & & & \\ a_{m1} & \cdots & \cdots & a_{mn} \end{bmatrix}$$

and also $A_{ij} = a_{ij}$.

DEFINITION 2.5

Two matrices are said to be *equal* if and only if they are of the same order and corresponding elements are equal. □

The 'if and only if' in the above definition means that:

(*a*) if $\mathbf{A}$ and $\mathbf{B}$ are matrices of the same order and $A_{ij} = B_{ij}$ for all appropriate $i$ and $j$ then $\mathbf{A} = \mathbf{B}$; and

(*b*) if $\mathbf{A}$ and $\mathbf{B}$ are matrices such that $\mathbf{A} = \mathbf{B}$ then they are of the same order and $A_{ij} = B_{ij}$ for all appropriate $i$ and $j$.

*Examples*

2.1.7 If

$$\mathbf{A} = \begin{bmatrix} 1 & 2 \\ 7 & 8 \end{bmatrix} \quad \text{and} \quad \mathbf{B} = \begin{bmatrix} 1 & 7 \\ 2 & 8 \end{bmatrix}$$

then, despite the fact that **A** and **B** are arrays of the same quantities, $\mathbf{A} \neq \mathbf{B}$. (For equality *corresponding* elements must be equal.)

2.1.8 If

$$\mathbf{A} = \begin{bmatrix} 1 & 2 \\ 7 & 8 \end{bmatrix} \quad \text{and} \quad \mathbf{B} = \mathbf{A}$$

then we must have

$$\mathbf{B} = \begin{bmatrix} 1 & 2 \\ 7 & 8 \end{bmatrix}.$$

2.1.9 If **A** is a square matrix of order 3 such that $A_{ij} = \cos^2 i + \sin^2 j$, and $\mathbf{B} = \mathbf{A}$ show that $B_{22} = 1$.

Since **A** is of order 3 x 3 then so is **B** (by Definition 2.5). Hence it is meaningful to talk of $B_{22}$. Then

$$\begin{aligned} B_{22} &= A_{22} \quad \text{(again by Definition 2.5)} \\ &= \cos^2 2 + \sin^2 2 \\ &= 1. \end{aligned}$$

(N.B. We *cannot* say that $B_{44} = A_{44} = \cos^2 4 + \sin^2 4 = 1$ since $A_{44}$ and hence $B_{44}$ are undefined.)

*Exercises*

2.1.1 If **A** is a matrix of order 3 x 4 such that $A_{ij} = |i - j|$ write down the matrix **A**.

2.1.2 If

$$\mathbf{A} = \begin{bmatrix} -1 & 2 & 5 \\ 8 & 9 & -30 \\ 2 & -7 & 81 \end{bmatrix},$$

$$\mathbf{B} = \begin{bmatrix} x^2 + y^2 & 2xy \\ -2xy & x^2 \\ y^2 & x^2 - y^2 \end{bmatrix},$$

$$\mathbf{C} = \begin{bmatrix} e^\theta \\ \cosh\theta \\ \sinh\theta \end{bmatrix}, \quad \text{and} \quad \mathbf{u} = [\alpha \quad \beta \quad \gamma],$$

write down the following matrix elements:

$$A_{13}, A_{22}, A_{31}, B_{12}, B_{13}, B_{32}, C_{11}, C_{31}, u_2, u_3.$$

2.1.3 If

$$\mathbf{A} = \begin{bmatrix} x+y & 3 \\ -2 & x-y \end{bmatrix},$$

$$\mathbf{B} = \begin{bmatrix} 3 & 4 \\ -3 & 1 \end{bmatrix},$$

and

$$\mathbf{C} = \begin{bmatrix} 3 & 3 \\ -2 & 1 \end{bmatrix},$$

find, if possible, the values of $x$ and $y$ such that: (i) $\mathbf{A} = \mathbf{B}$; and (ii) $\mathbf{A} = \mathbf{C}$.

## 2.2 Basic Matrix Operations

In this section we define the basic operations involving matrices, such as addition and multiplication, and investigate some of the consequences of these definitions.

DEFINITION 2.6 (Multiplication of a matrix by a number)

If $\mathbf{A}$ is a matrix and $c$ a number then the product of $\mathbf{A}$ with $c$ is defined to the matrix of the same order as $\mathbf{A}$ obtained from $\mathbf{A}$ by multiplying every element of $\mathbf{A}$ by $c$. If we denote the product by $c\mathbf{A}$ then in terms of matrix elements we have

$$(c\mathbf{A})_{ij} = cA_{ij}.$$ □

The product as defined is independent of order; if we use the same notation for product as in ordinary algebra (x) this means that if $c$ is a number and $\mathbf{A}$ is a matrix then $c \times \mathbf{A} = \mathbf{A} \times c$. Again as in ordinary algebra we often simply omit the multiplication sign (as indicated in the definition) and we have $c\mathbf{A} = \mathbf{A}c$.

*Examples*

2.2.1 If

$$\mathbf{A} = \begin{bmatrix} 2 & 4 \\ 9 & 1 \end{bmatrix}$$

obtain $3\mathbf{A}$ and $\mathbf{A}5$. We have

$$3\mathbf{A} = 3 \times \begin{bmatrix} 2 & 4 \\ 9 & 1 \end{bmatrix} = \begin{bmatrix} 3 \times 2 & 3 \times 4 \\ 3 \times 9 & 3 \times 1 \end{bmatrix} = \begin{bmatrix} 6 & 12 \\ 27 & 3 \end{bmatrix},$$

and

$$\mathbf{A}5 = \mathbf{A} \times 5 = 5 \times \mathbf{A} = 5 \times \begin{bmatrix} 2 & 4 \\ 9 & 1 \end{bmatrix} = \begin{bmatrix} 10 & 20 \\ 45 & 5 \end{bmatrix}.$$

2.2.2 If

$$\mathbf{A} = \begin{bmatrix} \alpha & \beta \\ \gamma & \delta \end{bmatrix}, \qquad \mathbf{X} = \begin{bmatrix} p & q \\ r & s \end{bmatrix}$$

and $\mathbf{A} = -\lambda\mathbf{X}$, obtain the relationships between the matrix elements. $\mathbf{A} = -\lambda\mathbf{X}$ means that

$$\begin{bmatrix} \alpha & \beta \\ \gamma & \delta \end{bmatrix} = (-\lambda) \times \begin{bmatrix} p & q \\ r & s \end{bmatrix} = \begin{bmatrix} -\lambda p & -\lambda q \\ -\lambda r & -\lambda s \end{bmatrix}$$

and hence $\alpha = -\lambda p$, $\beta = -\lambda q$, $\gamma = -\lambda r$ and $\delta = -\lambda s$.

DEFINITION 2.7 (Addition of two matrices)

If **A** and **B** are two matrices, their sum is defined if and only if they are of the same order. In that case the sum is defined to be the matrix of the same order as **A** and **B** obtained by adding corresponding elements of **A** and **B**; it is denoted by $\mathbf{A} + \mathbf{B}$. In terms of elements this means that $(\mathbf{A} + \mathbf{B})_{ij} = A_{ij} + B_{ij}$. □

DEFINITION 2.8 (Subtraction of two matrices)

If **A** and **B** are two matrices their difference is defined if and only if they are of the same order. In that case we may utilize Definitions 2.6 and 2.7; denoting the difference by $\mathbf{A} - \mathbf{B}$ we may define it by $\mathbf{A} - \mathbf{B} = \mathbf{A} + (-1) \times \mathbf{B}$. □

It follows that in terms of elements we have

$$\begin{aligned} (\mathbf{A} - \mathbf{B})_{ij} &= A_{ij} + (-1 \times \mathbf{B})_{ij} && \text{(by Definition 2.7)} \\ &= A_{ij} - B_{ij} && \text{(by Definition 2.6)} \end{aligned}$$

which is consistent with what we intuitively expect.

*Examples*

2.2.3 If

$$\mathbf{A} = \begin{bmatrix} 3 & 6 \\ 2 & 4 \\ 1 & 2 \end{bmatrix} \quad \text{and} \quad \mathbf{B} = \begin{bmatrix} -1 & 2 \\ 7 & 4 \\ 0 & 5 \end{bmatrix}$$

evaluate $\mathbf{A} + \mathbf{B}$, $\mathbf{A} + 2\mathbf{B}$ and $-\mathbf{A} + 3\mathbf{B}$.

Since **A** and **B** are of the same order (3 x 2) it is possible to add them. Then we have

$$\mathbf{A} + \mathbf{B} = \begin{bmatrix} 3 & 6 \\ 2 & 4 \\ 1 & 2 \end{bmatrix} + \begin{bmatrix} -1 & 2 \\ 7 & 4 \\ 0 & 5 \end{bmatrix} = \begin{bmatrix} 3-1 & 6+2 \\ 2+7 & 4+4 \\ 1+0 & 2+5 \end{bmatrix} = \begin{bmatrix} 2 & 8 \\ 9 & 8 \\ 1 & 7 \end{bmatrix},$$

$$\mathbf{A} + 2\mathbf{B} = \begin{bmatrix} 3 & 6 \\ 2 & 4 \\ 1 & 2 \end{bmatrix} + \begin{bmatrix} -2 & 4 \\ 14 & 8 \\ 0 & 10 \end{bmatrix} = \begin{bmatrix} 1 & 10 \\ 16 & 12 \\ 1 & 12 \end{bmatrix},$$

and

$$-\mathbf{A} + 3\mathbf{B} = \begin{bmatrix} -3 & -6 \\ -2 & -4 \\ -1 & -2 \end{bmatrix} + \begin{bmatrix} -3 & 6 \\ 21 & 12 \\ 0 & 15 \end{bmatrix} = \begin{bmatrix} -6 & 0 \\ 19 & 8 \\ -1 & 13 \end{bmatrix}.$$

2.2.4 If

$$\mathbf{X} = \begin{bmatrix} \alpha \\ \beta \end{bmatrix}, \quad \mathbf{Y} = \begin{bmatrix} \beta \\ \alpha \end{bmatrix}, \quad \mathbf{Z} = \begin{bmatrix} 4 \\ 3 \end{bmatrix},$$

and $\mathbf{X} + 2\mathbf{Y} = \mathbf{Z}$, find $\alpha$ and $\beta$.
Since $\mathbf{X} + 2\mathbf{Y} = \mathbf{Z}$ we have

$$\begin{bmatrix} \alpha \\ \beta \end{bmatrix} + 2\begin{bmatrix} \beta \\ \alpha \end{bmatrix} = \begin{bmatrix} 4 \\ 3 \end{bmatrix}$$

so that

$$\begin{bmatrix} \alpha + 2\beta \\ \beta + 2\alpha \end{bmatrix} = \begin{bmatrix} 4 \\ 3 \end{bmatrix}.$$

Hence

$$\alpha + 2\beta = 4,$$

and

$$\beta + 2\alpha = 3.$$

Solving these simultaneous equations in the usual way gives

$$\alpha = 2/3 \quad \text{and} \quad \beta = 5/3.$$

In the algebra of real and complex numbers, the order in which two numbers occur in addition does not matter; in other words, $a + b = b + a$. This holds true for matrix addition as well, and constitutes the so-called *commutative law* for matrix addition. The relevant result is proved in the following theorem.

THEOREM 2.1

If $\mathbf{A}$ and $\mathbf{B}$ are matrices of the same order then $\mathbf{A} + \mathbf{B} = \mathbf{B} + \mathbf{A}$.

*Proof*

Since $\mathbf{A}$ and $\mathbf{B}$ are of the same order, their sum exists and is also of the same order. Also, from Definition 2.7 we have

$$(\mathbf{A} + \mathbf{B})_{ij} = A_{ij} + B_{ij}$$

and

$$\begin{aligned} (\mathbf{B} + \mathbf{A})_{ij} &= B_{ij} + A_{ij} \\ &= A_{ij} + B_{ij} \end{aligned}$$

(since these quantities are numbers, and so the order of addition does not matter). Thus $(\mathbf{A} + \mathbf{B})_{ij} = (\mathbf{B} + \mathbf{A})_{ij}$ and hence from Definition 2.5 we have $\mathbf{A} + \mathbf{B} = \mathbf{B} + \mathbf{A}$. ■

So far we have only presented rules for adding two matrices. The sum of three matrices **A**, **B** and **C**, all of the same order, could thus be obtained by either

(*a*) adding **A** and **B** first to obtain a new matrix, and then adding **C**; or
(*b*) adding **B** and **C** to obtain a new matrix, and then adding **A**.

These two ways are equivalent to forming $(\mathbf{A} + \mathbf{B}) + \mathbf{C}$ and $\mathbf{A} + (\mathbf{B} + \mathbf{C})$ where the brackets indicate the operation to be performed first. That the two ways give the same result is the content of the following theorem, the so-called *associative law* of matrix addition.

THEOREM 2.2

If **A**, **B** and **C** are matrices of the same order then

$$(\mathbf{A} + \mathbf{B}) + \mathbf{C} = \mathbf{A} + (\mathbf{B} + \mathbf{C}).$$

*Proof*

Let us denote $(\mathbf{A} + \mathbf{B}) + \mathbf{C}$ by **L** and $\mathbf{A} + (\mathbf{B} + \mathbf{C})$ by **R**. We then prove the result by showing that $L_{ij} = R_{ij}$. (Both **L** and **R** are obviously of the same order as **A**, **B** and **C**.) Now $\mathbf{L} = (\mathbf{A} + \mathbf{B}) + \mathbf{C}$ and hence

$$\begin{aligned} L_{ij} &= (\mathbf{A} + \mathbf{B})_{ij} + C_{ij} && \text{(by Definition 2.7)} \\ &= A_{ij} + B_{ij} + C_{ij} && \text{(again by Definition 2.7).} \end{aligned}$$

Also $\mathbf{R} = \mathbf{A} + (\mathbf{B} + \mathbf{C})$ so that

$$\begin{aligned} R_{ij} &= A_{ij} + (\mathbf{B} + \mathbf{C})_{ij} \\ &= A_{ij} + B_{ij} + C_{ij}. \end{aligned}$$

Thus the required result is proved.

From the above theorem it follows that the brackets in $\mathbf{A} + (\mathbf{B} + \mathbf{C})$ and $(\mathbf{A} + \mathbf{B}) + \mathbf{C}$ are unnecessary and we may write $\mathbf{A} + \mathbf{B} + \mathbf{C}$ with no ambiguity in result.

*Example*

2.2.5 If

$$\mathbf{A} = \begin{bmatrix} 3 & 4 \\ 6 & 2 \end{bmatrix}, \quad \mathbf{B} = \begin{bmatrix} -2 & 1 \\ 4 & 8 \end{bmatrix}$$

and

$$\mathbf{C} = \begin{bmatrix} 6 & 7 \\ 4 & 1 \end{bmatrix},$$

find $\mathbf{A} + \mathbf{B} + \mathbf{C}$ and $\mathbf{A} + 3\mathbf{B} - 2\mathbf{C}$.

We have

$$\mathbf{A}+\mathbf{B}+\mathbf{C}=\begin{bmatrix}3 & 4\\ 6 & 2\end{bmatrix}+\begin{bmatrix}-2 & 1\\ 4 & 8\end{bmatrix}+\begin{bmatrix}6 & 7\\ 4 & 1\end{bmatrix}$$

$$=\begin{bmatrix}3-2+6 & 4+1+7\\ 6+4+4 & 2+8+1\end{bmatrix}$$

$$=\begin{bmatrix}7 & 12\\ 14 & 11\end{bmatrix}$$

and

$$\mathbf{A}+3\mathbf{B}-2\mathbf{C}=\begin{bmatrix}3 & 4\\ 6 & 2\end{bmatrix}+\begin{bmatrix}-6 & 3\\ 12 & 24\end{bmatrix}-\begin{bmatrix}12 & 14\\ 8 & 2\end{bmatrix}$$

$$=\begin{bmatrix}3-6-12 & 4+3-14\\ 6+12-8 & 2+24-2\end{bmatrix}$$

$$=\begin{bmatrix}-15 & -7\\ 10 & 24\end{bmatrix}.$$

The definitions presented so far have been reasonably straightforward and intuitively acceptable. We are now going to introduce the idea of the product of two matrices. At first sight the definition seems rather remote from usefulness, but in fact turns out to be just what is needed to simplify the rather tedious algebra which arises when linear systems are studied without the use of matrix theory.

DEFINITION 2.9 (Product of two matrices)

If **A** and **B** are two matrices the product of **A** with **B** is defined if and only if **A** has the same number of columns as **B** has rows; for example **A** can be of order $m \times n$ and **B** of order $n \times p$. In such a situation the product matrix is defined in such a way that it is of order $m \times p$ and is denoted by $\mathbf{A} \times \mathbf{B}$ or **AB**. The individual elements of **AB** are obtained from the following rule: the element in the $i$th row and $j$th column of **AB** is given by taking the sum of the products of corresponding elements from the $i$th row of **A** and the $j$th column of **B**. In other words

$$(\mathbf{AB})_{ij} = A_{i1}B_{1j} + A_{i2}B_{2j} + \ldots + A_{in}B_{nj}$$

$$= \sum_{r=1}^{n} A_{ir}B_{rj}.$$

(Note that the fact that **A** is of order $m \times n$ and **B** of order $n \times p$ guarantees that there is the same number of elements ($n$) in a row of **A** as there is in a column of **B**.) □

The above definition may be thought of as the 'row into column rule'; that is, the element $(\mathbf{AB})_{ij}$ is obtained by 'multiplying' the $i$th row of **A** into the $j$th column of **B**.

Notice also that the above definition gives the product of **A** with **B**, that is **AB**. Order is important, and we shall see that the product **BA** of **B** with **A** is in general different from the product **AB** of **A** with **B**. In fact the existence of **AB** does not even guarantee the existence of **BA**.

*Examples*

2.2.6 If

$$\mathbf{X} = \begin{bmatrix} 4 & 2 \\ -7 & 6 \end{bmatrix}, \qquad \mathbf{Y} = \begin{bmatrix} 1 & 3 \\ 8 & 15 \\ 19 & 12 \end{bmatrix}$$

and

$$\mathbf{Z} = \begin{bmatrix} 1 & 10 & 8 \\ 9 & 3 & 4 \\ 8 & 2 & 1 \\ -7 & 6 & 3 \end{bmatrix}$$

determine (i) which of the following products are defined and (ii) the order of those which are defined:

**XY, YX, YZ, ZY.**

We make use of the general result contained in Definition 2.9 that **AB** is defined if **A** is of order $m \times n$ and **B** is of order $n \times p$ (for some $m, n, p$); then also by Definition 2.9 **AB** is of order $m \times p$.

**XY**: **X** is of order 2 x 2 and **Y** is of order 3 x 2; hence **XY** is *not* defined.

**YX**: **Y** is of order 3 x 2 and **X** is of order 2 x 2; hence **YX** *is* defined and is of order 3 x 2.

**YZ**: **Y** is of order 3 x 2 and **Z** is of order 4 x 3; hence **YZ** is *not* defined.

**ZY**: **Z** is of order 4 x 3 and **Y** is of order 3 x 2; hence **ZY** *is* defined and is of order 4 x 2.

2.2.7 If

$$\mathbf{P} = \begin{bmatrix} a & b & c \\ d & e & f \end{bmatrix} \quad \text{and} \quad \mathbf{R} = \begin{bmatrix} p & s \\ q & t \\ r & u \end{bmatrix}$$

evaluate $(\mathbf{PR})_{11}$ and $(\mathbf{RP})_{23}$. First we verify that **PR** and **RP** exist;

**PR**: **P** is of order 2 x 3 and **R** is of order 3 x 2; hence **PR** exists and is of order 2 x 2.

**RP**: **R** is of order 3 x 2 and **P** is of order 2 x 3; hence **RP** exists and is of order 3 x 3.

Hence it is meaningful to talk of both $(\mathbf{PR})_{11}$ and $(\mathbf{RP})_{23}$. From Definition 2.9 we have

$$(\mathbf{PR})_{11} = P_{11}R_{11} + P_{12}R_{21} + P_{13}R_{31} \qquad \text{('row 1 of } \mathbf{P} \text{ into column 1 of } \mathbf{R}\text{')}$$
$$= ap + bq + cr$$

and

$$(\mathbf{RP})_{23} = R_{21}P_{13} + R_{22}P_{23} \qquad \text{('row 2 of } \mathbf{R} \text{ into column 3 of } \mathbf{P}\text{')}$$
$$= qc + tf.$$

2.2.8 For the matrices **P** and **R** in Example 2.2.7 determine **PR** and **RP**. We have

$$\mathbf{PR} = \begin{bmatrix} a & b & c \\ d & e & f \end{bmatrix} \begin{bmatrix} p & s \\ q & t \\ r & u \end{bmatrix}$$

$$= \begin{bmatrix} ap + bq + cr & as + bt + cu \\ dp + eq + fr & ds + et + fu \end{bmatrix},$$

(making use of the 'row into column rule' for the calculation of all elements) and

$$\mathbf{RP} = \begin{bmatrix} p & s \\ q & t \\ r & u \end{bmatrix} \begin{bmatrix} a & b & c \\ d & e & f \end{bmatrix}$$

$$= \begin{bmatrix} pa + sd & pb + se & pc + sf \\ qa + td & qb + te & qc + tf \\ ra + ud & rb + ue & rc + uf \end{bmatrix}$$

(again using the 'row into column rule').

We note again that in this example **PR** and **RP** are both defined, but are of different orders.

2.2.9 If

$$\mathbf{L} = \begin{bmatrix} 1 & -2 \\ 5 & 0 \end{bmatrix} \quad \text{and} \quad \mathbf{M} = \begin{bmatrix} 2 & 7 \\ -1 & 0 \end{bmatrix}$$

determine **LM** and **ML**.

Since **L** and **M** are both of order 2 x 2, both **LM** and **ML** exist and are of order 2 x 2. We have

$$\mathbf{LM} = \begin{bmatrix} 1 & -2 \\ 5 & 0 \end{bmatrix} \begin{bmatrix} 2 & 7 \\ -1 & 0 \end{bmatrix}$$

$$= \begin{bmatrix} 1 \times 2 + (-2) \times (-1) & 1 \times 7 + (-2) \times 0 \\ 5 \times 2 + 0 \times (-1) & 5 \times 7 + 0 \times 0 \end{bmatrix}$$

$$= \begin{bmatrix} 4 & 7 \\ 10 & 35 \end{bmatrix},$$

and

$$\mathbf{ML} = \begin{bmatrix} 2 & 7 \\ -1 & 0 \end{bmatrix} \begin{bmatrix} 1 & -2 \\ 5 & 0 \end{bmatrix}$$

$$= \begin{bmatrix} 2 \times 1 + 7 \times 5 & 2 \times (-2) + 7 \times 0 \\ -1 \times 1 + 0 \times 5 & -1 \times (-2) + 0 \times 0 \end{bmatrix}$$

$$= \begin{bmatrix} 37 & -4 \\ -1 & 2 \end{bmatrix}.$$

In this example we note that **LM** and **ML** are of the same order, but **LM** $\neq$ **ML**.

2.2.10 If

$$\mathbf{G} = \begin{bmatrix} 2 & 0 \\ 0 & 3 \end{bmatrix} \quad \text{and} \quad \mathbf{H} = \begin{bmatrix} 1 & 0 \\ 0 & 2 \end{bmatrix}$$

determine **GH** and **HG**. We have

$$\mathbf{GH} = \begin{bmatrix} 2 & 0 \\ 0 & 3 \end{bmatrix} \begin{bmatrix} 1 & 0 \\ 0 & 2 \end{bmatrix}$$

$$= \begin{bmatrix} 2 \times 1 + 0 \times 0 & 2 \times 0 + 0 \times 2 \\ 0 \times 1 + 3 \times 0 & 0 \times 0 + 3 \times 2 \end{bmatrix}$$

$$= \begin{bmatrix} 2 & 0 \\ 0 & 6 \end{bmatrix}$$

and

$$\mathbf{HG} = \begin{bmatrix} 1 & 0 \\ 0 & 2 \end{bmatrix} \begin{bmatrix} 2 & 0 \\ 0 & 3 \end{bmatrix}$$

$$= \begin{bmatrix} 1 \times 2 + 0 \times 0 & 1 \times 0 + 0 \times 3 \\ 0 \times 2 + 2 \times 0 & 0 \times 0 + 2 \times 3 \end{bmatrix}$$

$$= \begin{bmatrix} 2 & 0 \\ 0 & 6 \end{bmatrix}.$$

We note that in this example **GH** = **HG**.

We have noted both in general terms and in the above examples that **AB** and **BA** are in general different. Some results about the connection between **AB** and **BA** are contained in the following theorem.

THEOREM 2.3

(i) If **A** and **B** are matrices then the products **AB** and **BA** will both exist only if **A** is of order $m \times n$ and **B** is of order $n \times m$ for some $m, n$.

(ii) If the condition in (i) above is satisfied **AB** is square of order $m$ and **BA** is square of order $n$.

(iii) If **A** and **B** are both square of order $n$ then so are both **AB** and **BA**. However, in general we shall still have **AB** $\neq$ **BA**.

*Proof*

(i) Suppose that **A** is of order $m \times n$, **B** is of order $r \times s$ and that both **AB** and **BA** exist.

**AB**: **A** is of order $m \times n$; **B** is of order $r \times s$; hence we must have $r = n$ (by Definition 2.9)

**BA**: **B** is of order $r \times s$; **A** is of order $m \times n$; hence we must have $s = m$ (by Definition 2.9).

Thus $r = n$ and $s = m$ so that **A** is of order $m \times n$ and **B** is of order $n \times m$, as required.

(ii) **A** is of order $m \times n$ and **B** is of order $n \times m$; hence by Definition 2.9 **AB** must be of order $m \times m$ and **BA** must be of order $n \times n$.

(iii) That **AB** and **BA** are both square of order $n$ follows from (ii) above by choosing $m = n$; that **AB** $\neq$ **BA** in general is shown by the counter-example in Example 2.2.9 above.

In the algebra of real and complex numbers we know that multiplication is *commutative.* That is to say, if $a$ and $b$ are numbers then $ab = ba$—order of multiplication is unimportant. Theorem 2.3 shows that in general matrix multiplication is not commutative. That is, in general **AB** $\neq$ **BA** for matrices **A** and **B**. If, however, **A** and **B** are matrices such that **AB** = **BA** then **A** and **B** are said to *commute* or to be *commutative.* Theorem 2.3 shows that if **A** and **B** commute then it is necessary that they should be square and of the same order. Example 2.2.10 above shows that commutative matrices do in fact exist.

It is now natural to ask whether or not matrix multiplication obeys any of the other laws of ordinary algebra. These are the *associative* laws saying that the order in which multiplications are performed is unimportant (i.e. $(ab)c = a(bc)$) and the law of *distribution of multiplication over addition* which says that $a(b + c) = ab + ac$. In fact both these laws hold for matrix multiplication and are proved in the following theorems.

THEOREM 2.4

If **A**, **B** and **C** are matrices such that one of the products **(AB)C** and **A(BC)** exists, then so does the other and in fact both are equal (i.e. **(AB)C** = **A(BC)**).

*Proof*

We first show that if **(AB)C** exists then so does **A(BC)**. Suppose that **A** is of order $m \times n$, **B** is of order $p \times q$ and **C** is of order $r \times s$. Then we must have:

**AB**: **A** is of order $m \times n$; **B** is of order $p \times q$; hence $n = p$ and **AB** is of order $m \times q$.

**(AB)C**: **AB** is of order $m \times q$; **C** is of order $r \times s$; hence $q = r$ and **(AB)C** is of order $m \times s$.

Thus we have shown that the existence of **(AB)C** implies that

**A** is of order $m \times n$
**B** is of order $n \times r$
**C** is of order $r \times s$

and

**(AB)C** is of order $m \times s$.

We now show that these results guarantee the existence of **A(BC)**:

**BC**: **B** is of order $n \times r$; **C** is of order $r \times s$; hence **BC** exists and is of order $n \times s$.

**A(BC)**: **A** is of order $m \times n$; **BC** is of order $n \times s$; hence **A(BC)** exists and is of order $m \times s$.

Thus we have shown that the existence of **(AB)C** guarantees the existence of **A(BC)** and that both are of the same order; similarly we could show that the existence of **A(BC)** would guarantee the existence of **(AB)C** of the same order.

We must now show that **(AB)C = A(BC)**. We do this by proving that the elements in the $i$th row and $j$th column of both products are equal. Then by Definition 2.5 the matrices must be equal. Remembering that

**A** is of order $m \times n$
**B** is of order $n \times q$
**C** is of order $q \times s$
**AB** is of order $m \times q$

and **BC** is of order $n \times s$

we have

$$\{(\mathbf{AB})\mathbf{C}\}_{ij} = \sum_{k=1}^{q} (\mathbf{AB})_{ik} C_{kj} \qquad \text{(by Definition 2.9)}$$

$$= \sum_{k=1}^{q} \left( \sum_{l=1}^{n} A_{il} B_{lk} \right) C_{kj} \qquad \text{(again by Definition 2.9)}$$

$$= \sum_{k=1}^{q} \sum_{l=1}^{n} A_{il} B_{lk} C_{kj}.$$

Also we have

$$\{\mathbf{A}(\mathbf{BC})\}_{ij} = \sum_{k=1}^{n} \mathbf{A}_{ik} (\mathbf{BC})_{kj} \qquad \text{(by Definition 2.9)}$$

$$= \sum_{k=1}^{n} A_{ik} \left( \sum_{l=1}^{q} B_{kl} C_{lj} \right) \qquad \text{(again by Definition 2.9)}$$

$$= \sum_{k=1}^{n} \sum_{l=1}^{q} A_{ik} B_{kl} C_{lj}$$

$$= \sum_{l=1}^{n} \sum_{k=1}^{q} A_{il} B_{lk} C_{kj} \qquad \text{(interchanging the dummy summation indices } k \text{ and } l\text{)}$$

$$= \sum_{k=1}^{q} \sum_{l=1}^{n} A_{il} B_{lk} C_{kj}.$$

Hence $\{(\mathbf{AB})\mathbf{C}\}_{ij} = \{\mathbf{A}(\mathbf{BC})\}_{ij}$ and so the required result is proved. ∎

Note that as a result of the above theorem the brackets in **(AB)C** and **A(BC)** are redundant; it is meaningful to write **ABC** without specifying in which of the two possible ways the product is formed.

THEOREM 2.5

If **A**, **B** and **C** are matrices such that **AB**, **AC** and **B** + **C** all exist then **A(B + C)** also exists and **A(B + C) = AB + AC**.

*Proof*

Since **B** + **C** exists, **B** and **C** must be of the same order (see Definition 2.7). Suppose that **A** is of order $m \times n$; then for **AB** (and **AC**) to exist **B** and **C** must be of order $n \times p$. So therefore **B** + **C** is also of order $n \times p$ and hence **A(B + C)** is of the same order ($m \times p$) as **AB** and **AC**.

To prove equality of matrices we prove equality of elements. We have

$$
\begin{aligned}
\{\mathbf{A}(\mathbf{B}+\mathbf{C})\}_{ij} &= \sum_{k=1}^{n} A_{ik}(\mathbf{B}+\mathbf{C})_{kj} && \text{(by Definition 2.9)}\\
&= \sum_{k=1}^{n} A_{ik}(B_{kj}+C_{kj}) && \text{(by Definition 2.7)}\\
&= \sum_{k=1}^{n} A_{ik}B_{kj} + \sum_{k=1}^{n} A_{ik}C_{kj}
\end{aligned}
$$

and

$$
\begin{aligned}
(\mathbf{AB}+\mathbf{AC})_{ij} &= (\mathbf{AB})_{ij} + (\mathbf{AC})_{ij} && \text{(by Definition 2.7)}\\
&= \sum_{k=1}^{n} A_{ik}B_{kj} + \sum_{k=1}^{n} A_{ik}C_{kj} && \text{(by Definition 2.9).}
\end{aligned}
$$

Hence we have $\{\mathbf{A}(\mathbf{B}+\mathbf{C})\}_{ij} = (\mathbf{AB}+\mathbf{AC})_{ij}$ and so it follows that **A(B + C) = AB + AC**.

Strictly speaking, the above theorem shows that matrix multiplication is *left* distributive over addition (i.e. **A(B + C) = AB + AC**). Matrix multiplication is also *right* distributive over addition as the next theorem shows (i.e. **(A + B)C = AC + BC**). To say that matrix multiplication is distributive over addition means that it is both left distributive and right distributive.

THEOREM 2.6

If **A**, **B** and **C** are matrices such that **AC**, **BC** and **A** + **B** all exist then **(A + B)C** also exists and **(A + B)C = AC + BC**.

*Proof*

This is exactly similar to the proof of Theorem 2.5, so is not given here.

The associativity of multiplication (Theorem 2.4) means that products of several matrices may be taken without the necessity of inserting brackets to specify the order in which the products are formed. Thus, provided the matrices are of suitable orders it is meaningful to talk of the matrix product **ABC . . . LMN**. The condition on the orders is that if **A** is of order $m \times n$, **B** must be of order $n \times p$, **C** of order $p \times q$ and so on.

We might ask under what conditions it is possible to multiply a matrix by itself. Considering a matrix **A**, the existence of $\mathbf{A} \times \mathbf{A}$ means that the number of columns of **A** (the first matrix in the product) must equal the number of rows of **A** (the second matrix in the product)—see Definition 2.9. It follows that **A** must be square and hence for square matrices (and only for these) the product $\mathbf{A} \times \mathbf{A}$ is defined. It then follows that for square matrices we can define *powers* of a matrix:

$$\mathbf{A}^2 = \mathbf{A} \times \mathbf{A}$$

$$\mathbf{A}^3 = \mathbf{A} \times \mathbf{A} \times \mathbf{A}$$

$$\mathbf{A}^n = \underbrace{\mathbf{A} \times \mathbf{A} \times \mathbf{A} \ldots\ldots \times \mathbf{A}}_{n \text{ terms in all.}}$$

We have now seen that many of the laws of ordinary algebra hold also for matrix algebra (an important exception being the commutativity of multiplication). It follows that much of the manipulation to which we are accustomed can be carried over to matrix algebra, provided we remember that in general $\mathbf{AB} \neq \mathbf{BA}$. We illustrate this in the following examples.

*Examples*

2.2.11 Show that $(\mathbf{A} + \mathbf{B})^2 = \mathbf{A}^2 + \mathbf{AB} + \mathbf{BA} + \mathbf{B}^2$.

We have

$$\begin{aligned}(\mathbf{A} + \mathbf{B})^2 &= (\mathbf{A} + \mathbf{B})(\mathbf{A} + \mathbf{B}) \\ &= \mathbf{A}(\mathbf{A} + \mathbf{B}) + \mathbf{B}(\mathbf{A} + \mathbf{B}) \qquad \text{(by Theorem 2.6)} \\ &= \mathbf{A}^2 + \mathbf{AB} + \mathbf{BA} + \mathbf{B}^2 \qquad \text{(by Theorem 2.5).}\end{aligned}$$

(N.B. In general $\mathbf{BA} \neq \mathbf{AB}$ so we cannot say $(\mathbf{A} + \mathbf{B})^2 = \mathbf{A}^2 + 2\mathbf{AB} + \mathbf{B}^2$ unless **A** and **B** commute.)

2.2.12 Under what circumstances is $\mathbf{A}^2 - \mathbf{B}^2 = (\mathbf{A} - \mathbf{B})(\mathbf{A} + \mathbf{B})$?

We have

$$\begin{aligned}(\mathbf{A} - \mathbf{B})(\mathbf{A} + \mathbf{B}) &= \mathbf{A}(\mathbf{A} + \mathbf{B}) - \mathbf{B}(\mathbf{A} + \mathbf{B}) \\ &= \mathbf{A}^2 + \mathbf{AB} - \mathbf{BA} - \mathbf{B}^2 \\ &= \mathbf{A}^2 - \mathbf{B}^2 + \mathbf{AB} - \mathbf{BA}\end{aligned}$$

so $\quad (\mathbf{A} - \mathbf{B})(\mathbf{A} + \mathbf{B}) = \mathbf{A}^2 - \mathbf{B}^2$

only if $\mathbf{AB} = \mathbf{BA}$, i.e. if **A** and **B** commute.

*Exercises*

2.2.1 If

$$\mathbf{A} = \begin{bmatrix} 2 & 3 & 5 \\ 3 & 2 & 1 \\ 1 & 5 & 2 \end{bmatrix} \quad \text{and} \quad \mathbf{B} = \begin{bmatrix} -1 & 2 & 0 \\ 1 & -2 & 3 \\ 4 & 1 & -1 \end{bmatrix}$$

find (i) $\mathbf{A} + 2\mathbf{B}$ (ii) $\mathbf{AB}$ (iii) $\mathbf{BA}$.

2.2.2 Evaluate the following matrix products

(i) $\begin{bmatrix} 1 & 3 & 2 \\ 2 & 6 & 1 \end{bmatrix} \begin{bmatrix} 1 & 4 & 2 \\ -3 & 1 & 8 \\ 7 & 4 & 5 \end{bmatrix}$

(ii) $[3 \quad 5 \quad 6] \begin{bmatrix} 4 \\ 2 \\ 1 \end{bmatrix}$

(iii) $\begin{bmatrix} 4 \\ 2 \\ 1 \end{bmatrix} [3 \quad 5 \quad 6]$ .

2.2.3 Find all the values of $x$ such that

$$[x \quad 4 \quad 1] \begin{bmatrix} 2 & 1 & 0 \\ 1 & 0 & 2 \\ 0 & 2 & 4 \end{bmatrix} \begin{bmatrix} x \\ 4 \\ 1 \end{bmatrix} = 0.$$

2.2.4 Find numbers $a$, $b$, $c$, $d$ such that

$$\begin{bmatrix} 2a & 1 & 5 \\ 3 & 2 & -b \\ 3c & 4 & 1 \end{bmatrix} + \begin{bmatrix} 2 & 3 & 1 \\ 3 & d & 4 \\ 3 & 2 & 5 \end{bmatrix} = \begin{bmatrix} 6 & 4 & 6 \\ 6 & -1 & -2 \\ 3 & 6 & 6 \end{bmatrix}.$$

2.2.5 If

$$\mathbf{R}(\theta) = \begin{bmatrix} \cos\theta & \sin\theta \\ -\sin\theta & \cos\theta \end{bmatrix}$$

prove that $\mathbf{R}(\theta)\mathbf{R}(\phi) = \mathbf{R}(\theta + \phi)$. Deduce that $\mathbf{R}(2\theta) = \{\mathbf{R}(\theta)\}^2$ and write down a simple expression for $\{\mathbf{R}(\theta)\}^n$.

## 2.3 Null and Unit Matrices

DEFINITION 2.10

A matrix whose elements are all zero is known as a *null* or *zero* matrix. □

The usual notation for a null matrix is somewhat slipshod; strictly speaking we should denote a null matrix of order $m \times n$ by some symbol such as $\mathbf{0}^{(mn)}$, indicating the order of the matrix. But in fact we generally drop all reference to the order from the notation, leaving it to be understood by the context in which the null matrix appears, and denote a null matrix by $\mathbf{0}$; again boldface type is used to distinguish the fact that it is a matrix we are dealing with.

Thus, if we write $\mathbf{A} = \mathbf{0}$ and we know that $\mathbf{A}$ is square of order 2 we have

$$\mathbf{A} = \begin{bmatrix} 0 & 0 \\ 0 & 0 \end{bmatrix},$$

while if we write $\mathbf{B} = \mathbf{0}$ and we know that $\mathbf{B}$ is of order $3 \times 2$ then

$$\mathbf{B} = \begin{bmatrix} 0 & 0 \\ 0 & 0 \\ 0 & 0 \end{bmatrix}.$$

There are several properties of the zero matrix analogous to the zero of ordinary algebra. These we give in the following theorem.

THEOREM 2.7

If the zero matrices below are of suitable order then

(i) $\mathbf{A} + \mathbf{0} = \mathbf{0} + \mathbf{A} = \mathbf{A}$

(ii) $\mathbf{A} \times \mathbf{0} = \mathbf{0} \times \mathbf{A} = \mathbf{0}$

(iii) $\mathbf{A} = \mathbf{B}$ if and only if $\mathbf{A} - \mathbf{B} = \mathbf{0}$.

*Proof*

The proofs are trivial; the results should be immediately obvious to the reader. If this is not the case then he can write them out in detail as an exercise. ■

There is, however, one property of the ordinary algebraic zero which does not carry over to matrix algebra. In ordinary algebra we know that if $ab = 0$ then either $a = 0$ or $b = 0$ (or both). In matrix algebra, however, we can have $\mathbf{AB} = \mathbf{0}$ without either $\mathbf{A} = \mathbf{0}$ or $\mathbf{B} = \mathbf{0}$. That this is so is shown in Example 2.3.1 below. It is an important fact and must be constantly borne in mind.

*Examples*

2.3.1 If

$$\mathbf{A} = \begin{bmatrix} 1 & 0 \\ 0 & 0 \end{bmatrix} \quad \text{and} \quad \mathbf{B} = \begin{bmatrix} 0 & 0 \\ 1 & 1 \end{bmatrix}$$

show that $\mathbf{AB} = \mathbf{0}$. We have

$$\mathbf{AB} = \begin{bmatrix} 1 & 0 \\ 0 & 0 \end{bmatrix} \begin{bmatrix} 0 & 0 \\ 1 & 1 \end{bmatrix}$$

$$= \begin{bmatrix} 1 \times 0 + 0 \times 1 & 1 \times 0 + 0 \times 1 \\ 0 \times 0 + 0 \times 1 & 0 \times 0 + 0 \times 1 \end{bmatrix}$$

$$= \begin{bmatrix} 0 & 0 \\ 0 & 0 \end{bmatrix}$$

$$= \mathbf{0}.$$

2.3.2 Find all matrices $\mathbf{A}$ of order $2 \times 2$ such that $\mathbf{A}^2 = \mathbf{0}$. Let

$$\mathbf{A} = \begin{bmatrix} p & q \\ r & s \end{bmatrix}.$$

Then $\mathbf{A}^2 = \mathbf{0}$ means that

$$\begin{bmatrix} p & q \\ r & s \end{bmatrix} \begin{bmatrix} p & q \\ r & s \end{bmatrix} = \begin{bmatrix} 0 & 0 \\ 0 & 0 \end{bmatrix}$$

and carrying out the matrix multiplication gives

$$\begin{bmatrix} p^2 + qr & pq + qs \\ rp + sr & rq + s^2 \end{bmatrix} = \begin{bmatrix} 0 & 0 \\ 0 & 0 \end{bmatrix}.$$

This means that we have the four equations

$$p^2 + qr = 0 \qquad \text{(i)}$$
$$q(p + s) = 0 \qquad \text{(ii)}$$
$$r(p + s) = 0 \qquad \text{(iii)}$$
$$s^2 + qr = 0 \qquad \text{(iv).}$$

From (ii) either $q = 0$ or $p + s = 0$. If $q = 0$, then from (i) $p = 0$, from (iv) $s = 0$ and then (iii) is satisfied irrespective of $r$. If $p + s = 0$, then (iii) is also satisfied; (i) and (iv) will then be satisfied by taking $p = -s = \pm\sqrt{(-qr)}$. We note that this latter case includes the former as a special case.

Thus matrices satisfying $\mathbf{A}^2 = \mathbf{0}$ must be of the form

$$\mathbf{A} = \begin{bmatrix} \pm\sqrt{(-qr)} & q \\ r & \pm\sqrt{(-qr)} \end{bmatrix}$$

where $q$ and $r$ are arbitrary.

The two 'special' numbers in ordinary algebra are 0 and 1. We have introduced the null matrix which is the matrix analogue of 0; we now introduce the unit matrix which is the matrix analogue of 1 inasmuch as it possesses in matrix algebra some of the properties of 1 in ordinary algebra.

DEFINITION 2.11

A square matrix of order $n$ which has all its diagonal elements equal to 1 and all its other elements equal to 0 is called a *unit matrix* of order $n$. □

Again the notation is slipshod; we use the symbol **I** to denote a unit matrix of any order, the actual order being obvious (hopefully) from the context.

From the definition we have

$$I_{ij} = \begin{cases} 1 & \text{if } i = j \\ 0 & \text{if } i \neq j \end{cases}.$$

In particular the unit matrix of order 3 is

$$\mathbf{I} = \begin{bmatrix} 1 & 0 & 0 \\ 0 & 1 & 0 \\ 0 & 0 & 1 \end{bmatrix}.$$

The property of unity in ordinary algebra that the unit matrix reflects in matrix algebra is the multiplicative property $1 \times a = a \times 1 = a$. The corresponding matrix result is contained in the following theorem.

THEOREM 2.8

Suppose **I** is the unit matrix of order $n$.

(i) If **A** is of order $n \times m$ then **IA** = **A**.
(ii) If **A** is of order $m \times n$ then **AI** = **A**.
(iii) If **A** is square of order $n$ then **IA** = **AI** = **A**.

*Proof*

(i) Since **I** is of order $n \times n$ and **A** is of order $n \times m$, the product **IA** exists and is of order $n \times m$ (Definition 2.9). Thus **IA** is of the same order as **A**. To prove equality we prove equality of elements. Now,

$$(\mathbf{IA})_{ij} = \sum_{k=1}^{n} I_{ik} A_{kj}$$

But

$$I_{ik} = \begin{cases} 0 & \text{for } k \neq i \\ 1 & \text{for } k = i \end{cases}$$

and so the sum over $k$ on the right-hand side above gives only one term—the one for which $k = i$. Thus

$$(\mathbf{IA})_{ij} = I_{ii} A_{ij} = A_{ij}$$

and hence

$$\mathbf{IA} = \mathbf{A}.$$

(ii), (iii). These proofs are exactly similar to (i) above, and so will not be given here. ■

The introduction of the unit matrix enables us to extend the idea of powers of a matrix; so far $\mathbf{A}^n$ is defined for positive integer $n$ only. We extend the definition to $n = 0$ by defining $\mathbf{A}^0 = \mathbf{I}$. (This is consistent with $a^0 = 1$ in ordinary algebra.)

We can now consider matrix *polynomials*; a polynomial of degree $n$ in the matrix $\mathbf{A}$ is an expression of the form

$$P(\mathbf{A}) = c_n\mathbf{A}^n + c_{n-1}\mathbf{A}^{n-1} + \ldots + c_r\mathbf{A}^r + \ldots + c_1\mathbf{A} + c_0\mathbf{I}$$

$$= \sum_{r=0}^{n} c_r\mathbf{A}^r$$

where $c_0, c_1, \ldots, c_n$ are numbers. Of course, for this to be well defined, $\mathbf{A}$ must be square.

*Examples*

2.3.3 Expand $(\mathbf{A} + \mathbf{I})(\mathbf{B} + \mathbf{I})$.

We have

$$\begin{aligned}(\mathbf{A} + \mathbf{I})(\mathbf{B} + \mathbf{I}) &= \mathbf{A}(\mathbf{B} + \mathbf{I}) + \mathbf{I}(\mathbf{B} + \mathbf{I})\\ &= \mathbf{AB} + \mathbf{AI} + (\mathbf{B} + \mathbf{I})\\ &= \mathbf{AB} + \mathbf{A} + \mathbf{B} + \mathbf{I}.\end{aligned}$$

2.3.4 If

$$\mathbf{A} = \begin{bmatrix}\cos\theta & \sin\theta\\ -\sin\theta & \cos\theta\end{bmatrix} \quad\text{and}\quad \mathbf{B} = \begin{bmatrix}\cos\theta & -\sin\theta\\ \sin\theta & \cos\theta\end{bmatrix}$$

show that $\mathbf{AB} = \mathbf{I}$.

We have

$$\mathbf{AB} = \begin{bmatrix}\cos\theta & \sin\theta\\ -\sin\theta & \cos\theta\end{bmatrix}\begin{bmatrix}\cos\theta & -\sin\theta\\ \sin\theta & \cos\theta\end{bmatrix}$$

$$= \begin{bmatrix}\cos^2\theta + \sin^2\theta & -\cos\theta\sin\theta + \sin\theta\cos\theta\\ -\sin\theta\cos\theta + \cos\theta\sin\theta & \sin^2\theta + \cos^2\theta\end{bmatrix}$$

$$= \begin{bmatrix}1 & 0\\ 0 & 1\end{bmatrix}$$

$$= \mathbf{I}.$$

2.3.5 If

$$\mathbf{A} = \begin{bmatrix}\lambda_1 & 0 & 0\\ 0 & \lambda_2 & 0\\ 0 & 0 & \lambda_3\end{bmatrix}$$

and $F(x)$ is a polynomial in $x$, show that

(i) $\mathbf{A}^r = \begin{bmatrix}\lambda_1^r & 0 & 0\\ 0 & \lambda_2^r & 0\\ 0 & 0 & \lambda_3^r\end{bmatrix}$ (where $r$ is a non-negative integer);

$$\text{(ii)}\ F(\mathbf{A}) = \begin{bmatrix} F(\lambda_1) & 0 & 0 \\ 0 & F(\lambda_2) & 0 \\ 0 & 0 & F(\lambda_3) \end{bmatrix}.$$

(i) We shall prove this result by induction. We first assume the result to be true for $r = R$ where $R$ is some non-negative integer; then we prove it true for $r = R + 1$; then we prove it true for $r = 0$, and hence it will be true for all non-negative integers $r$. Assuming true for $r = R$ we have

$$\mathbf{A}^R = \begin{bmatrix} \lambda_1^R & 0 & 0 \\ 0 & \lambda_2^R & 0 \\ 0 & 0 & \lambda_3^R \end{bmatrix}$$

and then

$$\mathbf{A}^{R+1} = \mathbf{A} \times \mathbf{A}^R = \begin{bmatrix} \lambda_1 & 0 & 0 \\ 0 & \lambda_2 & 0 \\ 0 & 0 & \lambda_3 \end{bmatrix} \begin{bmatrix} \lambda_1^R & 0 & 0 \\ 0 & \lambda_2^R & 0 \\ 0 & 0 & \lambda_3^R \end{bmatrix}$$

$$= \begin{bmatrix} \lambda_1 \times \lambda_1^R + 0 \times 0 + 0 \times 0 & \lambda_1 \times 0 + 0 \times \lambda_2^R + 0 \times 0 & \lambda_1 \times 0 + 0 \times 0 + 0 \times \lambda_3^R \\ 0 \times \lambda_1^R + \lambda_2 \times 0 + 0 \times 0 & 0 \times 0 + \lambda_2 \times \lambda_2^R + 0 \times 0 & 0 \times 0 + \lambda_2 \times 0 + 0 \times \lambda_3^R \\ 0 \times \lambda_1^R + 0 \times 0 + \lambda_3 \times 0 & 0 \times 0 + 0 \times \lambda_2^R + \lambda_3 \times 0 & 0 \times 0 + 0 \times 0 + \lambda_3 \times \lambda_3^R \end{bmatrix}$$

$$= \begin{bmatrix} \lambda_1^{R+1} & 0 & 0 \\ 0 & \lambda_2^{R+1} & 0 \\ 0 & 0 & \lambda_3^{R+1} \end{bmatrix}.$$

Thus the result is true for $r = R + 1$. But when $r = 0$ we have

$$\mathbf{A}^0 = \mathbf{I} \qquad \text{and} \qquad \begin{bmatrix} \lambda_1^0 & 0 & 0 \\ 0 & \lambda_2^0 & 0 \\ 0 & 0 & \lambda_3^0 \end{bmatrix} = \mathbf{I}$$

and hence the result is true for $r = 0$. Thus, by the principle of induction, the result must be true for all positive integers $r$.

(ii) Since $F(x)$ is a polynomial in $x$ we have

$$F(x) = \sum_{r=0}^{n} c_r x^r \qquad \text{(for some } n\text{)}$$

and similarly the matrix polynomial $F(\mathbf{A})$ is given by

$$F(\mathbf{A}) = \sum_{r=0}^{n} c_r \mathbf{A}^r.$$

But

$$\mathbf{A}^r = \begin{bmatrix} \lambda_1^r & 0 & 0 \\ 0 & \lambda_2^r & 0 \\ 0 & 0 & \lambda_3^r \end{bmatrix}$$

and hence

$$F(\mathbf{A}) = \sum_{r=0}^{n} c_r \begin{bmatrix} \lambda_1^r & 0 & 0 \\ 0 & \lambda_2^r & 0 \\ 0 & 0 & \lambda_3^r \end{bmatrix}$$

$$= \sum_{r=0}^{n} \begin{bmatrix} c_r\lambda_1^r & 0 & 0 \\ 0 & c_r\lambda_2^r & 0 \\ 0 & 0 & c_r\lambda_3^r \end{bmatrix}$$

$$= \begin{bmatrix} \sum_{r=0}^{n} c_r\lambda_1^r & 0 & 0 \\ 0 & \sum_{r=0}^{n} c_r\lambda_2^r & 0 \\ 0 & 0 & \sum_{r=0}^{n} c_r\lambda_3^r \end{bmatrix}$$ (since to add matrices we just add corresponding elements)

$$= \begin{bmatrix} F(\lambda_1) & 0 & 0 \\ 0 & F(\lambda_2) & 0 \\ 0 & 0 & F(\lambda_3) \end{bmatrix},$$ as required.

*Exercises*

2.3.1 If

$$\mathbf{A} = \begin{bmatrix} 1 & 1 & -1 \\ 0 & 1 & 0 \\ 1 & 0 & 1 \end{bmatrix} \quad \text{and} \quad \mathbf{B} = \begin{bmatrix} 0{\cdot}5 & -0{\cdot}5 & 0{\cdot}5 \\ 0 & 1 & 0 \\ -0{\cdot}5 & 0{\cdot}5 & 0{\cdot}5 \end{bmatrix}$$

show that $\mathbf{AB} = \mathbf{BA} = \mathbf{I}$.

2.3.2 Expand $(\mathbf{A} + \mathbf{I})^3$ and write down the general expression for $(\mathbf{A} + \mathbf{I})^n$ where $n$ is a positive integer. Compare with the binomial expansion of $(a + 1)^n$. Does a similar analogue exist between $(a + b)^n$ in ordinary algebra and $(\mathbf{A} + \mathbf{B})^n$ in matrix algebra?

## 2.4 Transposition

DEFINITION 2.12

If $\mathbf{A}$ is a matrix of order $m \times n$ then the *transpose* of $\mathbf{A}$, denoted by $\mathbf{A}^{\mathrm{T}}$, is the matrix of order $n \times m$ such that $(\mathbf{A}^{\mathrm{T}})_{ij} = A_{ji}$. □

The above definition means that the element appearing in the $i$th row and $j$th column of $\mathbf{A}^T$ is the same as the element in the $j$th row and $i$th column of $\mathbf{A}$. This implies that $\mathbf{A}^T$ is obtained from $\mathbf{A}$ merely by interchanging rows and columns. It follows immediately from this that $(\mathbf{A}^T)^T = \mathbf{A}$; this just means that transposing twice brings us back to the original matrix. It also follows that if $\mathbf{A}$ and $\mathbf{B}$ are of the same order then $(\mathbf{A} + \mathbf{B})^T = \mathbf{A}^T + \mathbf{B}^T$.

*Examples*

2.4.1 If

$$\mathbf{A} = \begin{bmatrix} 3 & 7 & -2 \\ 6 & 8 & 15 \end{bmatrix} \quad \text{find } \mathbf{A}^T.$$

Interchanging rows and columns gives

$$\mathbf{A}^T = \begin{bmatrix} 3 & 6 \\ 7 & 8 \\ -2 & 15 \end{bmatrix}.$$

2.4.2 If

$$\mathbf{A} = \begin{bmatrix} p \\ q \\ r \end{bmatrix} \quad \text{and} \quad \mathbf{B} = [x \quad y \quad z]$$

evaluate $(\mathbf{AB})^T$ and $\mathbf{B}^T\mathbf{A}^T$. We have

$$\mathbf{AB} = \begin{bmatrix} p \\ q \\ r \end{bmatrix} [x \quad y \quad z]$$

$$= \begin{bmatrix} px & py & pz \\ qx & qy & qz \\ rx & ry & rz \end{bmatrix},$$

and so

$$(\mathbf{AB})^T = \begin{bmatrix} px & qx & rx \\ py & qy & ry \\ pz & qz & rz \end{bmatrix}.$$

Also

$$\mathbf{A}^T = [p \quad q \quad r] \quad \text{and} \quad \mathbf{B}^T = \begin{bmatrix} x \\ y \\ z \end{bmatrix}$$

so

$$\mathbf{B}^{\mathrm{T}}\mathbf{A}^{\mathrm{T}} = \begin{bmatrix} x \\ y \\ z \end{bmatrix} [p \quad q \quad r]$$

$$= \begin{bmatrix} xp & xq & xr \\ yp & yq & yr \\ zp & zq & zr \end{bmatrix}.$$

We see that for the particular case of Example 2.4.2 above that $(\mathbf{AB})^{\mathrm{T}} = \mathbf{B}^{\mathrm{T}}\mathbf{A}^{\mathrm{T}}$. That this is a general result is proved in the following theorem.

THEOREM 2.9

If $\mathbf{A}$ and $\mathbf{B}$ are such that $\mathbf{AB}$ exists then $(\mathbf{AB})^{\mathrm{T}} = \mathbf{B}^{\mathrm{T}}\mathbf{A}^{\mathrm{T}}$.

*Proof*

Let $\mathbf{A}$ be of order $m \times n$ and $\mathbf{B}$ of order $n \times p$ so that their product exists and is of order $m \times p$. Then $(\mathbf{AB})^{\mathrm{T}}$ is of order $p \times m$.

Also $\mathbf{B}^{\mathrm{T}}$ is of order $p \times n$ and $\mathbf{A}^{\mathrm{T}}$ is of order $n \times m$. Thus the product $\mathbf{B}^{\mathrm{T}}\mathbf{A}^{\mathrm{T}}$ exists and is of order $p \times m$. So $(\mathbf{AB})^{\mathrm{T}}$ and $\mathbf{B}^{\mathrm{T}}\mathbf{A}^{\mathrm{T}}$ are matrices of the same order; we now show they are equal by proving the equality of individual elements. We have

$$(\mathbf{AB})^{\mathrm{T}}_{ij} = (\mathbf{AB})_{ji} \qquad \text{(by Definition 2.12)}$$

$$= \sum_{k=1}^{n} A_{jk}B_{ki} \qquad \text{(by Definition 2.9).}$$

Also

$$(\mathbf{B}^{\mathrm{T}}\mathbf{A}^{\mathrm{T}})_{ij} = \sum_{k=1}^{n} (\mathbf{B}^{\mathrm{T}})_{ik}(\mathbf{A}^{\mathrm{T}})_{kj}$$

$$= \sum_{k=1}^{n} B_{ki}A_{jk}$$

$$= \sum_{k=1}^{n} A_{jk}B_{ki}.$$

Thus $(\mathbf{AB})^{\mathrm{T}}_{ij} = (\mathbf{B}^{\mathrm{T}}\mathbf{A}^{\mathrm{T}})_{ij}$ and so the required result is proved.

Theorem 2.9 can be extended to the product of more than two matrices. This extension is the content of the following theorem.

THEOREM 2.10

Provided the product $\mathbf{A}_1\mathbf{A}_2 \ldots \mathbf{A}_n$ exists then $(\mathbf{A}_1\mathbf{A}_2 \ldots \mathbf{A}_n)^{\mathrm{T}} = \mathbf{A}_n^{\mathrm{T}} \ldots \mathbf{A}_2^{\mathrm{T}}\mathbf{A}_1^{\mathrm{T}}$.

*Proof*

$$\begin{aligned}(\mathbf{A}_1\mathbf{A}_2 \cdots \mathbf{A}_n)^{\mathrm{T}} &= \{(\mathbf{A}_1)(\mathbf{A}_2 \ldots \mathbf{A}_n)\}^{\mathrm{T}} \\ &= (\mathbf{A}_2\mathbf{A}_3 \ldots \mathbf{A}_n)^{\mathrm{T}}\mathbf{A}_1^{\mathrm{T}} \qquad \text{(by Theorem 2.9)} \\ &= \{(\mathbf{A}_2)(\mathbf{A}_3 \ldots \mathbf{A}_n)\}^{\mathrm{T}}\mathbf{A}_1^{\mathrm{T}} \\ &= (\mathbf{A}_3 \ldots \mathbf{A}_n)^{\mathrm{T}}\mathbf{A}_2^{\mathrm{T}}\mathbf{A}_1^{\mathrm{T}} \qquad \text{(again by Theorem 2.9).}\end{aligned}$$

Repeating this process $(n - 1)$ times in all obviously leads to the required result. (Alternatively we could prove the result by induction on $n$, the number of matrices in the product.)

In the applications of matrix algebra it often happens that simplifications occur if the matrices involved are of some special type. We shall meet various special types throughout this book; the first of these we now introduce in the following definitions.

DEFINITION 2.13

A matrix $\mathbf{A}$ is said to be *symmetric* if $\mathbf{A}^{\mathrm{T}} = \mathbf{A}$.

DEFINITION 2.14

A matrix $\mathbf{A}$ is said to be *anti-symmetric* (or *skew-symmetric*) if $\mathbf{A}^{\mathrm{T}} = -\mathbf{A}$.

If $\mathbf{A}$ is of order $m \times n$ we know that $\mathbf{A}^{\mathrm{T}}$ is of order $n \times m$ (Definition 2.12). It hence follows that if $\mathbf{A}$ is either symmetric or anti-symmetric it must necessarily be square. It is also easy to show that an anti-symmetric matrix must have its diagonal elements equal to zero. For suppose that $\mathbf{A}$ is anti-symmetric so that

$$\mathbf{A}^{\mathrm{T}} = -\mathbf{A}.$$

Then

$$(\mathbf{A}^{\mathrm{T}})_{ii} = -A_{ii}$$

gives

$$A_{ii} = -A_{ii} \qquad \text{(by Definition 2.12)}$$

and hence

$$A_{ii} = 0.$$

*Example*

2.4.3 State whether each of the following matrices is symmetric, anti-symmetric or neither.

$$\text{(i)} \begin{bmatrix} a & h & g \\ h & b & f \\ g & f & c \end{bmatrix} \quad \text{(ii)} \begin{bmatrix} 0 & h & g \\ -h & 0 & f \\ -g & -f & 0 \end{bmatrix} \quad \text{(iii)} \begin{bmatrix} 1 & h & g \\ -h & 1 & f \\ -g & -f & 1 \end{bmatrix} \quad \text{(iv)} \begin{bmatrix} 1 & 1 \\ 1 & 1 \\ 1 & 1 \end{bmatrix}.$$

(i) Symmetric (ii) Anti-symmetric (iii) Neither (iv) Neither.

Since symmetric and anti-symmetric matrices have special properties it is sometimes useful to work as much as possible in terms of such matrices. The following theorem shows how a general square matrix can be replaced by a combination of symmetric and anti-symmetric matrices.

THEOREM 2.11

Any square matrix $\mathbf{A}$ can be written in the form $\mathbf{A} = \mathbf{B} + \mathbf{C}$ where $\mathbf{B}$ is symmetric and $\mathbf{C}$ is anti-symmetric; indeed $\mathbf{B} = \frac{1}{2}(\mathbf{A} + \mathbf{A}^T)$ and $\mathbf{C} = \frac{1}{2}(\mathbf{A} - \mathbf{A}^T)$.

*Proof*

We first prove two lemmas.

Lemma 1: $\mathbf{A} + \mathbf{A}^T$ is symmetric. For

$$(\mathbf{A} + \mathbf{A}^T)^T = \mathbf{A}^T + (\mathbf{A}^T)^T = \mathbf{A}^T + \mathbf{A} = \mathbf{A} + \mathbf{A}^T.$$

Lemma 2: $\mathbf{A} - \mathbf{A}^T$ is anti-symmetric. For

$$(\mathbf{A} - \mathbf{A}^T)^T = \mathbf{A}^T - (\mathbf{A}^T)^T = \mathbf{A}^T - \mathbf{A} = -(\mathbf{A} - \mathbf{A}^T).$$

We now let $\mathbf{B} = \frac{1}{2}(\mathbf{A} + \mathbf{A}^T)$ so that $\mathbf{B}$ is symmetric and $\mathbf{C} = \frac{1}{2}(\mathbf{A} - \mathbf{A}^T)$ so that $\mathbf{C}$ is anti-symmetric. Then

$$\begin{aligned}\mathbf{B} + \mathbf{C} &= \tfrac{1}{2}(\mathbf{A} + \mathbf{A}^T) + \tfrac{1}{2}(\mathbf{A} - \mathbf{A}^T)\\ &= \tfrac{1}{2}\mathbf{A} + \tfrac{1}{2}\mathbf{A}^T + \tfrac{1}{2}\mathbf{A} - \tfrac{1}{2}\mathbf{A}^T\\ &= \mathbf{A}\end{aligned}$$

and hence the result is proved. ■

*Examples*

2.4.4 If

$$\mathbf{A} = \begin{bmatrix} 1 & -7 & 2 \\ -3 & 4 & 7 \\ 8 & 1 & 5 \end{bmatrix}$$

write $\mathbf{A}$ in the form $\mathbf{B} + \mathbf{C}$ where $\mathbf{B}$ is symmetric and $\mathbf{C}$ is anti-symmetric.

We make use of Theorem 2.11. We have

$$\mathbf{A}^T = \begin{bmatrix} 1 & -3 & 8 \\ -7 & 4 & 1 \\ 2 & 7 & 5 \end{bmatrix}$$

and hence

$$\mathbf{A} + \mathbf{A}^T = \begin{bmatrix} 2 & -10 & 10 \\ -10 & 8 & 8 \\ 10 & 8 & 10 \end{bmatrix} \quad \text{and} \quad \mathbf{A} - \mathbf{A}^T = \begin{bmatrix} 0 & -4 & -6 \\ 4 & 0 & 6 \\ 6 & -6 & 0 \end{bmatrix}.$$

Thus

$$\mathbf{A} = \begin{bmatrix} 1 & -5 & 5 \\ -5 & 4 & 4 \\ 5 & 4 & 5 \end{bmatrix} + \begin{bmatrix} 0 & -2 & -3 \\ 2 & 0 & 3 \\ 3 & -3 & 0 \end{bmatrix}.$$

2.4.5 If **A** and **B** commute, do $\mathbf{A}^T$ and $\mathbf{B}^T$? Yes. For $\mathbf{AB} = \mathbf{BA}$ gives

$$(\mathbf{AB})^T = (\mathbf{BA})^T$$

and hence

$$\mathbf{B}^T\mathbf{A}^T = \mathbf{A}^T\mathbf{B}^T \quad \text{using Theorem 2.9.}$$

2.4.6 If **A** is symmetric show that $\mathbf{P}^T\mathbf{AP}$ is also symmetric where **P** is any matrix such that the product exists. We have

$$\begin{aligned} (\mathbf{P}^T\mathbf{AP})^T &= \mathbf{P}^T\mathbf{A}^T(\mathbf{P}^T)^T && \text{(by Theorem 2.10)} \\ &= \mathbf{P}^T\mathbf{A}^T\mathbf{P} \\ &= \mathbf{P}^T\mathbf{AP} && \text{(since } \mathbf{A} \text{ is symmetric).} \end{aligned}$$

Thus $\mathbf{P}^T\mathbf{AP}$ is equal to its own transpose and so (by Definition 2.13) is symmetric.

2.4.7 If **P** and **Q** are symmetric show that **PQ** is not necessarily symmetric. We have

$$\begin{aligned} (\mathbf{PQ})^T &= \mathbf{Q}^T\mathbf{P}^T \\ &= \mathbf{QP} \\ &= \mathbf{PQ} \end{aligned}$$

if and only if **P** and **Q** commute.

## 2.5 Partitioned Matrices

A matrix may be considered to be made up of 'submatrices'; for example, if

$$\mathbf{A} = \begin{bmatrix} a & h & g \\ h & b & f \\ g & f & c \end{bmatrix}$$

we may write

$$\mathbf{A} = \begin{bmatrix} \mathbf{A}_1 & \mathbf{A}_2 \\ \mathbf{A}_3 & \mathbf{A}_4 \end{bmatrix}$$

with

$$\mathbf{A}_1 = \begin{bmatrix} a & h \\ h & b \end{bmatrix} \qquad \mathbf{A}_2 = \begin{bmatrix} g \\ f \end{bmatrix}$$

$$\mathbf{A}_3 = [g \quad f] \qquad \mathbf{A}_4 = [c].$$

In such a situation the original matrix is said to have been *partitioned* into submatrices as indicated by the dotted lines:

$$\mathbf{A} = \left[\begin{array}{cc|c} a & h & g \\ h & b & f \\ \hline g & f & c \end{array}\right].$$

If two partitioned matrices are multiplied together, it may be shown that the result may be obtained by treating all the submatrices as if they were matrix elements and using the usual row into column rule of multiplication. This is only valid of course, if the partitioning is such that all matrix products involved exist. A typical situation is illustrated by the following theorem.

THEOREM 2.12

Suppose that **A** is a matrix of order $m \times n$ and **B** is a matrix of order $n \times p$ so that the product **AB** exists and is of order $m \times p$. Then if partitioning takes place as follows

$$\mathbf{A} = \left[\begin{array}{c|c} \alpha_{11} & \alpha_{12} \\ \hline \alpha_{21} & \alpha_{22} \end{array}\right] \begin{array}{l} \} \ t \text{ rows} \\ \\ \end{array}$$

$s$ columns

$$\mathbf{B} = \left[\begin{array}{c|c} \beta_{11} & \beta_{12} \\ \hline \beta_{21} & \beta_{22} \end{array}\right] \begin{array}{l} \} \ s \text{ rows} \\ \\ \end{array}$$

$u$ columns

we shall have

$$\mathbf{AB} = \begin{bmatrix} \alpha_{11}\beta_{11} + \alpha_{12}\beta_{21} & \alpha_{11}\beta_{12} + \alpha_{12}\beta_{22} \\ \alpha_{21}\beta_{11} + \alpha_{22}\beta_{21} & \alpha_{21}\beta_{12} + \alpha_{22}\beta_{22} \end{bmatrix} \begin{array}{l} \} \ t \text{ rows} \\ \\ \end{array}.$$

$u$ columns

*Proof*

The proof of the general result is not particularly difficult, but is rather complicated by the number of subscripts and attention to detail which is involved. We shall prove a special case and leave it to the reader either to prove the general case as an exercise or else to accept the result as an intuitively obvious generalization from the special case.

Let us consider, then, the product **AB** where

$$\mathbf{A} = \begin{bmatrix} a_{11} & a_{12} & a_{13} \\ a_{21} & a_{22} & a_{23} \\ a_{31} & a_{32} & a_{33} \end{bmatrix} \quad \text{and} \quad \mathbf{B} = \begin{bmatrix} b_{11} & b_{12} & b_{13} \\ b_{21} & b_{22} & b_{23} \\ b_{31} & b_{32} & b_{33} \end{bmatrix}.$$

Suppose **A** and **B** are partitioned as follows

$$\mathbf{A} = \left[\begin{array}{c|cc} a_{11} & a_{12} & a_{13} \\ \hline a_{21} & a_{22} & a_{23} \\ a_{31} & a_{32} & a_{33} \end{array}\right] \quad \text{and} \quad \mathbf{B} = \left[\begin{array}{c|cc} b_{11} & b_{12} & b_{13} \\ \hline b_{21} & b_{22} & b_{23} \\ b_{31} & b_{32} & b_{33} \end{array}\right]$$

so that

$$\boldsymbol{\alpha}_{11} = [a_{11}] \qquad \boldsymbol{\alpha}_{12} = [a_{12} \quad a_{13}]$$

$$\boldsymbol{\alpha}_{21} = \begin{bmatrix} a_{21} \\ a_{31} \end{bmatrix} \qquad \boldsymbol{\alpha}_{11} = \begin{bmatrix} a_{22} & a_{23} \\ a_{32} & a_{33} \end{bmatrix}$$

and

$$\boldsymbol{\beta}_{11} = [b_{11}] \qquad \boldsymbol{\beta}_{12} = [b_{12} \quad b_{13}]$$

$$\boldsymbol{\beta}_{21} = \begin{bmatrix} b_{21} \\ b_{31} \end{bmatrix} \qquad \boldsymbol{\beta}_{22} = \begin{bmatrix} b_{22} & b_{23} \\ b_{32} & b_{33} \end{bmatrix}$$

Then, by ordinary matrix multiplication, we have

$$\mathbf{AB} = \begin{bmatrix} a_{11} & a_{12} & a_{13} \\ a_{21} & a_{22} & a_{23} \\ a_{31} & a_{32} & a_{33} \end{bmatrix} \begin{bmatrix} b_{11} & b_{12} & b_{13} \\ b_{21} & b_{22} & b_{23} \\ b_{31} & b_{32} & b_{33} \end{bmatrix}$$

$$= \begin{bmatrix} a_{11}b_{11} + a_{12}b_{21} + a_{13}b_{31} & a_{11}b_{12} + a_{12}b_{22} + a_{13}b_{32} & a_{11}b_{13} + a_{12}b_{23} + a_{13}b_{33} \\ a_{21}b_{11} + a_{22}b_{21} + a_{23}b_{31} & a_{21}b_{12} + a_{22}b_{22} + a_{23}b_{32} & a_{21}b_{13} + a_{22}b_{23} + a_{23}b_{33} \\ a_{31}b_{11} + a_{32}b_{21} + a_{33}b_{31} & a_{31}b_{12} + a_{32}b_{22} + a_{33}b_{32} & a_{31}b_{13} + a_{32}b_{23} + a_{33}b_{33} \end{bmatrix}.$$

Let us now consider the quantities appearing in the partitioned matrix in the statement of the theorem. We have

$$\boldsymbol{\alpha}_{11}\boldsymbol{\beta}_{11} + \boldsymbol{\alpha}_{12}\boldsymbol{\beta}_{21} = [a_{11}]\,[b_{11}] + [a_{12}\,a_{13}] \begin{bmatrix} b_{21} \\ b_{31} \end{bmatrix}$$

$$= [a_{11}b_{11} + a_{12}b_{21} + a_{13}b_{31}]$$

$$\boldsymbol{\alpha}_{11}\boldsymbol{\beta}_{12} + \boldsymbol{\alpha}_{12}\boldsymbol{\beta}_{22} = [a_{11}]\,[b_{12}\,b_{13}] + [a_{12} \; a_{13}] \begin{bmatrix} b_{22} & b_{23} \\ b_{32} & b_{33} \end{bmatrix}$$

$$= [a_{11}b_{12} \quad a_{11}b_{13}] + [a_{12}b_{22} + a_{13}b_{32} \quad a_{12}b_{23} + a_{13}b_{33}]$$

$$= [a_{11}b_{12} + a_{12}b_{22} + a_{13}b_{32} \quad a_{11}b_{13} + a_{12}b_{23} + a_{13}b_{33}]$$

and similarly

$$\boldsymbol{\alpha}_{21}\boldsymbol{\beta}_{11} + \boldsymbol{\alpha}_{22}\boldsymbol{\beta}_{21} = \begin{bmatrix} a_{21}b_{11} + a_{22}b_{21} + a_{23}b_{31} \\ a_{31}b_{11} + a_{32}b_{21} + a_{33}b_{31} \end{bmatrix}$$

and

$$\boldsymbol{\alpha}_{21}\boldsymbol{\beta}_{12} + \boldsymbol{\alpha}_{22}\boldsymbol{\beta}_{22} = \begin{bmatrix} a_{21}b_{12} + a_{22}b_{22} + a_{23}b_{32} & a_{21}b_{13} + a_{22}b_{23} + a_{23}b_{33} \\ a_{31}b_{12} + a_{32}b_{22} + a_{33}b_{32} & a_{31}b_{13} + a_{32}b_{23} + a_{33}b_{33} \end{bmatrix}.$$

Thus we see that

$$\left[\begin{array}{c|c} \boldsymbol{\alpha}_{11}\boldsymbol{\beta}_{11} + \boldsymbol{\alpha}_{12}\boldsymbol{\beta}_{21} & \boldsymbol{\alpha}_{11}\boldsymbol{\beta}_{12} + \boldsymbol{\alpha}_{12}\boldsymbol{\beta}_{22} \\ \hline \boldsymbol{\alpha}_{21}\boldsymbol{\beta}_{21} + \boldsymbol{\alpha}_{22}\boldsymbol{\beta}_{22} & \boldsymbol{\alpha}_{21}\boldsymbol{\beta}_{12} + \boldsymbol{\alpha}_{22}\boldsymbol{\beta}_{22} \end{array}\right]$$

$$= \left[\begin{array}{c|cc} a_{11}b_{11} + a_{12}b_{21} + a_{13}b_{31} & a_{11}b_{12} + a_{12}b_{22} + a_{13}b_{32} & a_{11}b_{13} + a_{12}b_{23} + a_{13}b_{33} \\ \hline a_{21}b_{11} + a_{22}b_{21} + a_{23}b_{31} & a_{21}b_{12} + a_{22}b_{22} + a_{23}b_{32} & a_{21}b_{13} + a_{22}b_{23} + a_{23}b_{33} \\ a_{31}b_{11} + a_{32}b_{21} + a_{33}b_{31} & a_{31}b_{12} + a_{32}b_{22} + a_{33}b_{32} & a_{31}b_{13} + a_{32}b_{23} + a_{33}b_{33} \end{array}\right]$$

$$= \mathbf{AB},$$

which proves the required result for this special case. ■

The above theorem restricts itself to the situation in which the matrices are partitioned into four submatrices; the result can be extended to more general partitioning and we have the following theorem which we state without proof.

THEOREM 2.13

Suppose that **A** and **B** are matrices such that the product **AB** exists and that **A** is partitioned into submatrices $\boldsymbol{\alpha}_{ij}$ and **B** into submatrices $\boldsymbol{\beta}_{ij}$ so that

$$\mathbf{A} = \begin{bmatrix} \boldsymbol{\alpha}_{11} & \cdots & \boldsymbol{\alpha}_{1n} \\ \vdots & & \\ \boldsymbol{\alpha}_{m1} & & \boldsymbol{\alpha}_{mn} \end{bmatrix} \quad \text{and} \quad \mathbf{B} = \begin{bmatrix} \boldsymbol{\beta}_{11} & \cdots & \boldsymbol{\beta}_{1p} \\ \vdots & & \\ \boldsymbol{\beta}_{n1} & & \boldsymbol{\beta}_{np} \end{bmatrix}$$

(where, of course, all submatrices appearing in the same row of the partitioning have the same number of rows, and all submatrices appearing in the same column have the same number of columns). Then the product **AB** may be obtained by multiplying the partitioned matrices as if $\boldsymbol{\alpha}_{ij}$ and $\boldsymbol{\beta}_{ij}$ were matrix elements. This result will hold true for matrices and partitionings such that all matrix products involved exist. ■

*Examples*

2.5.1 If

$$\mathbf{A} = \begin{bmatrix} 1 & 5 & 7 & 8 \\ -3 & 2 & 0 & 4 \\ 0 & 0 & 1 & 5 \\ 0 & 0 & -3 & 2 \end{bmatrix}$$

use partitioning to evaluate $\mathbf{A}^2$. We have

$$\mathbf{A} = \begin{bmatrix} \boldsymbol{\alpha} & \boldsymbol{\beta} \\ \mathbf{0} & \boldsymbol{\alpha} \end{bmatrix}$$

where

$$\boldsymbol{\alpha} = \begin{bmatrix} 1 & 5 \\ -3 & 2 \end{bmatrix} \quad \text{and} \quad \boldsymbol{\beta} = \begin{bmatrix} 7 & 8 \\ 0 & 4 \end{bmatrix}.$$

Hence

$$\mathbf{A}^2 = \begin{bmatrix} \boldsymbol{\alpha} & \boldsymbol{\beta} \\ \mathbf{0} & \boldsymbol{\alpha} \end{bmatrix} \begin{bmatrix} \boldsymbol{\alpha} & \boldsymbol{\beta} \\ \mathbf{0} & \boldsymbol{\alpha} \end{bmatrix}$$

$$= \begin{bmatrix} \boldsymbol{\alpha}^2 & \boldsymbol{\alpha\beta} + \boldsymbol{\beta\alpha} \\ \mathbf{0} & \boldsymbol{\alpha}^2 \end{bmatrix}.$$

(We notice here how the use of partitioning gives us information about the *structure* of $\mathbf{A}^2$—the bottom left-hand 'corner' consists of zeros and top left and bottom right 'corners' are equal.) But

$$\boldsymbol{\alpha}^2 = \begin{bmatrix} 1 & 5 \\ -3 & 2 \end{bmatrix} \begin{bmatrix} 1 & 5 \\ -3 & 2 \end{bmatrix} = \begin{bmatrix} 1-15 & 5+10 \\ -3-6 & -15+4 \end{bmatrix} = \begin{bmatrix} -14 & 15 \\ -9 & -11 \end{bmatrix},$$

$$\boldsymbol{\alpha\beta} = \begin{bmatrix} 1 & 5 \\ -3 & 2 \end{bmatrix} \begin{bmatrix} 7 & 8 \\ 0 & 4 \end{bmatrix} = \begin{bmatrix} 7 & 8+20 \\ -21 & -24+8 \end{bmatrix} = \begin{bmatrix} 7 & 28 \\ -21 & -16 \end{bmatrix},$$

and

$$\boldsymbol{\beta\alpha} = \begin{bmatrix} 7 & 8 \\ 0 & 4 \end{bmatrix} \begin{bmatrix} 1 & 5 \\ -3 & 2 \end{bmatrix} = \begin{bmatrix} 7-24 & 35+16 \\ -12 & 8 \end{bmatrix} = \begin{bmatrix} -17 & 51 \\ -12 & 8 \end{bmatrix},$$

and hence

$$\mathbf{A}^2 = \begin{bmatrix} -14 & 15 & -10 & 79 \\ -9 & -11 & -33 & -8 \\ 0 & 0 & -14 & 15 \\ 0 & 0 & -9 & -11 \end{bmatrix}.$$

2.5.2 If $\mathbf{A}$ is a $4 \times 4$ matrix such that when partitioned into four $2 \times 2$ matrices it has the form

$$\mathbf{A} = \begin{bmatrix} \mathbf{X} & -\mathbf{X} \\ -\mathbf{X} & \mathbf{X} \end{bmatrix}$$

show that

$$\mathbf{A}^n = 2^{n-1} \begin{bmatrix} \mathbf{X}^n & -\mathbf{X}^n \\ -\mathbf{X}^n & \mathbf{X}^n \end{bmatrix}.$$

We use induction. Assuming the result true for $n = N$, say, we have

$$\mathbf{A}^N = 2^{N-1}\begin{bmatrix} \mathbf{X}^N & -\mathbf{X}^N \\ -\mathbf{X}^N & \mathbf{X}^N \end{bmatrix}.$$

Then

$$\mathbf{A}^{N+1} = 2^{N-1}\begin{bmatrix} \mathbf{X} & -\mathbf{X} \\ -\mathbf{X} & \mathbf{X} \end{bmatrix}\begin{bmatrix} \mathbf{X}^N & -\mathbf{X}^N \\ -\mathbf{X}^N & \mathbf{X}^N \end{bmatrix}$$

$$= 2^{N-1}\begin{bmatrix} \mathbf{X}^{N+1} + \mathbf{X}^{N+1} & -\mathbf{X}^{N+1} - \mathbf{X}^{N+1} \\ -\mathbf{X}^{N+1} - \mathbf{X}^{N+1} & \mathbf{X}^{N+1} + \mathbf{X}^{N+1} \end{bmatrix}$$

$$= 2^{N}\begin{bmatrix} \mathbf{X}^{N+1} & -\mathbf{X}^{N+1} \\ -\mathbf{X}^{N+1} & \mathbf{X}^{N+1} \end{bmatrix}.$$

Thus, if the result is true for $n = N$ it is also true for $n = N + 1$. But it is obviously true for $n = 1$; hence it is true for all positive integral $n$.

*Exercises*

2.5.1 Use a suitable partitioning to evaluate $\mathbf{A}^2$ when

$$\mathbf{A} = \begin{bmatrix} 0 & 0 & 1 & 8 \\ 0 & 0 & 3 & 2 \\ 7 & 9 & 1 & 0 \\ -8 & 6 & 0 & 1 \end{bmatrix}.$$

2.5.2 If $\mathbf{X} = [\mathbf{x}_1, \mathbf{x}_2, \mathbf{x}_3, \ldots \mathbf{x}_n]$ where the $\mathbf{x}_i$ are column vectors of order $n$ such that $\mathbf{A}\mathbf{x}_i = \lambda_i \mathbf{x}_i$ (with $\lambda_i$ a number) and

$$\mathbf{x}_i^{\mathrm{T}}\mathbf{x}_j = \begin{cases} 1 & \text{if } i = j \\ 0 & \text{if } i \neq j \end{cases}$$

show that

$$\mathbf{X}^{\mathrm{T}}\mathbf{A}\mathbf{X} = \begin{bmatrix} \lambda_1 & 0 & 0 & \cdots & & 0 \\ 0 & \lambda_2 & 0 & \cdots & & 0 \\ 0 & 0 & \lambda_3 & 0 & \cdots & 0 \\ \cdot & & & & & \\ \cdot & & & & & \\ \cdot & & & & & \\ 0 & 0 & \cdots & & & \lambda_n \end{bmatrix}.$$

2.5.3 Use partitioning to show that the product of any two matrices of the form

$$\begin{bmatrix} a & 0 & 0 \\ 0 & p & q \\ 0 & r & s \end{bmatrix}$$

is also a matrix of this form (it has the same elements equal to zero).

## 2.6 Diagonal and Triangular Matrices

There are some specially simple forms of matrices in which a number of the elements are zero. In this section we define some of these forms. Very often these forms arise naturally in the matrix formulation of scientific and engineering problems, but sometimes it is advantageous to write a general matrix in terms of several of these forms. We shall see how to do this in Theorems 2.14–2.16.

We emphasize here that this section is concerned with square matrices only.

DEFINITION 2.15

A square matrix **A** is said to be *diagonal* if its only non-zero elements occur on its principal diagonal (i.e. $A_{ij} = 0$ if $i \neq j$). □

DEFINITION 2.16

A square matrix **A** is said to be *lower triangular* if its only non-zero elements occur on and below its principal diagonal (i.e. $A_{ij} = 0$ if $j > i$). It is said to be *strictly lower triangular* if in addition the diagonal elements are zero (i.e. $A_{ij} = 0$ if $j \geqslant i$). It is said to be *unit lower triangular* if it is lower triangular with diagonal elements equal to unity. □

DEFINITION 2.17

A square matrix **A** is said to be *upper triangular* if its only non-zero elements occur on and above its principal diagonal (i.e. $A_{ij} = 0$ if $i > j$). It is said to be *strictly upper triangular* if in addition the diagonal elements are zero (i.e. $A_{ij} = 0$ if $i \geqslant j$). It is said to be *unit upper triangular* if it is upper triangular with diagonal elements equal to unity. □

*Example*

2.6.1 Classify the following matrices as either diagonal, lower triangular, strictly lower triangular, upper triangular or strictly upper triangular:

$$\text{(i)} \begin{bmatrix} 1 & 0 & 0 \\ 2 & 3 & 0 \\ -2 & 4 & 7 \end{bmatrix} \quad \text{(ii)} \begin{bmatrix} 1 & 3 & 2 \\ 0 & 1 & -4 \\ 0 & 0 & 1 \end{bmatrix} \quad \text{(iii)} \begin{bmatrix} a & 0 & 0 \\ 0 & b & 0 \\ 0 & 0 & c \end{bmatrix}$$

$$\text{(iv)} \begin{bmatrix} 0 & 0 & 0 \\ i & 0 & 0 \\ j & k & 0 \end{bmatrix}.$$

From the above definitions we have:

(i) lower triangular
(ii) (unit) upper triangular
(iii) diagonal (also lower triangular and upper triangular)
(iv) strictly lower triangular.

*Exercises*

2.6.1 If **D** is a diagonal matrix of order $n$ and **A** is square of the same order as **D**, show that **DA** is obtained from **A** by multiplying each element in the $i$th row of **A** by $D_{ii}$ $(1 \leqslant i \leqslant n)$, and that **AD** is obtained from **A** by multiplying each element in the $j$th column of **A** by $D_{jj}$ $(1 \leqslant j \leqslant n)$.

2.6.2 If $\mathbf{L}_1$ and $\mathbf{L}_2$ are two lower triangular matrices of the same order, determine whether or not $\mathbf{L}_1\mathbf{L}_2$ is also lower triangular.

2.6.3 If **L** is a lower triangular matrix and **U** an upper triangular matrix of the same order, determine whether either **LU** or **UL** has any special form.

It turns out that very often a problem involving a matrix **A** is particularly simple if **A** is triangular, and that a problem can be simplified by using triangular matrices as much as possible. One method of introducing triangular matrices is provided by the result of the following theorem.

THEOREM 2.14

Suppose that **A** is a square matrix of order $n$ and let $\Delta_p = \det(\mathbf{A}_p)$† where $\mathbf{A}_p$ is the square matrix of order $p$ formed by the first $p$ rows and columns of **A**. Then if $\Delta_p \neq 0$ $(p = 1,2, \ldots n)$ **A** can be expressed uniquely in the form $\mathbf{A} = \mathbf{LU}$ where **L** is a unit lower triangular matrix of order $n$ and **U** is an upper triangular matrix of order $n$.

*Proof*‡

We make use of induction and the idea of partitioning.

Suppose we have succeeded in the decomposition as far as the submatrix $\mathbf{A}_p$ consisting of the first $p$ rows and columns of **A** is concerned: that is, suppose we have **A** in the form

$$\mathbf{A} = \left[\begin{array}{c|c} \mathbf{L}_p\mathbf{U}_p & \mathbf{X}_p \\ \hline \underbrace{\mathbf{Y}_p}_{p \text{ columns}} & \mathbf{Z}_p \end{array}\right] \begin{array}{l} \}\, p \text{ rows} \\ \\ \end{array}.$$

We now wish to show that the decomposition can be extended to the next row and column, so we wish to prove that the square matrix of order $(p + 1)$ given by

$$\left[\begin{array}{c|c} \mathbf{L}_p\mathbf{U}_p & \mathbf{c}_{p+1} \\ \hline \underbrace{\mathbf{d}_{p+1}}_{p \text{ columns}} & \underbrace{a_{p+1,\,p+1}}_{1 \text{ column}} \end{array}\right] \begin{array}{l} \}\, p \text{ rows} \\ \}\, 1 \text{ row} \end{array}$$

can be written in the form $\mathbf{L}_{p+1}\mathbf{U}_{p+1}$. Now $a_{p+1,\,p+1}$ is, of course, just the element in the $(p + 1)$th row and column of **A**; $\mathbf{c}_{p+1}$ and $\mathbf{d}_{p+1}$ are respectively column and row vectors (portions of the $(p + 1)$th column and row of **A**).

† See Section 2.7 for the definition of det (**A**).
‡ This proof may be omitted on a first reading.

We shall show that the decomposition is possible. Let us choose

$$\mathbf{L}_{p+1} = \left[\begin{array}{c|c} \mathbf{L}_p & \mathbf{0} \\ \hline \mathbf{l}_{p+1} & 1 \end{array}\right]$$

and

$$\mathbf{U}_{p+1} = \left[\begin{array}{c|c} \mathbf{U}_p & \mathbf{u}_{p+1} \\ \hline \mathbf{0} & u_{p+1,\,p+1} \end{array}\right]$$

where $\mathbf{l}_{p+1}$ and $\mathbf{u}_{p+1}$ are, respectively, row and column vectors and $u_{p+1,\,p+1}$ is a single element; this choice preserves the required triangular properties of $\mathbf{L}$ and $\mathbf{U}$ (remembering that we require $\mathbf{L}$ to be *unit* triangular).

We now show that a suitable choice can be made so that

$$\begin{bmatrix} \mathbf{L}_p & \mathbf{0} \\ \mathbf{l}_{p+1} & 1 \end{bmatrix} \begin{bmatrix} \mathbf{U}_p & \mathbf{u}_{p+1} \\ \mathbf{0} & u_{p+1,\,p+1} \end{bmatrix} = \begin{bmatrix} \mathbf{L}_p\mathbf{U}_p & \mathbf{c}_{p+1} \\ \mathbf{d}_{p+1} & a_{p+1,\,p+1} \end{bmatrix}.$$

Multiplying out the left-hand side gives

$$\begin{bmatrix} \mathbf{L}_p\mathbf{U}_p & \mathbf{L}_p\mathbf{u}_{p+1} \\ \mathbf{l}_{p+1}\mathbf{U}_p & \mathbf{l}_{p+1}\mathbf{u}_{p+1} + u_{p+1,\,p+1} \end{bmatrix} = \begin{bmatrix} \mathbf{L}_p\mathbf{U}_p & \mathbf{c}_{p+1} \\ \mathbf{d}_{p+1} & a_{p+1,\,p+1} \end{bmatrix}$$

and hence equality is ensured provided we have

$$\mathbf{L}_p\mathbf{u}_{p+1} = \mathbf{c}_{p+1}$$
$$\mathbf{l}_{p+1}\mathbf{U}_p = \mathbf{d}_{p+1}$$
$$u_{p+1,\,p+1} = a_{p+1,\,p+1}.$$

Now we know that $\mathbf{A}_p = \mathbf{L}_p\mathbf{U}_p$ and that $\mathbf{A}_p$ is non-singular (since $\det(\mathbf{A}_p) \neq 0$—see Definition 2.19). Hence it follows by Theorem 2.18 that $\mathbf{L}_p$ and $\mathbf{U}_p$ are both non-singular, so that $\mathbf{L}_p^{-1}$ and $\mathbf{U}_p^{-1}$ exist. From the above equations we thus have

$$\mathbf{u}_{p+1} = \mathbf{L}_p^{-1}\mathbf{c}_{p+1}$$
$$\mathbf{l}_{p+1} = \mathbf{d}_{p+1}\mathbf{U}_p^{-1}$$
$$u_{p+1,\,p+1} = a_{p+1,\,p+1} - \mathbf{l}_{p+1}\mathbf{u}_{p+1}$$

which are sufficient to determine uniquely the unknown quantities.

Thus we have proved that: if $\mathbf{A}_p = \mathbf{L}_p\mathbf{U}_p$ and if $\mathbf{A}_p$ is non-singular then $\mathbf{A}_{p+1}$ can be written in the form $\mathbf{L}_{p+1}\mathbf{U}_{p+1}$.

But it is easy to show that $\mathbf{A}_2 = \mathbf{L}_2\mathbf{U}_2$

(for $\begin{bmatrix} a & b \\ c & d \end{bmatrix} = \begin{bmatrix} 1 & 0 \\ c/a & 1 \end{bmatrix} \begin{bmatrix} a & b \\ 0 & (ad-bc)/a \end{bmatrix}$—valid if $a \neq 0$).

Hence it follows that the result is true for all submatrices $\mathbf{A}_2, \mathbf{A}_3, \ldots, \mathbf{A}_n$ and hence for the complete matrix $\mathbf{A}$ (= $\mathbf{A}_n$). ∎

There are the following additional theorems on decomposition.

THEOREM 2.15

If **A** is a square matrix satisfying the same conditions as in Theorem 2.14 then **A** can be written in the form

$$\mathbf{A} = \mathbf{LU}$$

where **L** is a lower triangular matrix and **U** is a unit upper triangular matrix.

*Proof*

Exactly similar to that of Theorem 2.14. ■

THEOREM 2.16

If **A** is a square matrix satisfying the same conditions as those of Theorem 2.14 then **A** can be written in the form

$$\mathbf{A} = \mathbf{LDU}$$

where **L** is a unit lower triangular matrix, **D** is a diagonal matrix, and **U** is a unit upper triangular matrix.

*Proof*

By Theorem 2.14, all that we need to show is that if **U** is an upper triangular matrix then it can be written in the form $\mathbf{U} = \mathbf{DU}_1$, where $\mathbf{U}_1$ is a unit upper triangular matrix.

That this is possible is easily verified by noting that

$$\begin{bmatrix} u_{11} & u_{12} & u_{13} & \ldots & u_{1n} \\ 0 & u_{22} & u_{23} & \ldots & u_{2n} \\ 0 & 0 & u_{33} & \ldots & u_{3n} \\ \cdot & & & & \\ \cdot & & & & \\ \cdot & & & & \\ 0 & \ldots & \ldots & \ldots & u_{nn} \end{bmatrix}$$

$$= \begin{bmatrix} u_{11} & 0 & 0 & \ldots & 0 \\ 0 & u_{22} & 0 & \ldots & 0 \\ 0 & 0 & u_{33} & 0\ldots & 0 \\ \cdot & & & & \\ \cdot & & & & \\ \cdot & & & & \\ 0 & \ldots & \ldots & \ldots & u_{nn} \end{bmatrix} \begin{bmatrix} 1 & u_{12}/u_{11} & u_{13}/u_{11} & \ldots & u_{1n}/u_{11} \\ 0 & 1 & u_{23}/u_{22} & \ldots & u_{2n}/u_{22} \\ 0 & 0 & 1 & \ldots & u_{3n}/u_{33} \\ \cdot & & & & \\ \cdot & & & & \\ \cdot & & & & \\ 0 & 0 & \ldots & \ldots & 1 \end{bmatrix}.$$

The conditions of the theorem ensure that none of the $u_{ii}$ can be zero, and hence all the quotients above do in fact exist. ■

*Examples*

2.6.2 For each of the following matrices investigate whether or not triangular decomposition is possible:

$$\text{(i)} \begin{bmatrix} 1 & 2 & 4 \\ 2 & 6 & 6 \\ 3 & 6 & 9 \end{bmatrix} \quad \text{(ii)} \begin{bmatrix} 2 & 3 & 7 \\ 4 & 6 & 0 \\ 8 & 9 & 1 \end{bmatrix} \quad \text{(iii)} \begin{bmatrix} 0 & 1 \\ 4 & 7 \end{bmatrix} \quad \text{(iv)} \begin{bmatrix} a & b \\ -b & a \end{bmatrix}.$$

Using the notation of Theorem 2.14 we have

$$\text{(i)}\ \mathbf{A}_1 = [1], \qquad \Delta_1 = 1$$

$$\mathbf{A}_2 = \begin{bmatrix} 1 & 2 \\ 2 & 6 \end{bmatrix}, \qquad \Delta_2 = 2$$

$$\mathbf{A}_3 = \begin{bmatrix} 1 & 2 & 4 \\ 2 & 6 & 6 \\ 3 & 6 & 9 \end{bmatrix}, \qquad \Delta_3 = -6.$$

Thus $\Delta_p \neq 0$ ($p = 1,2,3$) and so decomposition is possible.

$$\text{(ii)}\ \mathbf{A}_1 = [2], \qquad \Delta_1 = 2$$

$$\mathbf{A}_2 = \begin{bmatrix} 2 & 3 \\ 4 & 6 \end{bmatrix}, \qquad \Delta_2 = 0.$$

Hence decomposition is not possible, since one of the $\Delta_p$ is zero.

$$\text{(iii)}\ \mathbf{A}_1 = [0], \qquad \Delta_1 = 0.$$

Hence decomposition is not possible.

$$\text{(iv)}\ \mathbf{A}_1 = [a], \qquad \Delta_1 = a$$

$$\mathbf{A}_2 = \begin{bmatrix} a & b \\ -b & a \end{bmatrix}, \qquad \Delta_2 = a^2 + b^2.$$

Hence $\Delta_p \neq 0$ ($p = 1,2$) provided $a$ is non-zero, and, subject to this condition decomposition is possible.

2.6.3 For the matrix in 2.6.2 (i) above, obtain the decomposition in the form **LU** with **U** unit upper triangular. Let

$$\mathbf{L} = \begin{bmatrix} l_{11} & 0 & 0 \\ l_{21} & l_{22} & 0 \\ l_{31} & l_{32} & l_{33} \end{bmatrix} \quad \text{and} \quad \mathbf{U} = \begin{bmatrix} 1 & u_{12} & u_{13} \\ 0 & 1 & u_{23} \\ 0 & 0 & 1 \end{bmatrix}.$$

Then we have

$$\mathbf{LU} = \begin{bmatrix} l_{11} & 0 & 0 \\ l_{21} & l_{22} & 0 \\ l_{31} & l_{32} & l_{33} \end{bmatrix} \begin{bmatrix} 1 & u_{12} & u_{13} \\ 0 & 1 & u_{23} \\ 0 & 0 & 1 \end{bmatrix}$$

$$= \begin{bmatrix} l_{11} & l_{11}u_{12} & l_{11}u_{13} \\ l_{21} & l_{21}u_{12} + l_{22} & l_{21}u_{13} + l_{22}u_{23} \\ l_{31} & l_{31}u_{12} + l_{32} & l_{31}u_{13} + l_{32}u_{23} + l_{33} \end{bmatrix},$$

and this matrix has to equal

$$\begin{bmatrix} 1 & 2 & 4 \\ 2 & 6 & 6 \\ 3 & 6 & 9 \end{bmatrix}.$$

Hence we have

$$\begin{array}{lll} l_{11} = 1 & l_{11}u_{12} = 2 & l_{11}u_{13} = 4 \\ l_{21} = 2 & l_{21}u_{12} + l_{22} = 6 & l_{21}u_{13} + l_{22}u_{23} = 6 \\ l_{31} = 3 & l_{31}u_{12} + l_{32} = 6 & l_{31}u_{13} + l_{32}u_{23} + l_{33} = 9. \end{array}$$

These equations give in turn $l_{11} = 1, l_{21} = 2, l_{31} = 3, u_{12} = 2, l_{22} = 2, l_{32} = 0, u_{13} = 4,$ $u_{23} = -1, l_{33} = -3$, and hence we have

$$\begin{bmatrix} 1 & 2 & 4 \\ 2 & 6 & 6 \\ 3 & 6 & 9 \end{bmatrix} = \begin{bmatrix} 1 & 0 & 0 \\ 2 & 2 & 0 \\ 3 & 0 & -3 \end{bmatrix} \begin{bmatrix} 1 & 2 & 4 \\ 0 & 1 & -1 \\ 0 & 0 & 1 \end{bmatrix}.$$

*Exercises*

2.6.4 For each of the following matrices determine whether or not a triangular decomposition is possible, and if it is write the matrix in the form **LDU** where **L** is unit lower triangular, **D** is diagonal and **U** is unit upper triangular:

(i) $\begin{bmatrix} 0 & 1 \\ 2 & 4 \end{bmatrix}$ (ii) $\begin{bmatrix} 1 & 0 \\ 4 & 2 \end{bmatrix}$ (iii) $\begin{bmatrix} 2 & 4 & 1 \\ -1 & 3 & -2 \\ 2 & -3 & 5 \end{bmatrix}$.

2.6.5 If **A** is symmetric and satisfies the conditions of Theorem 2.14 show that it is possible to decompose it into the form $\mathbf{A} = \mathbf{LDL}^T$ where **L** is unit lower triangular and **D** is diagonal. Deduce that it is possible to decompose **A** into the form $\mathbf{A} = \mathbf{LL}^T$ where **L** is lower triangular.

## 2.7 Determinants of Matrices

If **A** is a square matrix then it is possible to form the determinant of the array constituting **A** according to the definitions of Chapter 1. We shall denote this determinant by det (**A**). (An alternative notation sometimes used is $|\mathbf{A}|$.) Thus, if

$$\mathbf{A} = \begin{bmatrix} a_{11} & a_{12} & \cdots & a_{1n} \\ a_{21} & & & \\ a_{31} & & & \\ \cdot & & & \\ \cdot & & & \\ \cdot & & & \\ a_{n1} & & & a_{nn} \end{bmatrix}$$

then we have

$$\det(\mathbf{A}) = \begin{vmatrix} a_{11} & a_{12} & \cdots & a_{1n} \\ a_{21} & & & \cdot \\ \cdot & & & \cdot \\ \cdot & & & \cdot \\ \cdot & & & \cdot \\ a_{n1} & & & a_{nn} \end{vmatrix}.$$

The determinants of diagonal and triangular matrices are particularly easy to calculate, as is shown by the following theorem.

THEOREM 2.17

If **A** is an upper or lower triangular or diagonal matrix then det (**A**) is equal to the product of the diagonal elements of **A**.

*Proof*

Again we shall use induction; we give the proof for lower triangular **A**, but it is exactly similar for upper triangular, and diagonal is merely a special case of triangular so needs no separate proof.

Assume the result to be true for all matrices up to order $N$; then for order $N + 1$ we have

$$\det(\mathbf{A}) = \begin{vmatrix} a_{11} & 0 & 0 & \cdots\cdots & 0 \\ a_{21} & a_{22} & 0 & \cdots\cdots & 0 \\ a_{31} & a_{32} & a_{33} & 0 \quad \cdots & 0 \\ \cdot & & & & \\ \cdot & & & & \\ \cdot & & & & \\ a_{N+1,1} & \cdots & \cdots & \cdots & a_{N+1,N+1} \end{vmatrix}$$

$$= a_{11} \begin{vmatrix} a_{22} & 0 & \dots & 0 \\ a_{32} & a_{33} & \dots & 0 \\ \vdots & & & \\ a_{N+1,2} & \dots\dots & & a_{N+1,N+1} \end{vmatrix}$$

$= a_{11}(a_{22}\,a_{33} \dots a_{N+1,N+1})$

since the determinant is of a matrix of order $N$

$= a_{11}\,a_{22}\,a_{33} \dots a_{N+1,N+1}$

= product of diagonal elements of **A**.

Thus, if the result is true for matrices of order $N$ it is also true for matrices of order $(N + 1)$. But it is obviously true for $N = 2$

(by direct expansion $\begin{vmatrix} a_{11} & 0 \\ a_{21} & a_{22} \end{vmatrix} = a_{11}a_{22}$);

hence it is true for matrices of all orders. ■

The following theorem concerns the determinant of the product of two matrices. It is an example of the type of theorem for which the statement is extremely simple and easy to understand, but which is not particularly easy to prove.

THEOREM 2.18

If **A** and **B** are square matrices of the same order then det (**AB**) = det (**A**) × det (**B**).

*Proof*

As a lemma we first prove that if **L** is a unit lower triangular matrix then det (**LA**) = det (**A**) (in fact this is a special case of the theorem). To prove this, we let

$$\mathbf{L} = \begin{bmatrix} 1 & 0 & 0 & \dots\dots & 0 \\ l_{21} & 1 & 0 & \dots\dots & 0 \\ l_{31} & l_{32} & 1 & 0 \dots & 0 \\ \vdots & & & & \\ l_{n1} & l_{n2} & \dots & l_{n,n-1} & 1 \end{bmatrix}$$

and

$$\mathbf{A} = \begin{bmatrix} a_{11} & a_{12} & \dots & a_{1n} \\ a_{21} & & & \\ a_{31} & & & \\ \vdots & & & \\ a_{n1} & & & a_{nn} \end{bmatrix}$$

so that

$$\mathbf{LA} = \begin{bmatrix} a_{11} & a_{12} & \cdots & a_{1n} \\ l_{21}a_{11} + a_{21} & l_{21}a_{12} + a_{22} & & l_{21}a_{1n} + a_{2n} \\ l_{31}a_{11} + l_{32}a_{21} + a_{31} & l_{31}a_{12} + l_{32}a_{22} + a_{32} & & l_{31}a_{1n} + l_{32}a_{2n} + a_{3n} \\ \vdots & & & \vdots \\ l_{n1}a_{11} + l_{n2}a_{21} + \ldots + l_{n,n-1}a_{n-1,1} + a_{n1} & l_{n1}a_{12} + l_{n2}a_{22} + \ldots + a_{n2} & & l_{n1}a_{1n} + l_{n2}a_{2n} + l_{n3}a_{3n} + \ldots + l_{n,n-1}a_{n-1,n} + a_{nn} \end{bmatrix}.$$

But this matrix is just formed from the original matrix **A** by adding suitable multiples of rows from **A** to the original row to produce the new matrix and by Theorem 1.15 the value of the determinant is unaltered by operations of this type. It follows that det (**LA**) = det (**A**) and so the lemma is proved.

Now consider the product of the partitioned matrices

$$\begin{bmatrix} \mathbf{I} & \mathbf{0} \\ \mathbf{A} & \mathbf{I} \end{bmatrix} \quad \text{and} \quad \begin{bmatrix} \mathbf{B} & -\mathbf{I} \\ \mathbf{0} & \mathbf{A} \end{bmatrix}$$

where **I** is the unit matrix and **0** the zero matrix of the same order as **A** and **B**. Carrying out the multiplication we have

$$\begin{bmatrix} \mathbf{I} & \mathbf{0} \\ \mathbf{A} & \mathbf{I} \end{bmatrix} \begin{bmatrix} \mathbf{B} & -\mathbf{I} \\ \mathbf{0} & \mathbf{A} \end{bmatrix} = \begin{bmatrix} \mathbf{IB} + \mathbf{00} & \mathbf{I}(-\mathbf{I}) + \mathbf{0A} \\ \mathbf{AB} + \mathbf{I0} & \mathbf{A}(-\mathbf{I}) + \mathbf{IA} \end{bmatrix} = \begin{bmatrix} \mathbf{B} & -\mathbf{I} \\ \mathbf{AB} & \mathbf{0} \end{bmatrix}.$$

But

$$\begin{bmatrix} \mathbf{I} & \mathbf{0} \\ \mathbf{A} & \mathbf{I} \end{bmatrix}$$

is a unit lower triangular matrix; hence by the lemma we have

$$\begin{vmatrix} \mathbf{B} & -\mathbf{I} \\ \mathbf{0} & \mathbf{A} \end{vmatrix} = \begin{vmatrix} \mathbf{B} & -\mathbf{I} \\ \mathbf{AB} & \mathbf{0} \end{vmatrix}.$$

But if we expand the determinant on the left-hand side above by the Laplace expansion (Theorem 1.18) we obtain det (**A**) det (**B**) while if we expand the determinant on the right-hand side in the same way we obtain det (**AB**). Thus we have det (**AB**) = det (**A**) det (**B**) and the required result is proved. ■

N.B. It is important when using the above theorem to remember that **A** and **B** must be square and of the same order. For it is possible that **AB** is square so that det (**AB**) exists when **A** and **B** are rectangular so that det (**A**) and det (**B**) are undefined.

*Examples*

2.7.1 Evaluate

$$\begin{vmatrix} 1 & 2 & 4 \\ 2 & 6 & 6 \\ 3 & 6 & 9 \end{vmatrix}$$

using the result of Example 2.6.3. We have

$$\begin{bmatrix} 1 & 2 & 4 \\ 2 & 6 & 6 \\ 3 & 6 & 9 \end{bmatrix} = \begin{bmatrix} 1 & 0 & 0 \\ 2 & 2 & 0 \\ 3 & 0 & -3 \end{bmatrix} \begin{bmatrix} 1 & 2 & 4 \\ 0 & 1 & -1 \\ 0 & 0 & 1 \end{bmatrix}$$

and hence, by Theorem 2.18

$$\begin{vmatrix} 1 & 2 & 4 \\ 2 & 6 & 6 \\ 3 & 6 & 9 \end{vmatrix} = \begin{vmatrix} 1 & 0 & 0 \\ 2 & 2 & 0 \\ 3 & 0 & -3 \end{vmatrix} \begin{vmatrix} 1 & 2 & 4 \\ 0 & 1 & -1 \\ 0 & 0 & 1 \end{vmatrix}$$

$$= (1 \times 2 \times -3)(1 \times 1 \times 1) \qquad \text{by Theorem 2.17}$$

$$= -6.$$

2.7.2 If

$$\mathbf{A} = \begin{bmatrix} \cos\theta & \sin\theta & 0 \\ -\sin\theta & \cos\theta & 0 \\ 0 & 0 & 1 \end{bmatrix}$$

$$\mathbf{B} = \begin{bmatrix} 1 & 0 & 0 \\ 0 & \cos\phi & \sin\phi \\ 0 & -\sin\phi & \cos\phi \end{bmatrix}$$

and

$$\mathbf{C} = \begin{bmatrix} \cos\psi & \sin\psi & 0 \\ -\sin\psi & \cos\psi & 0 \\ 0 & 0 & 1 \end{bmatrix}$$

evaluate det (**ABC**).

We have, for det (**A**), by expanding down the third column,

$$\det(\mathbf{A}) = \cos^2\theta - (-\sin\theta)\sin\theta$$
$$= \cos^2\theta + \sin^2\theta$$
$$= 1.$$

Similarly

$$\det(\mathbf{B}) = \det(\mathbf{C}) = 1$$

and hence

det $(\mathbf{ABC}) = \det(\mathbf{A}) \times \det(\mathbf{B}) \times \det(\mathbf{C}) = 1$.

*Exercises*

2.7.1 Use triangular decomposition to evaluate det $(\mathbf{A})$ when

$$\mathbf{A} = \begin{bmatrix} 1 & 2 & 3 \\ 2 & 5 & 8 \\ 3 & 8 & 14 \end{bmatrix}.$$

2.7.2 (i) Show that det $(\mathbf{A}^T) = \det(\mathbf{A})$. (ii) If $\mathbf{AA}^T = \mathbf{I}$ show that det $(\mathbf{A}) = \pm 1$.

## 2.8 The Inverse of a Matrix

In the algebra of real and complex numbers, the existence of an inverse is examined at an early stage; given a number $x$ we seek a number $y$ with the properties $xy = yx = 1$. We then denote this number by $x^{-1}$ so that we have $xx^{-1} = x^{-1}x = 1$ and then all the laws of indices will extend to negative powers, provided we define $x^{-n} = (x^{-1})^n$.

We have already introduced the idea of the unit matrix $\mathbf{I}$. It is thus natural to ask whether for a given matrix $\mathbf{A}$ there exists another matrix $\mathbf{B}$ with the property $\mathbf{AB} = \mathbf{BA} = \mathbf{I}$. This matrix would then be denoted by $\mathbf{A}^{-1}$ so that we should have $\mathbf{AA}^{-1} = \mathbf{A}^{-1}\mathbf{A} = \mathbf{I}$ and the laws of indices for matrices would extend to negative powers provided we define $\mathbf{A}^{-n}$ as $(\mathbf{A}^{-1})^n$.

We are thus led to the following formal definition.

DEFINITION 2.18

The *inverse* of a given matrix $\mathbf{A}$ is a matrix $\mathbf{A}^{-1}$ such that $\mathbf{AA}^{-1} = \mathbf{A}^{-1}\mathbf{A} = \mathbf{I}$. (Note that we have as yet no guarantee that inverse matrices exist.)

We see immediately that if $\mathbf{AA}^{-1} = \mathbf{A}^{-1}\mathbf{A}$ then $\mathbf{A}$ and $\mathbf{A}^{-1}$ must be square and of the same order. Hence, if an inverse does exist, it will be for square matrices only, and in such a case it will be square of the same order as the original matrix.

We may also show that if an inverse exists then it is unique; for suppose that $\mathbf{B}$ and $\mathbf{C}$ are two different inverses for $\mathbf{A}$ so that $\mathbf{AB} = \mathbf{BA} = \mathbf{I}$ and $\mathbf{AC} = \mathbf{CA} = \mathbf{I}$. Then we have

$$\begin{aligned} \mathbf{B} &= \mathbf{IB} \\ &= \mathbf{CAB} \\ &= \mathbf{C}(\mathbf{AB}) \\ &= \mathbf{CI} \\ &= \mathbf{C} \end{aligned}$$

and hence the supposition that $\mathbf{B}$ was different from $\mathbf{C}$ has been proved false.

To investigate the existence of the inverse we first define the adjugate (or adjoint) matrix of a given square matrix $\mathbf{A}$.

DEFINITION 2.19

Suppose that **A** is a square matrix of order $n$; denote the cofactor of the element $a_{ij}$ of det (**A**) by $\alpha_{ij}$. Then the adjugate of **A**, denoted by adj (**A**), is the square matrix of order $n$ with elements defined by

$$\{\text{adj}(\mathbf{A})\}_{ij} = \alpha_{ji}.$$ □

(Concisely—the adjugate is the transpose of the matrix of cofactors.)

*Examples*

2.8.1 If

$$\mathbf{A} = \begin{bmatrix} 1 & 0 & -4 \\ 2 & 4 & 1 \\ 0 & 3 & 6 \end{bmatrix}$$

determine adj (**A**).

We first determine all the cofactors of det (**A**). Using the definition of cofactors (p. 6) we have

$$\alpha_{11} = \begin{vmatrix} 4 & 1 \\ 3 & 6 \end{vmatrix} = 21 \qquad \alpha_{12} = -\begin{vmatrix} 2 & 1 \\ 0 & 6 \end{vmatrix} = -12 \qquad \alpha_{13} = \begin{vmatrix} 2 & 4 \\ 0 & 3 \end{vmatrix} = 6$$

$$\alpha_{21} = -\begin{vmatrix} 0 & -4 \\ 3 & 6 \end{vmatrix} = -12 \quad \alpha_{22} = \begin{vmatrix} 1 & -4 \\ 0 & 6 \end{vmatrix} = 6 \qquad \alpha_{23} = -\begin{vmatrix} 1 & 0 \\ 0 & 3 \end{vmatrix} = -3$$

$$\alpha_{31} = \begin{vmatrix} 0 & -4 \\ 4 & 1 \end{vmatrix} = 16 \qquad \alpha_{32} = -\begin{vmatrix} 1 & -4 \\ 2 & 1 \end{vmatrix} = -9 \quad \alpha_{33} = \begin{vmatrix} 1 & 0 \\ 2 & 4 \end{vmatrix} = 4.$$

Thus

$$\begin{bmatrix} \alpha_{11} & \alpha_{12} & \alpha_{13} \\ \alpha_{21} & \alpha_{22} & \alpha_{23} \\ \alpha_{31} & \alpha_{32} & \alpha_{33} \end{bmatrix} = \begin{bmatrix} 21 & -12 & 6 \\ -12 & 6 & -3 \\ 16 & -9 & 4 \end{bmatrix},$$

and so

$$\text{adj}(\mathbf{A}) = \begin{bmatrix} \alpha_{11} & \alpha_{21} & \alpha_{31} \\ \alpha_{12} & \alpha_{22} & \alpha_{32} \\ \alpha_{13} & \alpha_{23} & \alpha_{33} \end{bmatrix} = \begin{bmatrix} 21 & -12 & 16 \\ -12 & 6 & -9 \\ 6 & -3 & 4 \end{bmatrix}.$$

We are now in a position to decide when a square matrix **A** has an inverse and, indeed, to present an explicit formula for the inverse.

THEOREM 2.19

If **A** is a square matrix of order $n$ such that det $(\mathbf{A}) \neq 0$ then the inverse exists and is given by

$$\mathbf{A}^{-1} = \frac{1}{\det(\mathbf{A})}\,\text{adj}(\mathbf{A}).$$

*Proof*

The result is proved if we show that

$$\mathbf{A}\,.\left(\frac{\text{adj}\,(\mathbf{A})}{\det\,(\mathbf{A})}\right) = \left(\frac{\text{adj}\,(\mathbf{A})}{\det\,(\mathbf{A})}\right).\,\mathbf{A} = \mathbf{I}.$$

Since adj $(\mathbf{A})$ is a square matrix of the same order as $\mathbf{A}$, then $\mathbf{A}\,.(\text{adj}\,(\mathbf{A})/\det\,(\mathbf{A}))$ is also a square matrix of the same order as $\mathbf{A}$ and we have

$$\left\{\mathbf{A}\,.\left(\frac{\text{adj}\,(\mathbf{A})}{\det\,(\mathbf{A})}\right)\right\}_{ij} = \frac{1}{\det\,(\mathbf{A})}\{\mathbf{A}\,.\,\text{adj}\,(\mathbf{A})\}_{ij}$$

$$= \frac{1}{\det\,(\mathbf{A})}\sum_{k=1}^{n} A_{ik}\,\text{adj}\,(\mathbf{A})_{kj} \qquad \text{(by Definition 2.9)}$$

$$= \frac{1}{\det\,(\mathbf{A})}\sum_{k=1}^{n} A_{ik}\alpha_{jk} \qquad \text{(by Definition 2.19).}$$

But

$$\sum_{k=1}^{n} A_{ik}\alpha_{jk} = \begin{cases} \det\,(\mathbf{A}) & \text{if } i = j \\ 0 & \text{if } i \neq j \end{cases} \qquad \text{(by Theorem 1.9)}$$

and hence

$$\left\{\mathbf{A}\,.\left(\frac{\text{adj}\,(\mathbf{A})}{\det\,(\mathbf{A})}\right)\right\}_{ij} = \begin{cases} 1 & \text{if } i = j \\ 0 & \text{if } i \neq j. \end{cases}$$

But these are just the elements of the unit matrix and hence we have

$$\mathbf{A}\,.\left(\frac{\text{adj}\,(\mathbf{A})}{\det\,(\mathbf{A})}\right) = \mathbf{I}.$$

In an exactly similar way we could prove that

$$\left(\frac{\text{adj}\,(\mathbf{A})}{\det\,(\mathbf{A})}\right).\,\mathbf{A} = \mathbf{I}$$

and thus the required result is proved. ■

We have just shown that if $\det\,(\mathbf{A}) \neq 0$, then a (unique) inverse exists. We now proceed to show that if $\det\,(\mathbf{A}) = 0$, then an inverse cannot exist.

THEOREM 2.20

If $\mathbf{A}$ is a square matrix such that $\det\,(\mathbf{A}) = 0$ then $\mathbf{A}^{-1}$ does not exist.

*Proof*

Suppose that $\mathbf{A}^{-1}$ does exist; then

$$\mathbf{A}\mathbf{A}^{-1} = \mathbf{I}$$

and hence

$$\det(\mathbf{A}\mathbf{A}^{-1}) = \det(\mathbf{I})$$

giving

$$\det(\mathbf{A}) \,.\, \det(\mathbf{A}^{-1}) = 1 \qquad \text{(using Theorem 2.18).}$$

But det $(\mathbf{A}) = 0$ and so we obtain $0 = 1$—an obvious contradiction—and hence the original assumption that $\mathbf{A}^{-1}$ exists must be false. ■

*Example*

2.8.2 If

$$\mathbf{A} = \begin{bmatrix} 1 & 0 & -4 \\ 2 & 4 & 1 \\ 0 & 3 & 6 \end{bmatrix}$$

evaluate $\mathbf{A}^{-1}$ if it exists.

We first evaluate det $(\mathbf{A})$; if det $(\mathbf{A}) \neq 0$ then we may deduce that $\mathbf{A}^{-1}$ exists and proceed to calculate it.

We have

$$\det(\mathbf{A}) = \begin{vmatrix} 1 & 0 & -4 \\ 2 & 4 & 1 \\ 0 & 3 & 6 \end{vmatrix}$$

$$= \begin{vmatrix} 1 & 0 & 0 \\ 2 & 4 & 9 \\ 0 & 3 & 6 \end{vmatrix}$$

$$= 24 - 27 = -3.$$

Thus $\mathbf{A}^{-1}$ exists and is given by

$$\mathbf{A}^{-1} = \text{adj}(\mathbf{A})/\det(\mathbf{A}).$$

We have already in a previous Example (2.8.1) obtained adj $(\mathbf{A})$:

$$\text{adj}(\mathbf{A}) = \begin{bmatrix} 21 & -12 & 16 \\ -12 & 6 & -9 \\ 6 & -3 & 4 \end{bmatrix}.$$

Thus

$$\mathbf{A}^{-1} = -\tfrac{1}{3} \begin{bmatrix} 21 & -12 & 16 \\ -12 & 6 & -9 \\ 6 & -3 & 4 \end{bmatrix} = \begin{bmatrix} -7 & 4 & -\frac{16}{3} \\ 4 & -2 & 3 \\ -2 & 1 & -\frac{4}{3} \end{bmatrix}.$$

We must emphasize that the main importance of Theorem 2.19 is that it provides a theoretical basis for the existence of the inverse. In general for matrices of order greater than about three, it does *not* provide a good means of actually calculating the inverse. Methods for performing this calculation will be discussed in Chapter 3.

In connection with the existence of an inverse, the terms singular and non-singular are often used. These are defined as follows.

DEFINITION 2.20

A square matrix is said to be *singular* if $\det(\mathbf{A}) = 0$; if $\det(\mathbf{A}) \neq 0$ it is said to be *non-singular.* □

Thus **A** possesses an inverse if and only if it is both square and non-singular.

*Exercises*

2.8.1 If

$$\mathbf{A} = \begin{bmatrix} a & b \\ c & a \end{bmatrix}$$

determine the conditions under which $\mathbf{A}^{-1}$ exists; if the inverse does exist use the result

$$\mathbf{A}^{-1} = \frac{\operatorname{adj}(\mathbf{A})}{\det(\mathbf{A})}$$

to obtain $\mathbf{A}^{-1}$ and verify that the matrix so obtained satisfies

$$\mathbf{A}\mathbf{A}^{-1} = \mathbf{I}.$$

2.8.2 If **A**, **B**, **C**, **D** are square matrices of the same order such that $\mathbf{ABC} = \mathbf{D}$, show that $\mathbf{B} = \mathbf{A}^{-1}\mathbf{D}\mathbf{C}^{-1}$ provided the appropriate inverses exist.

Hence find the matrix **A** which satisfies

$$\begin{bmatrix} 1 & 6 \\ 2 & 9 \end{bmatrix} \mathbf{A} \begin{bmatrix} 1 & -4 \\ -2 & 3 \end{bmatrix} = \begin{bmatrix} 8 & 6 \\ 4 & 3 \end{bmatrix}.$$

2.8.3 Find the inverse of

$$\mathbf{A} = \begin{bmatrix} 0 & 1 & 1 \\ 3 & 2 & -1 \\ 4 & 7 & 4 \end{bmatrix}.$$

2.8.4 Show that in general $(\mathbf{A} + \mathbf{B})^{-1} \neq \mathbf{A}^{-1} + \mathbf{B}^{-1}$.

There is a result concerning the inverse of a product rather similar to Theorem 2.10 concerning the transpose of a product. We give this result in the following theorem.

THEOREM 2.21

If **A**, **B**, **C**, . . . **M**, **N** are all non-singular matrices of the same order, then

$$(\mathbf{ABC} \ldots \mathbf{MN})^{-1} = \mathbf{N}^{-1}\mathbf{M}^{-1} \ldots \mathbf{C}^{-1}\mathbf{B}^{-1}\mathbf{A}^{-1}.$$

*Proof*

We first prove the result for two matrices; we thus wish to show that $(\mathbf{AB})^{-1} = \mathbf{B}^{-1}\mathbf{A}^{-1}$. This will be so if $(\mathbf{AB})(\mathbf{B}^{-1}\mathbf{A}^{-1}) = (\mathbf{B}^{-1}\mathbf{A}^{-1})(\mathbf{AB}) = \mathbf{I}$. But

$$\begin{aligned}(\mathbf{AB})(\mathbf{B}^{-1}\mathbf{A}^{-1}) &= \mathbf{A}(\mathbf{BB}^{-1})\mathbf{A}^{-1}\\ &= \mathbf{AIA}^{-1}\\ &= \mathbf{AA}^{-1}\\ &= \mathbf{I}\end{aligned}$$

and

$$\begin{aligned}(\mathbf{B}^{-1}\mathbf{A}^{-1})(\mathbf{AB}) &= \mathbf{B}^{-1}(\mathbf{A}^{-1}\mathbf{A})\mathbf{B}\\ &= \mathbf{B}^{-1}\mathbf{IB}\\ &= \mathbf{B}^{-1}\mathbf{B}\\ &= \mathbf{I}\end{aligned}$$

and hence the required result for two matrices is proved.

We thus have, for the general case,

$$\begin{aligned}(\mathbf{ABC}\ldots\mathbf{MN})^{-1} &= \{\mathbf{A}(\mathbf{BC}\ldots\mathbf{MN})\}^{-1}\\ &= (\mathbf{BC}\ldots\mathbf{MN})^{-1}\mathbf{A}^{-1} && \text{(by the result just proved for two matrices)}\\ &= \{\mathbf{B}(\mathbf{C}\ldots\mathbf{MN})\}^{-1}\mathbf{A}^{-1}\\ &= (\mathbf{C}\ldots\mathbf{MN})^{-1}\mathbf{B}^{-1}\mathbf{A}^{-1} && \text{(again using the result just proved).}\end{aligned}$$

Obviously this process can be repeated as many times as is necessary to give the required result. ■

The idea of the inverse can give us an insight into the fact already noted that if $\mathbf{AB} = \mathbf{0}$ we cannot in general deduce that either $\mathbf{A} = \mathbf{0}$ or $\mathbf{B} = \mathbf{0}$. However, if $\mathbf{A}$ is square and non-singular then we know that $\mathbf{A}^{-1}$ exists so we can premultiply both sides of $\mathbf{AB} = \mathbf{0}$ by $\mathbf{A}^{-1}$ to give

$$\mathbf{A}^{-1}(\mathbf{AB}) = \mathbf{A}^{-1}\mathbf{0}.$$

Thus $\mathbf{IB} = \mathbf{0}$ and so $\mathbf{B} = \mathbf{0}$.

Similarly if $\mathbf{B}$ is non-singular we may postmultiply $\mathbf{AB} = \mathbf{0}$ by $\mathbf{B}^{-1}$ to deduce that $\mathbf{A} = \mathbf{0}$.

In a similar manner we may deduce that if $\mathbf{AB} = \mathbf{AC}$ then $\mathbf{B} = \mathbf{C}$ if $\mathbf{A}$ is square and non-singular.

Sometimes the use of partitioning can help to find the inverse. For instance, suppose we have a matrix partitioned as shown, where $\mathbf{X}, \mathbf{Y}, \mathbf{Z}$ are square and non-singular:

$$\mathbf{A} = \begin{bmatrix}\mathbf{X} & \mathbf{Y}\\ \mathbf{0} & \mathbf{Z}\end{bmatrix}.$$

Let us try to find the inverse, partitioned in the same way; thus if

$$\mathbf{A}^{-1} = \begin{bmatrix}\mathbf{P} & \mathbf{Q}\\ \mathbf{R} & \mathbf{S}\end{bmatrix}$$

we must have

$$\begin{bmatrix} \mathbf{X} & \mathbf{Y} \\ \mathbf{0} & \mathbf{Z} \end{bmatrix} \begin{bmatrix} \mathbf{P} & \mathbf{Q} \\ \mathbf{R} & \mathbf{S} \end{bmatrix} = \begin{bmatrix} \mathbf{I} & \mathbf{0} \\ \mathbf{0} & \mathbf{I} \end{bmatrix}$$

where we have partitioned the unit matrix in the appropriate manner.

Carrying out the multiplication gives

$$\begin{bmatrix} \mathbf{XP} + \mathbf{YR} & \mathbf{XQ} + \mathbf{YS} \\ \mathbf{ZR} & \mathbf{ZS} \end{bmatrix} = \begin{bmatrix} \mathbf{I} & \mathbf{0} \\ \mathbf{0} & \mathbf{I} \end{bmatrix}$$

and hence

$$\mathbf{XP} + \mathbf{YR} = \mathbf{I} \tag{1}$$

$$\mathbf{XQ} + \mathbf{YS} = \mathbf{0} \tag{2}$$

$$\mathbf{ZR} = \mathbf{0} \tag{3}$$

$$\mathbf{ZS} = \mathbf{I}. \tag{4}$$

Since $\mathbf{Z}$ is non-singular we may deduce from (3) that $\mathbf{R} = \mathbf{0}$ and hence (1) reduces to $\mathbf{XP} = \mathbf{I}$ giving $\mathbf{P} = \mathbf{X}^{-1}$ (since $\mathbf{X}$ is non-singular). Also (4) gives $\mathbf{S} = \mathbf{Z}^{-1}$ and then (2) gives

$$\mathbf{XQ} = -\mathbf{YS} = -\mathbf{YZ}^{-1}$$

so that

$$\mathbf{Q} = -\mathbf{X}^{-1}\mathbf{YZ}^{-1}.$$

Hence we have

$$\mathbf{A}^{-1} = \begin{bmatrix} \mathbf{X}^{-1} & -\mathbf{X}^{-1}\mathbf{YZ}^{-1} \\ \mathbf{0} & \mathbf{Z}^{-1} \end{bmatrix}$$

and we have reduced the original problem to the easier problem of calculating inverses of lower order matrices.

*Examples*

2.8.3 Show that $\det(\mathbf{A}^{-1}) = 1/\det(\mathbf{A})$.

We have $\mathbf{A}^{-1}\mathbf{A} = \mathbf{I}$ and hence $\det(\mathbf{A}^{-1}\mathbf{A}) = \det(\mathbf{I}) = 1$.
But $\det(\mathbf{A}^{-1}\mathbf{A}) = \det(\mathbf{A}^{-1}) . \det(\mathbf{A})$ by Theorem 2.18 and hence $\det(\mathbf{A}^{-1}) \det(\mathbf{A}) = 1$ gives $\det(\mathbf{A}^{-1}) = 1/\det(\mathbf{A})$.

2.8.4 Suppose that $\mathbf{A}$ is a square non-singular matrix and that $\mathbf{B}_1, \mathbf{B}_2, \ldots \mathbf{B}_n \ldots$ is a sequence of square matrices such that $\mathbf{B}_{n+1} = \mathbf{B}_n(2\mathbf{I} - \mathbf{AB}_n)$. If $\mathbf{B}_n = \mathbf{A}^{-1} + \boldsymbol{\varepsilon}_n$ show that $\boldsymbol{\varepsilon}_{n+1} = -\boldsymbol{\varepsilon}_n\mathbf{A}\boldsymbol{\varepsilon}_n$ and explain how this result enables the sequence $\mathbf{B}_1, \mathbf{B}_2, \ldots$ to be used to obtain an approximation to $\mathbf{A}^{-1}$.

If $\mathbf{B}_n = \mathbf{A}^{-1} + \boldsymbol{\varepsilon}_n$ then $\mathbf{B}_{n+1} = \mathbf{A}^{-1} + \boldsymbol{\varepsilon}_{n+1}$ and substituting these expressions into $\mathbf{B}_{n+1} = \mathbf{B}_n(2\mathbf{I} - \mathbf{A}\mathbf{B}_n)$ gives

$$\begin{aligned}
\mathbf{A}^{-1} + \boldsymbol{\varepsilon}_{n+1} &= (\mathbf{A}^{-1} + \boldsymbol{\varepsilon}_n)\{2\mathbf{I} - \mathbf{A}(\mathbf{A}^{-1} + \boldsymbol{\varepsilon}_n)\} \\
&= (\mathbf{A}^{-1} + \boldsymbol{\varepsilon}_n)(2\mathbf{I} - \mathbf{I} - \mathbf{A}\boldsymbol{\varepsilon}_n) \\
&= (\mathbf{A}^{-1} + \boldsymbol{\varepsilon}_n)(\mathbf{I} - \mathbf{A}\boldsymbol{\varepsilon}_n) \\
&= \mathbf{A}^{-1} - \mathbf{A}^{-1}\mathbf{A}\boldsymbol{\varepsilon}_n + \boldsymbol{\varepsilon}_n - \boldsymbol{\varepsilon}_n\mathbf{A}\boldsymbol{\varepsilon}_n \\
&= \mathbf{A}^{-1} - \boldsymbol{\varepsilon}_n\mathbf{A}\boldsymbol{\varepsilon}_n.
\end{aligned}$$

Hence $\boldsymbol{\varepsilon}_{n+1} = -\boldsymbol{\varepsilon}_n\mathbf{A}\boldsymbol{\varepsilon}_n$, as required.

If we now suppose that $\mathbf{B}_n$ is an approximation to $\mathbf{A}^{-1}$ so that the elements of the matrix $\boldsymbol{\varepsilon}_n$ are small, then the elements of $\boldsymbol{\varepsilon}_{n+1}$ will be of the second order in small quantities, and thus will be smaller than those of $\boldsymbol{\varepsilon}_n$. Hence $\mathbf{B}_{n+1}$ will be a better approximation to $\mathbf{A}^{-1}$ than $\mathbf{B}_n$ is.

2.8.5 If two coordinate systems are related as shown in Figure 2.8.1 show that

$$\begin{bmatrix} x' \\ y' \end{bmatrix} = \mathbf{R}(\theta) \begin{bmatrix} x \\ y \end{bmatrix}$$

with

$$\mathbf{R}(\theta) = \begin{bmatrix} \cos\theta & \sin\theta \\ -\sin\theta & \cos\theta \end{bmatrix}.$$

Show also that

(i) $\mathbf{R}(\theta)\mathbf{R}(\phi) = \mathbf{R}(\theta + \phi)$

(ii) $\mathbf{R}(\theta)^{-1} = \mathbf{R}(-\theta)$

and interpret these results geometrically.

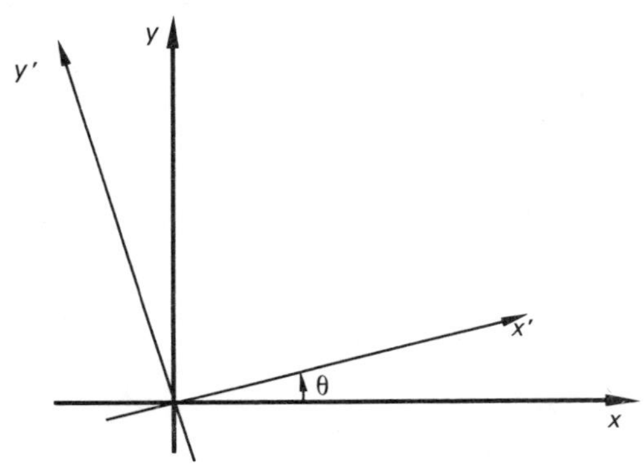


Figure 2.8.1

From Figure 2.8.2 we have

$$\begin{aligned}
x' &= \text{OM} \\
&= \text{OR} + \text{RM} \\
&= \text{ON}\cos\theta + \text{PN}\sin\theta \\
&= x\cos\theta + y\sin\theta.
\end{aligned}$$

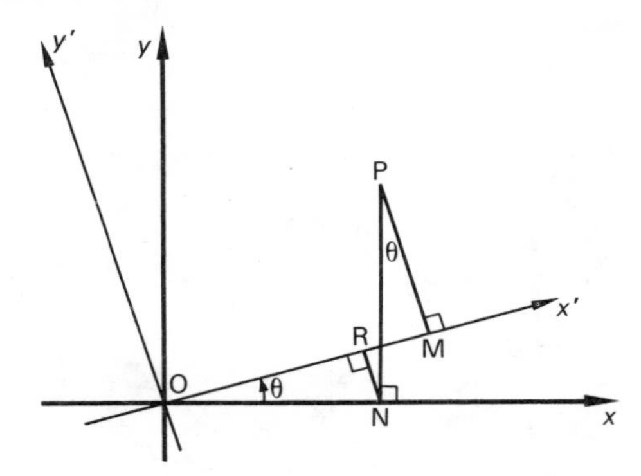


Figure 2.8.2

From Figure 2.8.3 we have

$$\begin{aligned} y' &= \mathrm{PM} \\ &= \mathrm{PQ} - \mathrm{MQ} \\ &= \mathrm{PQ} - \mathrm{NR} \\ &= \mathrm{PN} \cos\theta - \mathrm{ON} \sin\theta \\ &= y\cos\theta - x\sin\theta. \end{aligned}$$

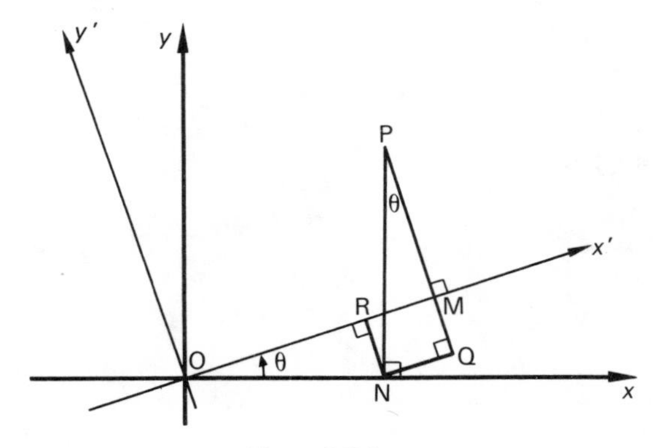


Figure 2.8.3

Thus we have

$$x' = x\cos\theta + y\sin\theta$$

$$y' = -x\sin\theta + y\cos\theta$$

which, in matrix notation is just

$$\begin{bmatrix} x' \\ y' \end{bmatrix} = \begin{bmatrix} \cos\theta & \sin\theta \\ -\sin\theta & \cos\theta \end{bmatrix} \begin{bmatrix} x \\ y \end{bmatrix}.$$

(i) We have

$$\mathbf{R}(\theta)\mathbf{R}(\phi) = \begin{bmatrix} \cos\theta & \sin\theta \\ -\sin\theta & \cos\theta \end{bmatrix} \begin{bmatrix} \cos\phi & \sin\phi \\ -\sin\phi & \cos\phi \end{bmatrix}$$

$$= \begin{bmatrix} \cos\theta\cos\phi - \sin\theta\sin\phi & \cos\theta\sin\phi + \sin\theta\cos\phi \\ -\sin\theta\cos\phi - \cos\theta\sin\phi & -\sin\theta\sin\phi + \cos\theta\cos\phi \end{bmatrix}$$

$$= \begin{bmatrix} \cos(\theta+\phi) & \sin(\theta+\phi) \\ -\sin(\theta+\phi) & \cos(\theta+\phi) \end{bmatrix}$$

$$= \mathbf{R}(\theta+\phi), \qquad \text{as required.}$$

(ii) The required result may be either proved directly from Theorem 2.19 or else we may use (i) above with $\phi = -\theta$.

Then we have

$$\mathbf{R}(\theta)\,.\,\mathbf{R}(-\theta) = \mathbf{R}(0) = \begin{bmatrix} 1 & 0 \\ 0 & 1 \end{bmatrix} = \mathbf{I}$$

and hence

$$\mathbf{R}(-\theta) = \{\mathbf{R}(\theta)\}^{-1}.$$

The geometrical interpretations are:

(i) Rotation $\theta$ followed by rotation $\phi$ is equivalent to rotation $(\theta + \phi)$.

(ii) Since

$$\begin{bmatrix} x' \\ y' \end{bmatrix} = \mathbf{R}(\theta) \begin{bmatrix} x \\ y \end{bmatrix}$$

implies that

$$\begin{bmatrix} x \\ y \end{bmatrix} = \{\mathbf{R}(\theta)\}^{-1} \begin{bmatrix} x' \\ y' \end{bmatrix}$$

the interpretation is that if the $x'y'$-axes are obtained from the $xy$-axes by a rotation of $\theta$, then the $xy$-axes are obtained from the $x'y'$-axes by a rotation of $-\theta$.

*Exercises*

2.8.5 If $\mathbf{D}$ is a diagonal matrix such that $D_{ii} = a_i \neq 0$ show that $\mathbf{D}^{-1}$ exists and is a diagonal matrix such that

$$(\mathbf{D}^{-1})_{ii} = 1/a_i.$$

2.8.6 Show that the inverse of a non-singular upper triangular matrix is also an upper triangular matrix. Show further that the inverse of a unit upper triangular matrix is a unit upper triangular matrix.

2.8.7 Show that $(\mathbf{A}^{-1})^{-1} = \mathbf{A}$.

2.8.8 If $\mathbf{A}$ is singular show that

$$\mathbf{A}\,\mathrm{adj}(\mathbf{A}) = (\mathrm{adj}(\mathbf{A}))\mathbf{A} = \mathbf{0}.$$

2.8.9 Find the inverse of

$$\begin{bmatrix} 1 & 0 & 3 & 4 \\ 0 & 1 & 2 & 3 \\ 0 & 0 & -1 & 2 \\ 0 & 0 & -2 & 3 \end{bmatrix}.$$

2.8.10 If $\mathbf{B} = \mathbf{P}^{-1}\mathbf{A}\mathbf{P}$ show that $\mathbf{B}^n = \mathbf{P}^{-1}\mathbf{A}^n\mathbf{P}$ and deduce that if $F(\mathbf{A})$ is a polynomial in the matrix $\mathbf{A}$ then

$$F(\mathbf{P}^{-1}\mathbf{A}\mathbf{P}) = \mathbf{P}^{-1}F(\mathbf{A})\mathbf{P}.$$

## 2.9 Special Matrices

In the applications of matrix theory it turns out that there are certain special types of matrix for which certain processes are particularly simple or which possess certain properties of interest both mathematically and physically. We have already met two such types of matrix—the symmetric and anti-symmetric. Before introducing more types we first define the complex and hermitian conjugates of a matrix.

DEFINITION 2.21

The *complex conjugate* of a matrix $\mathbf{A}$ is denoted by $\bar{\mathbf{A}}$ and is defined as the matrix of the same order as $\mathbf{A}$ such that

$$(\bar{\mathbf{A}})_{ij} = \overline{(A_{ij})}. \qquad \square$$

In other words, $\bar{\mathbf{A}}$ is obtained from $\mathbf{A}$ by taking the complex conjugate of all elements. The notation $\mathbf{A}^*$ is also used; this is sometimes more convenient since, for example, $(\mathbf{ABC}\ldots\mathbf{MN})^*$ is much neater than $\overline{(\mathbf{ABC}\ldots\mathbf{MN})}$.

*Example*

2.9.1 If

$$\mathbf{A} = \begin{bmatrix} 3+\mathrm{i} & 2+4\mathrm{i} \\ 5-7\mathrm{i} & 10\mathrm{i} \end{bmatrix}$$

where $\mathrm{i} = \sqrt{-1}$, obtain $\bar{\mathbf{A}}$.

From the definition we have, by taking the complex conjugate of each element,

$$\bar{\mathbf{A}} = \begin{bmatrix} 3-\mathrm{i} & 2-4\mathrm{i} \\ 5+7\mathrm{i} & -10\mathrm{i} \end{bmatrix}.$$

DEFINITION 2.22

The *hermitian conjugate* of a matrix $\mathbf{A}$ is defined to be the transpose of its complex conjugate and is denoted by $\mathbf{A}^+$. Thus $\mathbf{A}^+ = (\bar{\mathbf{A}})^{\mathrm{T}}$. $\square$

*Example*

2.9.2 If

$$\mathbf{A} = \begin{bmatrix} 3+i & 2+4i \\ 5-7i & 10i \end{bmatrix}$$

find $\mathbf{A}^+$.

We have

$$\bar{\mathbf{A}} = \begin{bmatrix} 3-i & 2-4i \\ 5+7i & -10i \end{bmatrix}$$

and hence

$$\mathbf{A}^+ = \begin{bmatrix} 3-i & 5+7i \\ 2-4i & -10i \end{bmatrix}.$$

We now proceed to define several types of matrix of special interest, namely orthogonal, hermitian and unitary matrices.

DEFINITION 2.23

A square matrix $\mathbf{A}$ is said to be *orthogonal* if it is such that $\mathbf{A}\mathbf{A}^T = \mathbf{A}^T\mathbf{A} = \mathbf{I}$. □

DEFINITION 2.24

A square matrix $\mathbf{A}$ is said to be *hermitian* if it is such that $\mathbf{A}^+ = \mathbf{A}$. □

DEFINITION 2.25

A square matrix $\mathbf{A}$ is said to be *unitary* if it is such that

$$\mathbf{A}\mathbf{A}^+ = \mathbf{A}^+\mathbf{A} = \mathbf{I}.$$ □

It follows immediately from these definitions that for an orthogonal matrix we have $\mathbf{A}^{-1} = \mathbf{A}^T$ and that for a unitary matrix $\mathbf{A}^{-1} = \mathbf{A}^+$.

It is also easy to see that a real unitary matrix is an orthogonal matrix. Thus all the properties of unitary matrices will also be properties of orthogonal matrices (since orthogonal matrices are just a subset of the set of unitary matrices).

*Examples*

2.9.3 If

$$\mathbf{A} = \frac{1}{\sqrt{2}} \begin{bmatrix} 1 & 1 & 0 \\ 0 & 0 & \sqrt{2} \\ 1 & -1 & 0 \end{bmatrix} \qquad \mathbf{B} = \begin{bmatrix} \cos\theta & i\sin\theta \\ -i\sin\theta & \cos\theta \end{bmatrix}$$

and

$$\mathbf{C} = \begin{bmatrix} \cos\theta & i\sin\theta \\ i\sin\theta & \cos\theta \end{bmatrix}$$

show that $\mathbf{A}$ is orthogonal, $\mathbf{B}$ is hermitian and $\mathbf{C}$ is unitary.

We have

$$\mathbf{A}^{\mathrm{T}} = \frac{1}{\sqrt{2}}\begin{bmatrix} 1 & 0 & 1 \\ 1 & 0 & -1 \\ 0 & \sqrt{2} & 0 \end{bmatrix}.$$

So

$$\mathbf{A}\mathbf{A}^{\mathrm{T}} = \tfrac{1}{2}\begin{bmatrix} 1 & 1 & 0 \\ 0 & 0 & \sqrt{2} \\ 1 & -1 & 0 \end{bmatrix}\begin{bmatrix} 1 & 0 & 1 \\ 1 & 0 & -1 \\ 0 & \sqrt{2} & 0 \end{bmatrix}$$

$$= \tfrac{1}{2}\begin{bmatrix} 2 & 0 & 0 \\ 0 & 2 & 0 \\ 0 & 0 & 2 \end{bmatrix}$$

$$= \mathbf{I},$$

and hence **A** is orthogonal. Also

$$\mathbf{B}^{+} = (\overline{\mathbf{B}})^{\mathrm{T}} = \begin{bmatrix} \cos\theta & -\mathrm{i}\sin\theta \\ \mathrm{i}\sin\theta & \cos\theta \end{bmatrix}^{\mathrm{T}}$$

$$= \begin{bmatrix} \cos\theta & \mathrm{i}\sin\theta \\ -\mathrm{i}\sin\theta & \cos\theta \end{bmatrix}$$

$$= \mathbf{B},$$

and hence **B** is hermitian. Similarly for **C** we obtain

$$\mathbf{C}^{+} = \begin{bmatrix} \cos\theta & -\mathrm{i}\sin\theta \\ -\mathrm{i}\sin\theta & \cos\theta \end{bmatrix}$$

and so

$$\mathbf{C}\mathbf{C}^{+} = \begin{bmatrix} \cos\theta & \mathrm{i}\sin\theta \\ \mathrm{i}\sin\theta & \cos\theta \end{bmatrix}\begin{bmatrix} \cos\theta & -\mathrm{i}\sin\theta \\ -\mathrm{i}\sin\theta & \cos\theta \end{bmatrix}$$

$$= \begin{bmatrix} 1 & 0 \\ 0 & 1 \end{bmatrix}$$

$$= \mathbf{I},$$

and hence **C** is unitary.

2.9.4 If **A** is an orthogonal matrix of order $n$ partitioned into its columns $\mathbf{v}_1$, $\mathbf{v}_2, \ldots \mathbf{v}_n$ so that

$$\mathbf{A} = [\mathbf{v}_1, \mathbf{v}_2, \ldots, \mathbf{v}_n]$$

show that

$$\mathbf{v}_i^{\mathrm{T}}\mathbf{v}_j = \begin{cases} 1 & \text{if } i = j \\ 0 & \text{if } i \neq j. \end{cases}$$

We have

$$\mathbf{A} = [\mathbf{v}_1, \mathbf{v}_2, \ldots \mathbf{v}_n]$$

so

$$\mathbf{A}^{\mathrm{T}} = \begin{bmatrix} \mathbf{v}_1^{\mathrm{T}} \\ \mathbf{v}_2^{\mathrm{T}} \\ \cdot \\ \cdot \\ \cdot \\ \mathbf{v}_n^{\mathrm{T}} \end{bmatrix}$$

and hence

$$\mathbf{A}^{\mathrm{T}}\mathbf{A} = \begin{bmatrix} \mathbf{v}_1^{\mathrm{T}} \\ \mathbf{v}_2^{\mathrm{T}} \\ \cdot \\ \cdot \\ \cdot \\ \mathbf{v}_n^{\mathrm{T}} \end{bmatrix} [\mathbf{v}_1, \mathbf{v}_2 \ldots \mathbf{v}_n]$$

$$= \begin{bmatrix} \mathbf{v}_1^{\mathrm{T}}\mathbf{v}_1 & \mathbf{v}_1^{\mathrm{T}}\mathbf{v}_2 & \ldots & \mathbf{v}_1^{\mathrm{T}}\mathbf{v}_n \\ \mathbf{v}_2^{\mathrm{T}}\mathbf{v}_1 & \mathbf{v}_2^{\mathrm{T}}\mathbf{v}_2 & \ldots & \mathbf{v}_2^{\mathrm{T}}\mathbf{v}_n \\ \cdot & & & \\ \cdot & & & \\ \cdot & & & \\ \mathbf{v}_n^{\mathrm{T}}\mathbf{v}_1 & \ldots\ldots & \ldots & \mathbf{v}_n^{\mathrm{T}}\mathbf{v}_n \end{bmatrix}.$$

So we see that $(\mathbf{A}^{\mathrm{T}}\mathbf{A})_{ij} = \mathbf{v}_i^{\mathrm{T}}\mathbf{v}_j$. But $\mathbf{A}^{\mathrm{T}}\mathbf{A} = \mathbf{I}$ (since $\mathbf{A}$ is orthogonal) so

$$(\mathbf{A}^{\mathrm{T}}\mathbf{A})_{ij} = \mathbf{I}_{ij} = \begin{cases} 1 & \text{if } i = j \\ 0 & \text{if } i \neq j \end{cases}$$

and hence the result follows.

2.9.5 If **A** and **B** are both orthogonal matrices of the same order show that **AB** is also orthogonal.

We have $\mathbf{A}^{-1} = \mathbf{A}^{\mathrm{T}}$ and $\mathbf{B}^{-1} = \mathbf{B}^{\mathrm{T}}$ and we wish to show that $(\mathbf{AB})^{-1} = (\mathbf{AB})^{\mathrm{T}}$. But

$$\begin{aligned} (\mathbf{AB})^{-1} &= \mathbf{B}^{-1}\mathbf{A}^{-1} && \text{by Theorem 2.21} \\ &= \mathbf{B}^{\mathrm{T}}\mathbf{A}^{\mathrm{T}} \\ &= (\mathbf{AB})^{\mathrm{T}} && \text{by Theorem 2.9} \end{aligned}$$

and hence the result is proved.

*Exercises*

2.9.1 Prove that the product of any number of unitary matrices of the same order is also a unitary matrix.

2.9.2 Show that

(i) $\det(\mathbf{A}) = \pm 1$ if $\mathbf{A}$ is orthogonal and real

(ii) $|\det(\mathbf{A})| = 1$ if $\mathbf{A}$ is unitary.

2.9.3 Prove that the product of two hermitian matrices of the same order is not in general hermitian.

2.9.4 Prove that the sum of two orthogonal matrices of the same order is not in general orthogonal.

2.9.5 Prove that the inverse of a non-singular hermitian matrix is hermitian.

2.9.6 If $\mathbf{B} = \mathbf{A}_1\mathbf{A}_2 \ldots \mathbf{A}_n$ show that $\mathbf{B}^+ = \mathbf{A}_n^+\mathbf{A}_{n-1}^+ \ldots \mathbf{A}_2^+\mathbf{A}_1^+$.

## 2.10 Elementary Transformations

We often wish to manipulate matrices in one of the following ways:

(1) Multiply all the elements in a particular row (or column) by a given number.

(2) Interchange the elements in a particular row (or column) with the corresponding elements in another row (or column).

(3) Add a multiple of the elements in some row (or column) to the corresponding element in some other row (or column).

Operations of the above type are called *elementary transformations* of the original matrix; corresponding to the three operations introduced above we talk about elementary transformations on rows (or columns) of type 1, type 2 or type 3. We shall show that all elementary transformations on rows may be accomplished by premultiplying the given matrix by some suitable matrix, and that all elementary transformations on columns may be accomplished by postmultiplying the given matrix by some suitable matrix. The following theorem gives the rules whereby these pre- and postmultiplying matrices may be constructed.

THEOREM 2.22

Any elementary transformation on rows may be effected by premultiplying the given matrix by a matrix which is constructed from the unit matrix by subjecting it to the desired elementary transformation. Similarly, an elementary transformation on columns is effected by postmultiplying by the unit matrix subjected to the required transformation.

*Proof*

We carry through the proof in detail only for the row transformations; for column transformations the proof is essentially similar.

Suppose the given matrix is $\mathbf{A}$ and is partitioned into rows $\mathbf{r}^{(1)}, \mathbf{r}^{(2)}, \ldots \mathbf{r}^{(m)}$ so that

$$\mathbf{A} = \begin{bmatrix} \mathbf{r}^{(1)} \\ \mathbf{r}^{(2)} \\ \cdot \\ \cdot \\ \cdot \\ \mathbf{r}^{(m)} \end{bmatrix}.$$

There is no need for the matrix to be square, so each row may consist of $n$ elements.

Suppose the unit matrix of order $m$ (which is conformable for premultiplication with $\mathbf{A}$) is also partitioned into rows $\mathbf{p}^{(1)}, \mathbf{p}^{(2)}, \ldots \mathbf{p}^{(m)}$ so that

$$\mathbf{I} = \begin{bmatrix} \mathbf{p}^{(1)} \\ \mathbf{p}^{(2)} \\ \cdot \\ \cdot \\ \cdot \\ \mathbf{p}^{(m)} \end{bmatrix} \quad \text{where } p_j^{(i)} = \begin{cases} 1 & \text{if } j = i \\ 0 & \text{if } j \neq i. \end{cases} \tag{2.10.1}$$

Then, of course, the product $\mathbf{IA}$ could be written in the form

$$\mathbf{IA} = \begin{bmatrix} p_1^{(1)}\mathbf{r}^{(1)} + p_2^{(1)}\mathbf{r}^{(2)} + p_3^{(1)}\mathbf{r}^{(3)} + \ldots + p_m^{(1)}\mathbf{r}^{(m)} \\ p_1^{(2)}\mathbf{r}^{(1)} + p_2^{(2)}\mathbf{r}^{(2)} + p_3^{(2)}\mathbf{r}^{(3)} + \ldots + p_m^{(2)}\mathbf{r}^{(m)} \\ \cdot \\ \cdot \\ \cdot \\ p_1^{(m)}\mathbf{r}^{(1)} + p_2^{(m)}\mathbf{r}^{(2)} + p_3^{(m)}\mathbf{r}^{(3)} + \ldots + p_m^{(m)}\mathbf{r}^{(m)} \end{bmatrix} = \begin{bmatrix} \mathbf{r}^{(1)} \\ \mathbf{r}^{(2)} \\ \cdot \\ \cdot \\ \cdot \\ \mathbf{r}^{(m)} \end{bmatrix}$$

by virtue of equation (2.10.1).

(i) We accomplish the proof of this part by showing that multiplication of the $i$th row of $\mathbf{A}$ by $\lambda$ is achieved by the premultiplication of $\mathbf{A}$ by

$$\begin{bmatrix} \mathbf{p}^{(1)} \\ \mathbf{p}^{(2)} \\ \cdot \\ \cdot \\ \cdot \\ \lambda\mathbf{p}^{(i)} \\ \cdot \\ \cdot \\ \cdot \\ \mathbf{p}^{(m)} \end{bmatrix}$$

which, of course, is just the unit matrix with *its* $i$th row multiplied by $\lambda$. But

$$\begin{bmatrix} \mathbf{p}^{(1)} \\ \mathbf{p}^{(2)} \\ \cdot \\ \cdot \\ \cdot \\ \lambda\mathbf{p}^{(i)} \\ \cdot \\ \cdot \\ \cdot \\ \mathbf{p}^{(m)} \end{bmatrix} \mathbf{A} = \begin{bmatrix} p_1^{(1)} & p_2^{(1)} & \dots & p_m^{(1)} \\ p_1^{(2)} & p_2^{(2)} & \dots & p_m^{(2)} \\ \cdot & & & \cdot \\ \cdot & & & \cdot \\ \cdot & & & \cdot \\ \lambda p_1^{(i)} & \lambda p_2^{(i)} & \dots & \lambda p_m^{(i)} \\ \cdot & & & \cdot \\ \cdot & & & \cdot \\ \cdot & & & \cdot \\ p_1^{(m)} & p_2^{(m)} & \dots & p_m^{(m)} \end{bmatrix} \begin{bmatrix} \mathbf{r}^{(1)} \\ \mathbf{r}^{(2)} \\ \cdot \\ \cdot \\ \cdot \\ \cdot \\ \cdot \\ \cdot \\ \cdot \\ \mathbf{r}^{(m)} \end{bmatrix}$$

$$= \begin{bmatrix} p_1^{(1)}\mathbf{r}^{(1)} + p_2^{(1)}\mathbf{r}^{(2)} + \dots + p_m^{(1)}\mathbf{r}^{(m)} \\ p_1^{(2)}\mathbf{r}^{(1)} + p_2^{(2)}\mathbf{r}^{(2)} + \dots + p_m^{(2)}\mathbf{r}^{(m)} \\ \cdot \\ \cdot \\ \cdot \\ \lambda p_1^{(i)}\mathbf{r}^{(1)} + \lambda p_2^{(i)}\mathbf{r}^{(2)} + \dots + \lambda p_m^{(i)}\mathbf{r}^{(m)} \\ \cdot \\ \cdot \\ \cdot \\ p_1^{(m)}\mathbf{r}^{(1)} + p_2^{(m)}\mathbf{r}^{(2)} + \dots + p_m^{(m)}\mathbf{r}^{(m)} \end{bmatrix} = \begin{bmatrix} \mathbf{r}^{(1)} \\ \mathbf{r}^{(2)} \\ \cdot \\ \cdot \\ \cdot \\ \mathbf{r}^{(i-1)} \\ \lambda\mathbf{r}^{(i)} \\ \mathbf{r}^{(i+1)} \\ \cdot \\ \cdot \\ \cdot \\ \mathbf{r}^{(m)} \end{bmatrix}$$

using equation (2.10.1). This proves the required result.

(ii) We now show that interchanging the $i$th and $j$th rows of $\mathbf{A}$ is accomplished if we premultiply $\mathbf{A}$ by the matrix

$$\begin{bmatrix} \mathbf{p}^{(1)} \\ \cdot \\ \cdot \\ \cdot \\ \mathbf{p}^{(i-1)} \\ \mathbf{p}^{(j)} \\ \mathbf{p}^{(i+1)} \\ \cdot \\ \cdot \\ \cdot \\ \mathbf{p}^{(j-1)} \\ \mathbf{p}^{(i)} \\ \mathbf{p}^{(j+1)} \\ \cdot \\ \cdot \\ \cdot \\ \mathbf{p}^{(m)} \end{bmatrix}$$

which is, of course, just the unit matrix with rows $i$ and $j$ interchanged. But

$$\begin{bmatrix} \mathbf{p}^{(1)} \\ \cdot \\ \cdot \\ \cdot \\ \mathbf{p}^{(i-1)} \\ \mathbf{p}^{(j)} \\ \mathbf{p}^{(i+1)} \\ \cdot \\ \cdot \\ \cdot \\ \mathbf{p}^{(j-1)} \\ \mathbf{p}^{(i)} \\ \mathbf{p}^{(j+1)} \\ \cdot \\ \cdot \\ \cdot \\ \mathbf{p}^{(m)} \end{bmatrix} \mathbf{A} = \begin{bmatrix} p_1^{(1)} & p_2^{(1)} & \ldots & p_m^{(1)} \\ \cdot & \cdot & & \cdot \\ \cdot & \cdot & & \cdot \\ \cdot & \cdot & & \cdot \\ p_1^{(i-1)} & p_2^{(i-1)} & \ldots & p_m^{(i-1)} \\ p_1^{(j)} & p_2^{(j)} & \ldots & p_m^{(j)} \\ p_1^{(i+1)} & p_2^{(i+1)} & \ldots & p_m^{(i+1)} \\ \cdot & \cdot & & \cdot \\ \cdot & \cdot & & \cdot \\ \cdot & \cdot & & \cdot \\ p_1^{(j-1)} & p_2^{(j-1)} & \ldots & p_m^{(j-1)} \\ p_1^{(i)} & p_2^{(i)} & \ldots & p_m^{(i)} \\ p_1^{(j+1)} & p_2^{(j+1)} & \ldots & p_m^{(j+1)} \\ \cdot & \cdot & & \cdot \\ \cdot & \cdot & & \cdot \\ \cdot & \cdot & & \cdot \\ p_1^{(m)} & p_2^{(m)} & \ldots & p_m^{(m)} \end{bmatrix} \begin{bmatrix} \mathbf{r}^{(1)} \\ \mathbf{r}^{(2)} \\ \cdot \\ \cdot \\ \cdot \\ \cdot \\ \cdot \\ \cdot \\ \cdot \\ \cdot \\ \cdot \\ \cdot \\ \cdot \\ \cdot \\ \cdot \\ \mathbf{r}^{(m)} \end{bmatrix}$$

$$= \begin{bmatrix} p_1^{(1)}\mathbf{r}^{(1)} + p_2^{(2)}\mathbf{r}^{(2)} + \ldots + p_m^{(1)}\mathbf{r}^{(m)} \\ \cdot \\ \cdot \\ \cdot \\ p_1^{(i-1)}\mathbf{r}^{(1)} + p_2^{(i-1)}\mathbf{r}^{(2)} + \ldots + p_m^{(i-1)}\mathbf{r}^{(m)} \\ p_1^{(j)}\mathbf{r}^{(1)} + p_2^{(j)}\mathbf{r}^{(2)} + \ldots + p_m^{(j)}\mathbf{r}^{(m)} \\ p_1^{(i+1)}\mathbf{r}^{(1)} + p_2^{(i+1)}\mathbf{r}^{(2)} + \ldots + p_m^{(i+1)}\mathbf{r}^{(m)} \\ \cdot \\ \cdot \\ \cdot \\ p_1^{(j-1)}\mathbf{r}^{(1)} + p_2^{(j-1)}\mathbf{r}^{(2)} + \ldots + p_m^{(j-1)}\mathbf{r}^{(m)} \\ p_1^{(i)}\mathbf{r}^{(1)} + p_2^{(i)}\mathbf{r}^{(2)} + \ldots + p_m^{(i)}\mathbf{r}^{(m)} \\ p_1^{(j+1)}\mathbf{r}^{(1)} + p_2^{(j+1)}\mathbf{r}^{(2)} + \ldots + p_m^{(j+1)}\mathbf{r}^{(m)} \\ \cdot \\ \cdot \\ \cdot \\ p_1^{(m)}\mathbf{r}^{(1)} + p_2^{(m)}\mathbf{r}^{(2)} + \ldots + p_m^{(m)}\mathbf{r}^{(m)} \end{bmatrix} = \begin{bmatrix} \mathbf{r}^{(1)} \\ \cdot \\ \cdot \\ \cdot \\ \mathbf{r}^{(i-1)} \\ \mathbf{r}^{(j)} \\ \mathbf{r}^{(i+1)} \\ \cdot \\ \cdot \\ \cdot \\ \mathbf{r}^{(j-1)} \\ \mathbf{r}^{(i)} \\ \mathbf{r}^{(j+1)} \\ \cdot \\ \cdot \\ \cdot \\ \mathbf{r}^{(m)} \end{bmatrix}$$

using equation (2.10.1). Thus rows $i$ and $j$ have been interchanged as required.

(iii) Finally, in exactly the same way as (ii) above we may show that adding a multiple $\lambda$ of row $j$ to row $i$ of matrix $\mathbf{A}$ is accomplished by premultiplying $\mathbf{A}$ by

$$\begin{bmatrix} \mathbf{p}^{(1)} \\ \mathbf{p}^{(2)} \\ \cdot \\ \cdot \\ \cdot \\ \mathbf{p}^{(i-1)} \\ \mathbf{p}^{(i)} + \lambda \mathbf{p}^{(j)} \\ \mathbf{p}^{(i+1)} \\ \cdot \\ \cdot \\ \cdot \\ \mathbf{p}^{(m)} \end{bmatrix}.$$

The reader may find it instructive to work through the details of the proof. ■

*Examples*

2.10.1 Construct the matrices which, when premultiplying a 4 x 4 matrix produce the following transformations:

(i) multiply all elements in the second row by 6;

(ii) interchange third and fourth rows; and

(iii) add twice the first row to the third row.

Using Theorem 2.22 we have immediately that the required matrices are given by

$$\text{(i)} \begin{bmatrix} 1 & 0 & 0 & 0 \\ 0 & 6 & 0 & 0 \\ 0 & 0 & 1 & 0 \\ 0 & 0 & 0 & 1 \end{bmatrix} \qquad \text{(ii)} \begin{bmatrix} 1 & 0 & 0 & 0 \\ 0 & 1 & 0 & 0 \\ 0 & 0 & 0 & 1 \\ 0 & 0 & 1 & 0 \end{bmatrix}$$

and

$$\text{(iii)} \begin{bmatrix} 1 & 0 & 0 & 0 \\ 0 & 1 & 0 & 0 \\ 2 & 0 & 1 & 0 \\ 0 & 0 & 0 & 1 \end{bmatrix}.$$

2.10.2 Construct the matrix which, when postmultiplying a 5 x 3 matrix interchanges columns 1 and 3 and verify directly by applying to the matrix

$$\mathbf{A} = \begin{bmatrix} 1 & 7 & -3 \\ 2 & 1 & 9 \\ 3 & 0 & 4 \\ 2 & 2 & 1 \\ 6 & 2 & 0 \end{bmatrix}.$$

By Theorem 2.22 the required matrix is obtained by interchanging columns 1 and 3 of the unit matrix of order 3 and so is given by

$$\mathbf{Q} = \begin{bmatrix} 0 & 0 & 1 \\ 0 & 1 & 0 \\ 1 & 0 & 0 \end{bmatrix}.$$

Then

$$\mathbf{AQ} = \begin{bmatrix} 1 & 7 & -3 \\ 2 & 1 & 9 \\ 3 & 0 & 4 \\ 2 & 2 & 1 \\ 6 & 2 & 0 \end{bmatrix} \begin{bmatrix} 0 & 0 & 1 \\ 0 & 1 & 0 \\ 1 & 0 & 0 \end{bmatrix}$$

$$= \begin{bmatrix} -3 & 7 & 1 \\ 9 & 1 & 2 \\ 4 & 0 & 3 \\ 1 & 2 & 2 \\ 0 & 2 & 6 \end{bmatrix}$$

using the 'row into column' rule for matrix multiplication. Thus the required interchange has been achieved.

2.10.3 If $\mathbf{P}$ is a matrix which produces an elementary transformation of type 2, show that $\mathbf{P}^2 = \mathbf{I}$.

Since $\mathbf{P}$ interchanges row $i$ with row $j$, $\mathbf{P}^2$ will interchange these rows twice and so just reproduce the original matrix. Thus multiplying by $\mathbf{P}$ twice is equivalent to leaving the matrix unaltered, and hence $\mathbf{P}^2 = \mathbf{I}$.

*Exercises*

2.10.1 Write down the matrices which when premultiplying a 4 x 3 matrix will

(i) interchange rows 2 and 4

(ii) multiply row 3 by 99

(iii) subtract row 3 from row 1.

2.10.2 Write down the matrix which when postmultiplying a 4 x 3 matrix will

(i) interchange columns 2 and 3

(ii) multiply column 2 by –3

(iii) add twice column 1 to column 2.

2.10.3 Investigate the possible values of det(**P**) where **P** is a matrix producing an elementary transformation.

## 2.11 Quadratic Forms and Coordinate Transformations

Very often in physical problems described by the $n$ coordinates $x_1, x_2, x_3, \ldots, x_n$ expressions such as

$$\sum_{i,\,j=1}^{n} a_{ij}\, x_{ij} x_j$$

arise. A common example of such an expression is the potential energy of some system displaced a small amount from its equilibrium configuration. Such expressions are important enough to deserve special study in their own right.

DEFINITION 2.26

The function of $n$ variables $x_1, x_2, \ldots, x_n$ defined as $\Sigma_{i,\,j=1}^{n} a_{ij}x_ix_j$ (where the $a_{ij}$ are independent of the $x_i$) is called a quadratic form in the $n$ variables. □

The matrix **A** (square of order $n$) defined by $\mathbf{A}_{ij} = a_{ij}$ is called the matrix of the quadratic form. If we introduce the column vector **x** defined by

$$\mathbf{x} = \begin{bmatrix} x_1 \\ x_2 \\ \cdot \\ \cdot \\ \cdot \\ x_n \end{bmatrix}$$

then we see immediately from the properties of matrix multiplication that

$$\sum_{i,\,j=1}^{n} a_{ij}x_ix_j = \mathbf{x}^{\mathrm{T}}\mathbf{A}\mathbf{x};$$

thus we have expressed quadratic forms in a matrix notation.

A given quadratic form does not have a unique matrix. Consider as an example the form given by

$$Q = x^2 + 5xy + 3y^2$$

–a quadratic form in the two variables $x$ and $y$. We can write

$$Q = \mathbf{u}^{\mathrm{T}}\,\mathbf{A}\mathbf{u} \qquad \text{where } \mathbf{u} = \begin{bmatrix} x \\ y \end{bmatrix}$$

in many different ways. This is because we can consider the $5xy$ to be included by contributions from either or both the $xy$ and $yx$ terms in the expansion. Thus, for example, we have

$$Q = [x \quad y]\begin{bmatrix}1 & 5\\ 0 & 3\end{bmatrix}\begin{bmatrix}x\\ y\end{bmatrix} = [x \quad y]\begin{bmatrix}1 & 0\\ 5 & 3\end{bmatrix}\begin{bmatrix}x\\ y\end{bmatrix}$$

$$= [x \quad y]\begin{bmatrix}1 & 4\\ 1 & 3\end{bmatrix}\begin{bmatrix}x\\ y\end{bmatrix} \qquad \text{and so on.}$$

Because of this it is possible to split up the $5xy$ equally between the contributions from the $xy$ term and the $yx$ term to obtain

$$Q = [x \quad y]\begin{bmatrix}1 & \frac{5}{2}\\ \frac{5}{2} & 3\end{bmatrix}\begin{bmatrix}x\\ y\end{bmatrix}.$$

Thus we have written the quadratic form as $Q = \mathbf{u}^{\mathrm{T}}\mathbf{A}\mathbf{u}$ with $\mathbf{A}$ symmetric. That this is possible in general is proved in the following theorem.

THEOREM 2.23

For any quadratic form $Q = \mathbf{x}^{\mathrm{T}}\mathbf{A}\mathbf{x}$ we have $Q = \mathbf{x}^{\mathrm{T}}\mathbf{B}\mathbf{x}$ where the symmetric matrix $\mathbf{B}$ is defined by $\mathbf{B} = \frac{1}{2}(\mathbf{A} + \mathbf{A}^{\mathrm{T}})$.

*Proof*

We have already noted (see Theorem 2.11) that

$$\begin{aligned}\mathbf{A} &= \tfrac{1}{2}(\mathbf{A} + \mathbf{A}^{\mathrm{T}}) + \tfrac{1}{2}(\mathbf{A} - \mathbf{A}^{\mathrm{T}})\\ &= \mathbf{B} + \mathbf{C},\end{aligned}$$

say, where $\mathbf{B} = \mathbf{B}^{\mathrm{T}}$ and $\mathbf{C} = -\mathbf{C}^{\mathrm{T}}$. Thus

$$\begin{aligned}\mathbf{x}^{\mathrm{T}}\mathbf{A}\mathbf{x} &= \mathbf{x}^{\mathrm{T}}(\mathbf{B} + \mathbf{C})\mathbf{x}\\ &= \mathbf{x}^{\mathrm{T}}\mathbf{B}\mathbf{x} + \mathbf{x}^{\mathrm{T}}\mathbf{C}\mathbf{x}.\end{aligned}$$

We now show that $\mathbf{x}^{\mathrm{T}}\mathbf{C}\mathbf{x} = \mathbf{0}$; we have

$$\begin{aligned}\mathbf{x}^{\mathrm{T}}\mathbf{C}\mathbf{x} &= (\mathbf{x}^{\mathrm{T}}\mathbf{C}\mathbf{x})^{\mathrm{T}} && \text{since } \mathbf{x}^{\mathrm{T}}\mathbf{C}\mathbf{x} \text{ is a number}^{\dagger}\\ &= \mathbf{x}^{\mathrm{T}}\mathbf{C}^{\mathrm{T}}(\mathbf{x}^{\mathrm{T}})^{\mathrm{T}} && \text{by Theorem 2.15}\\ &= \mathbf{x}^{\mathrm{T}}(-\mathbf{C})\mathbf{x}\\ &= -\mathbf{x}^{\mathrm{T}}\mathbf{C}\mathbf{x}.\end{aligned}$$

So, since $\mathbf{x}^{\mathrm{T}}\mathbf{C}\mathbf{x} = -\mathbf{x}^{\mathrm{T}}\mathbf{C}\mathbf{x}$ we have $\mathbf{x}^{\mathrm{T}}\mathbf{C}\mathbf{x} = \mathbf{0}$. Thus $\mathbf{x}^{\mathrm{T}}\mathbf{A}\mathbf{x} = \mathbf{x}^{\mathrm{T}}\mathbf{B}\mathbf{x}$ and the required result is proved. ∎

† More precisely $\mathbf{x}^{\mathrm{T}}\mathbf{C}\mathbf{x}$ is a square matrix of order 1; however we do not here draw any distinction between an array of one element and that element itself. In other words we consider that $[a] = a$.

The most general quadratic form in three variables is

$$Q = ax^2 + by^2 + cz^2 + 2fyz + 2gzx + 2hxy$$
$$= [x \quad y \quad z] \begin{bmatrix} a & h & g \\ h & b & f \\ g & f & c \end{bmatrix} \begin{bmatrix} x \\ y \\ z \end{bmatrix}.$$

If we denote this quadratic form by $\phi(x, y, z)$ then the equation $\phi(x, y, z) = 1$ describes some surface in three-dimensional space. This particular type of surface is an example of a class of surfaces called quadric surfaces; it turns out to be either a hyperboloid ellipsoid, or pair of planes, depending on the coefficients.

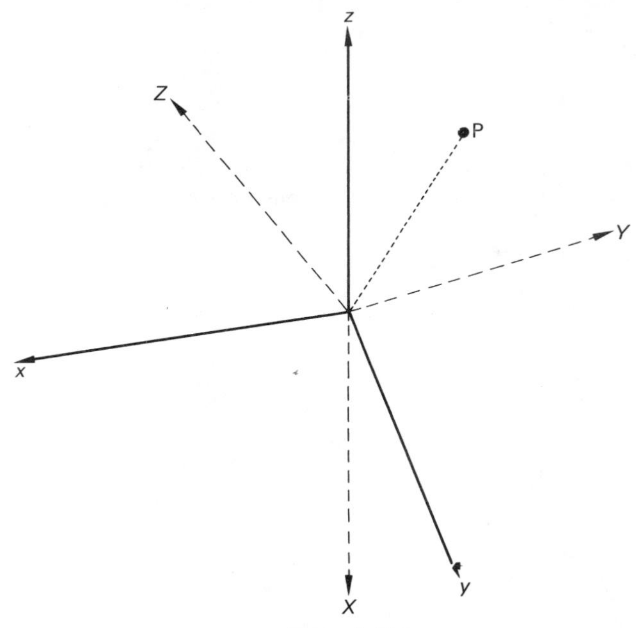


Figure 2.11.1

We look now and see how the equation is transformed when a new set of axes is introduced. Suppose a new set of rectangular Cartesian coordinates $(X, Y, Z)$ is introduced as shown in Figure 2.11.1. Then the coordinates in the two systems of any point are linearly related in some way so that:

$$X = t_{11}x + t_{12}y + t_{13}z$$
$$Y = t_{21}x + t_{22}y + t_{23}x$$
$$Z = t_{31}x + t_{32}y + t_{33}z$$

or, in matrix notation,

$$\begin{bmatrix} X \\ Y \\ Z \end{bmatrix} = \mathbf{T} \begin{bmatrix} x \\ y \\ z \end{bmatrix}$$

where the $t_{ij}$ depend on the actual axes chosen.

The point P is described by either the coordinates $(X, Y, Z)$ or $(x, y, z)$. However, we can obtain a condition on the coefficients $t_{ij}$ by noting that in the $XYZ$-system $OP^2 = X^2 + Y^2 + Z^2$ whereas in the $xyz$-system $OP^2 = x^2 + y^2 + z^2$. Now

$$X^2 + Y^2 + Z^2 = [X \quad Y \quad Z] \begin{bmatrix} X \\ Y \\ Z \end{bmatrix}$$

$$= [x \quad y \quad z] \mathbf{T}^{\mathrm{T}}\mathbf{T} \begin{bmatrix} x \\ y \\ z \end{bmatrix}$$

$$= x^2 + y^2 + z^2 \qquad \text{if and only if } \mathbf{T}^{\mathrm{T}}\mathbf{T} = \mathbf{I}.$$

Thus a necessary condition that $\mathbf{T}$ must satisfy is that $\mathbf{T}^{\mathrm{T}}\mathbf{T} = \mathbf{I}$. It may also be shown that if $\mathbf{T}^{\mathrm{T}}\mathbf{T} = \mathbf{I}$ then $\mathbf{T}$ will represent some transformation of the coordinate system. By Definition 2.23 this means that coordinate transformations are effected by orthogonal matrices.

Let us now write the quadratic form

$$Q = [x \quad y \quad z] \begin{bmatrix} a & h & g \\ h & b & f \\ g & f & c \end{bmatrix} \begin{bmatrix} x \\ y \\ z \end{bmatrix}$$

in terms of the new coordinates $X, Y, Z$. Writing

$$\mathbf{A} = \begin{bmatrix} a & h & g \\ h & b & f \\ g & f & c \end{bmatrix}$$

and noting that

$$\begin{bmatrix} X \\ Y \\ Z \end{bmatrix} = \mathbf{T} \begin{bmatrix} x \\ y \\ z \end{bmatrix}$$

gives

$$\begin{bmatrix} x \\ y \\ z \end{bmatrix} = \mathbf{T}^{-1} \begin{bmatrix} X \\ Y \\ Z \end{bmatrix}$$

we obtain

$$Q = [X \quad Y \quad Z]\,(\mathbf{T}^{-1})^{\mathrm{T}}\mathbf{A}\mathbf{T}^{-1}\begin{bmatrix} X \\ Y \\ Z \end{bmatrix}$$

$$= [X \quad Y \quad Z]\,\mathbf{T}\mathbf{A}\mathbf{T}^{\mathrm{T}}\begin{bmatrix} X \\ Y \\ Z \end{bmatrix}$$

since **T** is orthogonal. Thus, when the coordinate system is changed, the equation of the quadratic is changed from

$$[x \quad y \quad z]\,\mathbf{A}\begin{bmatrix} x \\ y \\ z \end{bmatrix} = 1,$$

to

$$[X \quad Y \quad Z]\,\mathbf{T}\mathbf{A}\mathbf{T}^{\mathrm{T}}\begin{bmatrix} X \\ Y \\ Z \end{bmatrix} = 1.$$

We shall see in a later chapter how a suitable choice of **T** enables us to simplify the form of the equation.

It is convenient to introduce at this point a definition which makes use of the idea of quadratic form.

DEFINITION 2.27

A real matrix **A** is said to be *positive definite* if $\mathbf{x}^{\mathrm{T}}\mathbf{A}\mathbf{x} > 0$ for all non-zero vectors **x**; similarly **A** is said to be *negative definite* if $\mathbf{x}^{\mathrm{T}}\mathbf{A}\mathbf{x} < 0$ for all non-zero vectors **x**. □

We shall see later that the investigation of quadratic forms is intimately connected with eigenvalue theory, and we shall return to this topic in Chapter 4.

*Examples*

2.11.1 If the $x_1x_2$- and $y_1y_2$-coordinate systems are related as shown in Figure 2.11.2, express the quadratic form $S = \alpha(x_1^2 + x_2^2) + 2\beta x_1 x_2$ in terms of the variables $y_1$ and $y_2$ and show that there is no term in $y_1y_2$ when $\theta = \pi/4$.

We have (see Example 2.8.5) that

$$\begin{bmatrix} y_1 \\ y_2 \end{bmatrix} = \mathbf{R}\begin{bmatrix} x_1 \\ x_2 \end{bmatrix} \qquad \text{where} \quad \mathbf{R} = \begin{bmatrix} \cos\theta & \sin\theta \\ -\sin\theta & \cos\theta \end{bmatrix}.$$

It may be directly verified that $\mathbf{R}\mathbf{R}^{\mathrm{T}} = \mathbf{I}$ thus proving the orthogonality of **R**—the two-dimensional analogue of **T**.

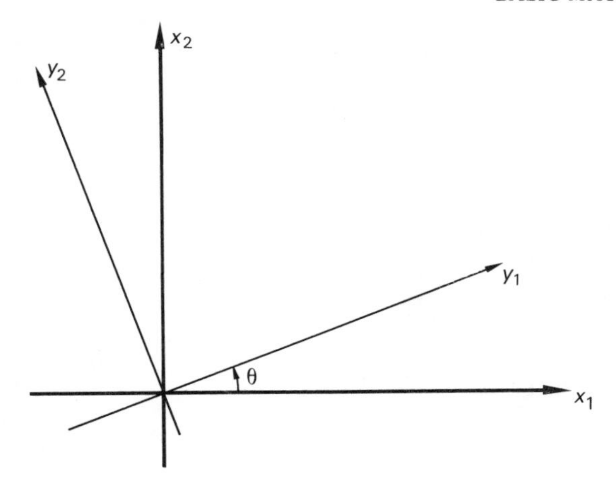


Figure 2.11.2

We thus have

$$S = [x_1 \quad x_2] \begin{bmatrix} \alpha & \beta \\ \beta & \alpha \end{bmatrix} \begin{bmatrix} x_1 \\ x_2 \end{bmatrix}$$

$$= [y_1 \quad y_2] \quad \mathbf{R} \begin{bmatrix} \alpha & \beta \\ \beta & \alpha \end{bmatrix} \mathbf{R}^{\mathrm{T}} \begin{bmatrix} y_1 \\ y_2 \end{bmatrix}.$$

But

$$\mathbf{R} \begin{bmatrix} \alpha & \beta \\ \beta & \alpha \end{bmatrix} \mathbf{R}^{\mathrm{T}} = \begin{bmatrix} \cos\theta & \sin\theta \\ -\sin\theta & \cos\theta \end{bmatrix} \begin{bmatrix} \alpha & \beta \\ \beta & \alpha \end{bmatrix} \begin{bmatrix} \cos\theta & -\sin\theta \\ \sin\theta & \cos\theta \end{bmatrix}$$

which reduces, after some manipulation, to

$$\begin{bmatrix} \alpha + \beta \sin 2\theta & \beta \cos 2\theta \\ \beta \cos 2\theta & \alpha - \beta \sin 2\theta \end{bmatrix}.$$

Thus

$$S = [y_1 \quad y_2] \begin{bmatrix} \alpha + \beta \sin 2\theta & \beta \cos 2\theta \\ \beta \cos 2\theta & \alpha - \beta \sin 2\theta \end{bmatrix} \begin{bmatrix} y_1 \\ y_2 \end{bmatrix};$$

there will be no terms in $y_1 y_2$ if the off-diagonal elements are zero. The condition for this is that $\cos 2\theta = 0$, giving $2\theta = \pi/2$ and hence $\theta = \pi/4$, as required.

2.11.2 Show that the matrix

$$\mathbf{A} = \begin{bmatrix} 1 & 1 & 0 \\ 1 & 2 & -1 \\ 0 & -1 & 1 \end{bmatrix}$$

is positive definite.

The quadratic form $\mathbf{x}^T\mathbf{A}\mathbf{x}$ is given by

$$\mathbf{x}^T\mathbf{A}\mathbf{x} = [x_1 \quad x_2 \quad x_3] \begin{bmatrix} 1 & 1 & 0 \\ 1 & 2 & -1 \\ 0 & -1 & 1 \end{bmatrix} \begin{bmatrix} x_1 \\ x_2 \\ x_3 \end{bmatrix}$$

$$= x_1^2 + 2x_2^2 + x_3^2 + 2x_1x_2 - 2x_2x_3$$

If we can write this expression as a sum of squares then it will be positive for all values of $x_1, x_2$ and $x_3$ and hence the result will be proved. But

$$\begin{aligned} &x_1^2 + 2x_2^2 + x_3^2 + 2x_1x_2 - 2x_2x_3 \\ &\quad = (x_1 + x_2)^2 + x_2^2 + x_3^2 - 2x_2x_3 \\ &\quad = (x_1 + x_2)^2 + (x_2 - x_3)^2 \end{aligned}$$

and hence the required property is proved.

*Exercises*

2.11.1 Show that the matrix

$$\begin{bmatrix} a & h \\ h & b \end{bmatrix}$$

is positive definite if and only if $ab - h^2 > 0$ and $a > 0$.

2.11.2 If $\mathbf{A} = \mathbf{P}^T\mathbf{P}$ where $\mathbf{P}$ is real show that $\mathbf{A}$ is positive definite provided $\mathbf{P}\mathbf{x} = \mathbf{O}$ implies that $\mathbf{x} = \mathbf{O}$.

2.11.3 If $\mathbf{D}$ is a diagonal matrix show that it is positive definite if all its diagonal elements are positive and negative definite if all its diagonal elements are negative. On the other hand, if some diagonal elements are positive and some negative, show that it is neither positive nor negative definite.

Deduce that $\mathbf{A}$ is positive definite if it is possible to find a non-singular matrix $\mathbf{P}$ such that $\mathbf{P}^T\mathbf{A}\mathbf{P}$ is diagonal with positive diagonal elements.

2.11.4 Determine whether or not the matrix

$$\begin{bmatrix} 1 & 1 & 1 \\ 1 & 1 & 1 \\ 1 & 1 & 0 \end{bmatrix}$$

is positive definite.

# 3. Systems of Simultaneous Linear Equations

## 3.1 Existence of Solutions

We have already briefly considered in Chapter 1 a system of $n$ linear equations in $n$ unknowns and seen how in certain circumstances Cramer's rule can lead to a solution. We wish now to consider the general problem of a system of $m$ equations in $n$ unknowns. We have to determine under what conditions solutions exist, whether or not these solutions are unique, and we must investigate methods for determining these solutions. If a solution exists, the equations are said to be consistent.

The general system which we shall be considering has the form

$$\begin{aligned} a_{11}x_1 + a_{12}x_2 + \ldots + a_{1n}x_n &= b_1 \\ a_{21}x_1 + a_{22}x_2 + \ldots + a_{2n}x_n &= b_2 \\ &\vdots \\ a_{m1}x_1 + a_{m2}x_2 + \ldots + a_{mn}x_n &= b_m \end{aligned} \qquad (3.1.1)$$

or, in matrix notation,

$$\mathbf{Ax} = \mathbf{b} \qquad (3.1.2)$$

where $\mathbf{A}$ is of the order $m \times n$, $\mathbf{x}$ is of order $n \times 1$, and $\mathbf{b}$ is of order $m \times 1$.

In order to obtain an idea of the types of situation which can arise in general, we shall consider some systems of equations involving the three unknowns $x, y$ and $z$.

(i) $m = 2, n = 3$

$$x + y + z = 0$$

$$x + y + z = 1.$$

These two equations obviously have no solution and are inconsistent.

(ii) $m = 2, n = 3$

$$x + y + z = 0$$

$$x + 2y + 3z = 1.$$

These two equations will not determine $x, y$ and $z$ *uniquely*; however, substituting $z = -(x + y)$ from the first into the second gives

$$x + 2y - 3x - 3y = 1$$

leading to

$$2x + y = -1$$

so we have

$$y = -1 - 2x$$

and

$$z = -x - y = -x + 1 + 2x = 1 + x.$$

Thus a solution is given by

$$x = k$$

$$y = -1 - 2k$$

$$z = 1 + k$$

for any value of $k$, and so in this case there exists an infinity of solutions.

(iii) $m = 3, n = 3$

$$x + y + z = 0$$

$$x + 2y + 3z = 4$$

$$2x + 3y + 4z = 6.$$

We see that adding the first and second equations gives $2x + 3y + 4z = 4$ which is inconsistent with the third equation. Hence the set of equations possesses no solution.

(iv) $m = 3, n = 3$

$$x + y + z = 0$$

$$x + 2y + 3z = 1$$

$$2x + 3y + 4z = 1.$$

Here the third equation is just the same as that obtained by adding the first two. Hence it gives no further information and we are back to case (ii) above with solution

$$x = k$$

$$y = -1 - 2k$$

$$z = 1 + k \qquad \text{for any value of } k.$$

Hence again we have an infinity of solutions.

(v) $m = 3, n = 3$

$$x + y + z = 6$$

$$x + y - z = 0$$

$$x - y + z = 2.$$

Solving by Cramer's rule or by systematic elimination gives $x = 1, y = 2, z = 3$ —a unique solution.

(vi) $m = 4, n = 3$

$$x + y + z = 6$$

$$x + y - z = 0$$

$$x - y + z = 2$$

$$x - y - z = 4.$$

The first three of these equations may be solved to give $x = 1, y = 2, z = 3$ as in (v) above. However these values do not satisfy the fourth equation, and hence the equations are inconsistent.

(vii) $m = 4, n = 3$

$$x + y + z = 0$$

$$x + 2y + 3z = 1$$

$$3x + 3y + 4z = 1$$

$$y + 2z = 1.$$

Here the third equation is obtained by adding the first two and the fourth equation by subtracting the first two. Thus the last two equations give no more information than the first two, and we are back to example (ii) with an infinity of solutions.

(viii) $m = 4, n = 3$

$$x + y + z = 6$$

$$x + y - z = 0$$

$$x - y + z = 2$$

$$x + y = 3.$$

This time the fourth equation is obtained by adding the first two. Thus we have the same amount of information as in case (v) above and thus we have the unique solution $x = 1, y = 2, z = 3$.

To sum up, we see that at least the following possibilities exist:

$m < n$ (*a*) inconsistent; (*b*) infinity of solutions

$m > n$ (*a*) inconsistent; (*b*) infinity of solutions; (*c*) unique solution.

We see that we have been unable to find a unique solution if $m < n$, while if $m > n$, solutions exist only if the 'extra' equations are obtainable from $n$ 'basic' equations— the equations are not all independent.

We now wish to show that these results hold true in general and we wish criteria which will enable us to determine the character of the solution of a particular set of simultaneous linear equations. We shall do this in Theorem 3.1, but first we need to introduce the concept of *rank.*

DEFINITION 3.1

If a matrix **A** is modified by the deletion of any number of rows and/or columns, the resulting rectangular array is said to be a *submatrix* of **A**. □

To illustrate the above definition, if we consider the matrix

$$\mathbf{A} = \begin{bmatrix} 3 & -9 & 2 \\ 6 & 1 & 0 \end{bmatrix}$$

the following are all submatrices:

$\begin{bmatrix} 3 & 2 \\ 6 & 0 \end{bmatrix}$ obtained by deleting the second column;

$\begin{bmatrix} 3 \\ 6 \end{bmatrix}$ obtained by deleting second and third columns;

$[-9]$ obtained by deleting the second row and first and third columns;

$\begin{bmatrix} 3 & -9 & 2 \\ 6 & 1 & 0 \end{bmatrix}$ obtained by deleting no rows and no columns.

(The last example shows that a matrix is a submatrix of itself.)

DEFINITION 3.2

If a matrix **A**, of order $m \times n$, is such that every square submatrix of order greater than $r$ has zero determinant while at least one square submatrix of order $r$ has determinant non-zero, then the matrix **A** is said to be of *rank* $r$. We sometimes use the notation $r(\mathbf{A})$ to mean the rank of **A**. □

Since we have to form *square* submatrices, we see that they cannot be of any order higher than the lesser of $m$ and $n$; in other words we must have $r \leqslant \min(m, n)$. A systematic way to determine rank would then be to evaluate the determinants of the highest possible order submatrices; then the next highest order and so on. Evaluation can stop as soon as a non-zero determinant is found; the rank is equal to the order of the corresponding submatrix.

*Example*

3.1.1 Determine the rank of

$$\mathbf{A} = \begin{bmatrix} 1 & 1 & 1 & 0 \\ 1 & 2 & 3 & 1 \\ 2 & 3 & 4 & 1 \end{bmatrix}.$$

**A** is of order 3 x 4 and hence the maximum possible value the rank can be is 3. It will have this value provided at least one square submatrix of order 3 has non-zero determinant. Otherwise the rank will be at most 2 and we shall have to investigate square submatrices of order 2.

There are four square submatrices of order 3, and these are as follows:

| Submatrix | Determinant (evaluated by the methods of Chapter 1) |
|---|---|
| $\begin{bmatrix} 1 & 1 & 0 \\ 2 & 3 & 1 \\ 3 & 4 & 1 \end{bmatrix}$ | 0 |
| $\begin{bmatrix} 1 & 1 & 0 \\ 1 & 3 & 1 \\ 2 & 4 & 1 \end{bmatrix}$ | 0 |
| $\begin{bmatrix} 1 & 1 & 0 \\ 1 & 2 & 1 \\ 2 & 3 & 1 \end{bmatrix}$ | 0 |
| $\begin{bmatrix} 1 & 1 & 1 \\ 1 & 2 & 3 \\ 2 & 3 & 4 \end{bmatrix}$ | 0 |

We must thus investigate submatrices of order 2:

| Submatrix | Determinant |
|---|---|
| $\begin{bmatrix} 1 & 1 \\ 1 & 2 \end{bmatrix}$ | 1 |

We have thus discovered a non-zero determinant, and so we can stop with the conclusion that $r(\mathbf{A}) = 2$.

DEFINITION 3.3

For the system of equations with matrix form $\mathbf{Ax} = \mathbf{b}$ (where **A** is of order $m \times n$, **x** is of order $n \times 1$ and **b** is of order $m \times 1$) we define the augmented matrix $\mathbf{A_b}$ to be the matrix of order $m \times (n + 1)$ obtained by conjoining the column **b** to the matrix **A**; in terms of partitioned matrices we have

$$\mathbf{A_b} = [\mathbf{A} \vdots \mathbf{b}]. \qquad \square$$

For example, if we consider the system of equations

$$x + y + z = 6$$

$$x + y - z = 0$$

$$x - y + z = 2$$

we have

$$\mathbf{A} = \begin{bmatrix} 1 & 1 & 1 \\ 1 & 1 & -1 \\ 1 & -1 & 1 \end{bmatrix}, \qquad \mathbf{b} = \begin{bmatrix} 6 \\ 0 \\ 2 \end{bmatrix} \qquad \text{and} \qquad \mathbf{A_b} = \begin{bmatrix} 1 & 1 & 1 & 6 \\ 1 & 1 & -1 & 0 \\ 1 & -1 & 1 & 2 \end{bmatrix}.$$

We are now in a position to state and prove the general result concerning the existence and uniqueness of solutions to the system of equations $\mathbf{Ax} = \mathbf{b}$.

THEOREM 3.1

The system of $m$ equations in $n$ unknowns, $\mathbf{Ax} = \mathbf{b}$, is consistent if and only if $r(\mathbf{A}) = r(\mathbf{A_b})$. This common value, if it exists, is called the rank of the system, and is denoted by $r$. In such a case the solution is unique if and only if $r = n$.

To help us in the proof of this theorem, we first introduce and prove the following lemma.

LEMMA

If a matrix $\mathbf{A}$ is subjected to a sequence of elementary transformations of type 1, 2 or 3 (as defined in Section 2.10) giving the matrix $\mathbf{B}$, then $r(\mathbf{B}) = r(\mathbf{A})$.

*Proof of Lemma*

Suppose $\mathbf{A}$ has rank $r$ (i.e. $r(\mathbf{A}) = r$); then we know from Definition 3.2 that all square submatrices of $\mathbf{A}$ of order $(r + 1)$ have determinant zero.

A submatrix of $\mathbf{B}$ of order $(r + 1)$ is obtained from the matrix $\mathbf{A}$ by an operation which ensures that its determinant differs from that of some submatrix of $\mathbf{A}$ of order $(r + 1)$ by at most a constant factor (we recall that elementary transformations involve either (i) multiplying all the elements in a row or column by a constant, (ii) interchanging rows or columns or (iii) adding multiples of rows (or columns) to a given row (or column) and the appropriate properties of determinants are given in Theorems 1.11, 1.13 and 1.15). Hence it follows that all submatrices of $\mathbf{B}$ of order $(r + 1)$ also have zero determinant, and hence $r(\mathbf{B}) \leqslant r(\mathbf{A})$.

However, if $\mathbf{B}$ is obtained from $\mathbf{A}$ by a sequence of elementary transformations, it follows also that $\mathbf{A}$ may be obtained from $\mathbf{B}$ by a sequence of different transformations. It thus follows from the argument in the preceding paragraph that $r(\mathbf{A}) \leqslant r(\mathbf{B})$.

Thus since we have both $r(\mathbf{B}) \leqslant r(\mathbf{A})$ and $r(\mathbf{B}) \geqslant r(\mathbf{A})$ we must have $r(\mathbf{B})\ r(\mathbf{A})$.

*Proof of Theorem*

The system of equations under consideration is

$$\begin{aligned} a_{11}x_1 + a_{12}x_2 + \ldots + a_{1n}x_n &= b_1 \\ a_{21}x_1 + a_{22}x_2 + \ldots + a_{2n}x_n &= b_2 \\ &\vdots \\ a_{m1}x_1 + a_{m2}x_2 + \ldots + a_{mn}x_n &= b_m \end{aligned} \tag{3.1.3}$$

which in matrix notation is $\mathbf{Ax} = \mathbf{b}$; the corresponding augmented matrix is $\mathbf{A_b} = [\mathbf{A}\ \mathbf{b}]$.

Suppose now that $x_{i_1}$ is some convenient variable which actually appears in one of the above equations; if we re-arrange the variables and equations so that $x_{i_1}$ appears with a non-zero coefficient as the first term of the first equation, then divide throughout this first equation by this non-zero coefficient and use the resulting equation to eliminate $x_{i_1}$ from all other equations we shall obtain a system of equations equivalent to (3.1.3) whose augmented matrix is obtained from the original augmented matrix by a sequence of elementary transformations.

We now continue in the above manner, choosing suitable variables $x_{i_2}, x_{i_3}, \ldots$ in turn, and eliminating the chosen variable from all the other equations. We are eventually forced to stop either when all $m$ of the equations have been used or else when the remaining equations contain no variables which can be used. In either case we end up with a system of equations of the form

$$\begin{aligned}
x_{i_1} \qquad\qquad &+ \alpha_{1,i_{r+1}} x_{i_{r+1}} + \alpha_{1,i_{r+2}} x_{i_{r+2}} + \ldots + \alpha_{1,i_n} x_{i_n} = \beta_1 \\
x_{i_2} \qquad &+ \alpha_{2,i_{r+1}} x_{i_{r+1}} + \alpha_{2,i_{r+2}} x_{i_{r+2}} + \ldots + \alpha_{2,i_n} x_{i_n} = \beta_2 \\
&\cdot \\
&\cdot \\
&\cdot \\
x_{i_r} &+ \alpha_{r,i_{r+1}} x_{i_{r+1}} + \alpha_{r,i_{r+2}} x_{i_{r+2}} + \ldots + \alpha_{r,i_n} x_{i_n} = \beta_r \\
&\qquad\qquad 0 = \beta_{r+1} \\
&\qquad\qquad 0 = \beta_{r+2} \\
&\qquad\qquad \cdot \\
&\qquad\qquad \cdot \\
&\qquad\qquad \cdot \\
&\qquad\qquad 0 = \beta_m .
\end{aligned} \tag{3.1.4}$$

The corresponding matrix form is

$$\begin{matrix} r\left\{ \right. \\ \\ m-r\left\{ \right. \end{matrix} \left[\begin{array}{c|c} \mathbf{I} & \alpha \\ \hline 0 & 0 \end{array}\right] \begin{bmatrix} x_{i_1} \\ x_{i_2} \\ \cdot \\ \cdot \\ \cdot \\ x_{i_n} \end{bmatrix} = \begin{bmatrix} \beta_1 \\ \beta_2 \\ \cdot \\ \cdot \\ \cdot \\ \beta_m \end{bmatrix}$$

$$\underbrace{\phantom{I}}_{r} \quad \underbrace{\phantom{\alpha}}_{n-r}$$

with augmented matrix

$$\begin{bmatrix} \mathbf{I} & \alpha & \mathbf{p} \\ \mathbf{0} & \mathbf{0} & \mathbf{q} \end{bmatrix} \quad \text{where} \quad \mathbf{p} = \begin{bmatrix} \beta_1 \\ \cdot \\ \cdot \\ \cdot \\ \beta_r \end{bmatrix} \quad \text{and} \quad \mathbf{q} = \begin{bmatrix} \beta_{r+1} \\ \cdot \\ \cdot \\ \cdot \\ \beta_m \end{bmatrix}.$$

Again, this augmented matrix has been obtained from the original augmented matrix by a sequence of elementary transformations and hence, by the lemma, has the same rank as the original augmented matrix.

Now, the equations (3.1.4) (and hence the original equations (3.1.3)) have solutions if and only if $\beta_{r+1} = \beta_{r+2} = \ldots = \beta_m = 0$ i.e. if and only if $\mathbf{q} = \mathbf{0}$. However, $\mathbf{q} = \mathbf{0}$ if and only if the rank of

$$\begin{bmatrix} \mathbf{I} & \alpha & \mathbf{p} \\ \mathbf{0} & \mathbf{0} & \mathbf{q} \end{bmatrix}$$

is $r$ (since det $(\mathbf{I}) \neq 0$ there is at least one square submatrix of order $r$ with non-zero determinant, and if $\mathbf{q} = \mathbf{0}$ all submatrices of order $(r + 1)$ and greater will contain a row of zeros and hence have zero determinant). Also of course the rank of

$$\begin{bmatrix} \mathbf{I} & \alpha \\ \mathbf{0} & \mathbf{0} \end{bmatrix}$$

is $r$. Thus the necessary and sufficient condition for the consistency of the system of equations (3.1.3) is that the ranks of the coefficient and augmented matrices should be the same. And since these matrices have been derived by a sequence of elementary transformations from the original matrices $\mathbf{A}$ and $\mathbf{A_b}$, this implies that the system of equations is consistent if and only if $r(\mathbf{A}) = r(\mathbf{A_b})$.

In the consistent case with rank $r\,(<n)$ it is obvious that $x_{i_{r+1}}, x_{i_{r+2}}, \ldots x_{i_n}$ may be chosen arbitrarily, and all the other variables $x_{i_1}, x_{i_2}, \ldots x_{i_r}$ expressed in terms of them (using the first $r$ of equations (3.1.4)); hence if $r < n$ there is an infinity of solutions involving $(n - r)$ arbitrary constants, while if $r = n$ there is a unique solution given by $x_{i_k} = \beta_k \; (1 \leqslant k \leqslant n)$.

This completes the proof of the required result. ■

*Example*

3.1.2 Apply Theorem 3.1 to the systems (i)–(v) above (pp 106–8).

(i) Here

$$m = 2, n = 3, \qquad \mathbf{A} = \begin{bmatrix} 1 & 1 & 1 \\ 1 & 1 & 1 \end{bmatrix}, \qquad \mathbf{b} = \begin{bmatrix} 0 \\ 1 \end{bmatrix}$$

and

$$\mathbf{A_b} = \begin{bmatrix} 1 & 1 & 1 & 0 \\ 1 & 1 & 1 & 1 \end{bmatrix}.$$

Thus we obtain $r(\mathbf{A}) = 1$, $r(\mathbf{A_b}) = 2$; hence $r(\mathbf{A}) \neq r(\mathbf{A_b})$ and so by Theorem 3.1 the equations are inconsistent.

(ii) Here

$$m = 2, n = 3, \qquad \mathbf{A} = \begin{bmatrix} 1 & 1 & 1 \\ 1 & 2 & 3 \end{bmatrix}, \qquad \mathbf{b} = \begin{bmatrix} 0 \\ 1 \end{bmatrix}$$

and

$$\mathbf{A_b} = \begin{bmatrix} 1 & 1 & 1 & 0 \\ 1 & 2 & 3 & 1 \end{bmatrix}.$$

In this case $r(\mathbf{A}) = 2$, $r(\mathbf{A_b}) = 2$ and so $r(\mathbf{A}) = r(\mathbf{A_b}) < n$; thus by Theorem 3.1 we have a consistent set of equations with an infinity of solutions.

(iii) Here

$$m = 3, n = 3, \qquad \mathbf{A} = \begin{bmatrix} 1 & 1 & 1 \\ 1 & 2 & 3 \\ 2 & 3 & 4 \end{bmatrix}, \qquad \mathbf{b} = \begin{bmatrix} 0 \\ 4 \\ 6 \end{bmatrix}$$

and

$$\mathbf{A_b} = \begin{bmatrix} 1 & 1 & 1 & 0 \\ 1 & 2 & 3 & 4 \\ 2 & 3 & 4 & 6 \end{bmatrix}.$$

In this case $r(\mathbf{A}) = 2$, $r(\mathbf{A_b}) = 3$; so $r(\mathbf{A}) \neq r(\mathbf{A_b})$ and the equations are inconsistent.

(iv) Here

$$m = 3, n = 3, \qquad \mathbf{A} = \begin{bmatrix} 1 & 1 & 1 \\ 1 & 2 & 3 \\ 2 & 3 & 4 \end{bmatrix}, \qquad \mathbf{b} = \begin{bmatrix} 0 \\ 1 \\ 1 \end{bmatrix}$$

and

$$\mathbf{A_b} = \begin{bmatrix} 1 & 1 & 1 & 0 \\ 1 & 2 & 3 & 1 \\ 2 & 3 & 4 & 1 \end{bmatrix}.$$

Hence $r(\mathbf{A}) = 2$, $r(\mathbf{A_b}) = 2$ and so $r(\mathbf{A}) = r(\mathbf{A_b}) < n$ showing by Theorem 3.1, that an infin of solutions exists.

(v) Here

$$m = 3, n = 3, \qquad \mathbf{A} = \begin{bmatrix} 1 & 1 & 1 \\ 1 & 1 & -1 \\ 1 & -1 & 1 \end{bmatrix}, \qquad \mathbf{b} = \begin{bmatrix} 6 \\ 0 \\ 2 \end{bmatrix}$$

and

$$\mathbf{A_b} = \begin{bmatrix} 1 & 1 & 1 & 6 \\ 1 & 1 & -1 & 0 \\ 1 & -1 & 1 & 2 \end{bmatrix}.$$

Here $r(\mathbf{A}) = 3$, $r(\mathbf{A_b}) = 3$ so $r(\mathbf{A}) = r(\mathbf{A_b}) = n$ and hence by Theorem 3.1 a unique solution exists.

A special case which often arises in practice is when we have a system of equations of the form (3.1.1) with all the $b_i$ zero. We distinguish this special case in the following definition.

DEFINITION 3.4

The system of equations $\mathbf{Ax} = \mathbf{0}$ is said to be a *homogeneous* system; the system $\mathbf{Ax} = \mathbf{b}$ where $\mathbf{b} \neq \mathbf{0}$ (i.e. at least one $b_i \neq 0$) is said to be an *inhomogeneous* system. □

Since $\mathbf{A0} = \mathbf{0}$ we see that the homogeneous system $\mathbf{Ax} = \mathbf{0}$ always possesses the solution $\mathbf{x} = \mathbf{0}$; we call this the *trivial solution*. However, we are often interested in the possible existence of a solution other than the trivial solution. No information is gained from Theorem 3.1 since it will give the result that the equations are always consistent (relevant for the trivial solution). The appropriate result of interest is contained in the following theorem.

THEOREM 3.2

The homogeneous system of $m$ equations in $n$ unknowns $\mathbf{Ax} = \mathbf{0}$ has a non-trivial solution if and only if $r(\mathbf{A}) < n$; in the important special case of $m = n$, this is equivalent to the condition $\det(\mathbf{A}) = 0$.

*Proof*

As in Theorem 3.1, the given set of equations may be replaced by the equivalent set

$$\begin{aligned}
x_{i_1} \qquad\qquad & + \alpha_{1,i_{r+1}} x_{i_{r+1}} + \alpha_{1,i_{r+2}} x_{i_{r+2}} + \ldots + \alpha_{1,i_n} x_{i_n} = 0 \\
x_{i_2} \qquad & + \alpha_{2,i_{r+1}} x_{i_{r+1}} + \alpha_{2,i_{r+2}} x_{i_{r+2}} + \ldots + \alpha_{2,i_n} x_{i_n} = 0 \\
& \vdots \\
x_{i_r} & + \alpha_{r,i_{r+1}} x_{i_{r+1}} + \alpha_{r,i_{r+2}} x_{i_{r+2}} + \ldots + \alpha_{r,i_n} x_{i_n} = 0
\end{aligned} \tag{3.1.5}$$

(the remaining $(m - r)$ equations in this case just being $0 = 0$). Again $r = r(\mathbf{A})$. If $r = n$, these equations just reduce to $x_{i_1} = x_{i_2} = \ldots = x_{i_n} = 0$, meaning that we just obtain the trivial solution. On the other hand, if $r < n$, the system (3.1.5) has a non-zero solution with $x_{i_{r+1}}, x_{i_{r+2}} \ldots x_{i_n}$ being chosen arbitrarily.

If $\mathbf{A}$ is square, the condition $r(\mathbf{A}) < n$ implies that $\det(\mathbf{A}) = 0$, thus proving the second part of the theorem. ■

We note that by the above theorem, the system $\mathbf{Ax} = \mathbf{0}$ posseses a non-trivial solution if and only if $\mathbf{A}^{-1}$ does not exist. We note also that in this case the solution is never unique; for if $\mathbf{x}$ satisfies $\mathbf{Ax} = \mathbf{0}$ then so does $c\mathbf{x}$ where $c$ is any constant. Thus at the most we can hope for the ratios of the $x_i$ to be unique.

*Exercise*

3.1.1 Determine whether or not solutions exist for the following sets of equations; if a solution exists determine whether or not it is unique.

(i)
$$\begin{aligned} x + y + z &= 0 \\ 3x + 2y - z &= -4 \\ x + 4y + 2z &= 1 \end{aligned}$$

(ii)
$$\begin{aligned} 3x + y + z &= 2 \\ -2x + y - z &= 1 \\ 12x - y + 5z &= 1 \end{aligned}$$

(iii)
$$\begin{aligned} x + y - z &= 1 \\ 3x + 2y - 4z &= 1 \\ -2x - 3y + z &= 1 \end{aligned}$$

(iv)
$$\begin{aligned} x + y + z &= 0 \\ x + 2y + z &= 0 \\ 3x + 2y - z &= -4 \\ x + 4y + 2z &= 1 \end{aligned}$$

(v)
$$\begin{aligned} x + y + z &= 0 \\ x + y + 2z &= 0 \\ 3x + 2y - z &= -4 \\ x + 4y + 2z &= 1. \end{aligned}$$

We have discussed above in some detail and generality the existence and uniqueness of a solution to the system of equations $\mathbf{Ax} = \mathbf{b}$. By far the most important situation in practice is the system of $n$ equations in $n$ unknowns with a unique solution. We shall devote the remainder of this chapter to a consideration of practical methods for solving such a system. A wide variety of methods exist, and we cannot hope to give here either a comprehensive survey of all available methods or even a detailed treatment of such methods as we do discuss; this is the province of a textbook on numerical analysis. What we shall do is present the more commonly used methods in such a way that the reader can either implement them as computer programs or else have an intelligent appreciation of programs which are available to him.

With the present day wide availability of digital computers, it is almost certain that in all real situations where the solution is needed to a system of linear equations, it will be found by computer means. It is thus essential that the user should have some knowledge of how the solutions are achieved; in particular he should be aware of situations in which the methods might fall, he should avoid the blind acceptance of numbers produced by a computer program as supposed answers to his problem, and he should be conscious of whether or not procedures he uses to check his answers are in fact valid. We discuss briefly some of these points in Section 3.4 after presenting methods of solution in Sections 3.2 and 3.3.

## 3.2 Direct Methods of Solution

By a direct method we mean a method which calculates the required solution without any intermediate approximations. This is opposed to an indirect or iterative method which starts from an initial approximation and proceeds by calculating a sequence of further approximations which eventually gives the required solution as accurately as is desired.

### METHOD 1: GAUSSIAN ELIMINATION

The basis of this method is to note that the following system of equations can be easily solved:

$$\begin{aligned}
\alpha_{11}x_1 + \alpha_{12}x_2 + \dots \qquad\qquad & + \alpha_{1n}x_n = \beta_1 \\
\alpha_{22}x_2 + \alpha_{23}x_3 + \dots & + \alpha_{2n}x_n = \beta_2 \\
\alpha_{33}x_3 + \dots & + \alpha_{3n}x_n = \beta_3 \\
& \;\vdots \\
\alpha_{n-1,n-1}x_{n-1} & + \alpha_{n-1,n}x_n = \beta_{n-1} \\
& \alpha_{nn}x_n = \beta_n.
\end{aligned} \tag{3.2.1}$$

The solution is accomplished by what we call 'back-substitution'. The $n$th equation gives $x_n$ directly ($\beta_n/\alpha_{nn}$); then the $(n-1)$th equation gives $x_{n-1}$ (since $x_n$ is known), the $(n-2)$th equation gives $x_{n-2}$ (since $x_n$ and $x_{n-1}$ are known) and so on until finally the first equation gives $x_1$. In general the $i$th equation gives $x_i$ in terms of $x_{i+1}$, $x_{i+2}, \ldots x_n$ which have already been calculated.

In matrix terms this means that the system of equations $\mathbf{Ax} = \mathbf{b}$ is easy to solve when $\mathbf{A}$ is an upper triangular matrix; a similar situation obviously arises when $\mathbf{A}$ is lower triangular.

We thus see that if we can transform the original set of equations

$$\begin{aligned} a_{11}x_1 + a_{12}x_2 + \ldots + a_{1n}x_n &= b_1 \\ a_{21}x_1 + a_{22}x_2 + \ldots + a_{2n}x_n &= b_2 \\ &\ \,\vdots \\ a_{n1}x_1 + a_{n2}x_2 + \ldots + a_{nn}x_n &= b_n \end{aligned} \tag{3.2.2}$$

into the form (3.2.1), then we shall be able to obtain a solution. The most straightforward way of doing this is to use the first equation to eliminate $x_1$ from the second and all subsequent equations, the new second equation to eliminate $x_2$ from the new third and all subsequent equations, and so on; the last step will be to use the $(n-1)$th equation to eliminate $x_{n-1}$ from the $n$th equation. The above procedure will work only if at each stage the $i$th equation does in fact contain $x_i$ with a non-zero coefficient so that the appropriate multiple of this equation may be subtracted from all subsequent equations to eliminate $x_i$ from these equations. If this condition is not satisfied then either it is possible to re-arrange the equations so as to obtain a non-zero coefficient for $x_i$ in the appropriate position or else all the coefficients are already zero and no elimination of $x_i$ is necessary.

*Example*

3.2.1 Use Gaussian elimination to solve the system of equations

$$\begin{aligned} x_1 + x_2 + x_3 - x_4 &= 2 \\ 4x_1 + 4x_2 + x_3 + x_4 &= 11 \\ x_1 - x_2 - x_3 + 2x_4 &= 0 \\ 2x_1 + x_2 + 2x_3 - 2x_4 &= 2. \end{aligned}$$

We use the first equation to eliminate $x_1$ from the subsequent equations by subtracting the appropriate multiple of the first equation from each of the others:

$$\begin{aligned} x_1 + x_2 + x_3 - x_4 &= 2 \\ -3x_3 + 5x_4 &= 3 \\ -2x_2 - 2x_3 + 3x_4 &= -2 \\ -x_2 \qquad\qquad &= -2. \end{aligned}$$

Our next step is to try to use the new second equation to eliminate $x_2$ from the subsequent equations. Unfortunately, here $x_2$ does not appear in the second equation, so we must re-arrange the equations in such a way that it does appear; a possible re-ordering would be

$$\begin{aligned} x_1 + x_2 + x_3 - x_4 &= 2 \\ -x_2 &= -2 \\ -3x_3 + 5x_4 &= 3 \\ -2x_2 - 2x_3 + 3x_4 &= -2. \end{aligned}$$

Eliminating $x_2$ now gives

$$\begin{aligned} x_1 + x_2 + x_3 - x_4 &= 2 \\ -x_2 &= -2 \\ -3x_3 + 5x_4 &= 3 \\ -2x_3 + 3x_4 &= 2. \end{aligned}$$

We now use the third equation to eliminate $x_3$ from the fourth equation and we obtain

$$\begin{aligned} x_1 + x_2 + x_3 - x_4 &= 2 \\ -x_2 &= -2 \\ -3x_3 + 5x_4 &= 3 \\ -\tfrac{1}{3}x_4 &= 0. \end{aligned}$$

We may now use these in turn, starting with the last to give:

$x_4 = 0$

$-3x_3 + 5 \times 0 = 3$ and hence $x_3 = -1$

$-x_2 = -2$ and hence $x_2 = 2$

$x_1 + 2 + (-1) - 0 = 2$ and hence $x_1 = 1$.

Thus the solution is $x_1 = 1, x_2 = 2, x_3 = -1, x_4 = 0$.

## METHOD 2: GAUSSIAN ELIMINATION WITH PARTIAL PIVOTING†

This is a modification of Method 1 above, designed to make it less susceptible to rounding error.

Suppose that at the stage when we are using the $i$th equation to eliminate $x_i$ from subsequent equations that the $i$th and subsequent equations have the form

$$\begin{aligned} c_{ii}x_i + c_{i,i+1}x_{i+1} + \ldots + c_{in}x_n &= d_i \\ &\vdots \\ c_{ji}x_i + c_{j,i+1}x_{i+1} + \ldots + c_{jn}x_n &= d_j \\ &\vdots \\ c_{ni}x_i + c_{n,i+1}x_{i+1} + \ldots + c_{nn}x_n &= d_n. \end{aligned}$$

† As we shall see, partial pivoting involves a possible interchange of equations (or equivalently rows in the matrix **A**); full pivoting would involve interchange of variables (or columns of the matrix) but is not discussed here.

Hence, to eliminate $x_i$ from the $j$th equation we have to subtract from this equation the $i$th equation multiplied by the factor $c_{ji}/c_{ii}$. Obviously this number can be very large if $c_{ii}$ is small; it makes sense to keep numbers of a reasonable size and avoid rounding error as much as possible by choosing the $c_{ii}$ to have the largest magnitude possible. This is achieved by re-ordering the equations if necessary. The $c_{ii}$ which is actually used for the elimination is called the *pivot* for this stage of the elimination.

*Example*

3.2.2 Use Gaussian elimination with partial pivoting to solve the system of equations

$$\begin{aligned} x_1 + x_2 + x_3 - x_4 &= 2 \\ 4x_1 + 4x_2 + x_3 + x_4 &= 11 \\ x_1 - x_2 - x_3 + 2x_4 &= 0 \\ 2x_1 + x_2 + 2x_3 - 2x_4 &= 2. \end{aligned}$$

This is the same set of equations as in the previous example. Now, however, we re-order at each stage so that the pivot is as large as possible. Thus the various steps in the solution are now as follows:

(i) Re-order to obtain

$$\begin{aligned} 4x_1 + 4x_2 + x_3 + x_4 &= 11 \\ x_1 + x_2 + x_3 - x_4 &= 2 \\ x_1 - x_2 - x_3 + 2x_4 &= 0 \\ 2x_1 + x_2 + 2x_3 - 2x_4 &= 2. \end{aligned}$$

(ii) Use the first equation to eliminate $x_1$ giving

$$\begin{aligned} 4x_1 + 4x_2 + x_3 + x_4 &= 11 \\ \tfrac{3}{4}x_3 - \tfrac{5}{4}x_4 &= -\tfrac{3}{4} \\ -2x_2 - \tfrac{5}{4}x_3 + \tfrac{7}{4}x_4 &= -\tfrac{11}{4} \\ -x_2 + \tfrac{3}{2}x_3 - \tfrac{5}{2}x_4 &= -\tfrac{7}{2}. \end{aligned}$$

(iii) Re-order to obtain

$$\begin{aligned} 4x_1 + 4x_2 + x_3 + x_4 &= 11 \\ -2x_2 - \tfrac{5}{4}x_3 + \tfrac{7}{4}x_4 &= -\tfrac{11}{4} \\ \tfrac{3}{4}x_3 - \tfrac{5}{4}x_4 &= -\tfrac{3}{4} \\ -x_2 + \tfrac{3}{2}x_3 - \tfrac{5}{2}x_4 &= -\tfrac{7}{2}. \end{aligned}$$

(iv) Use the second equation to eliminate $x_2$ giving

$$\begin{aligned} 4x_1 + 4x_2 + x_3 + x_4 &= 11 \\ -2x_2 - \tfrac{5}{4}x_3 + \tfrac{7}{4}x_4 &= -\tfrac{11}{4} \\ \tfrac{3}{4}x_3 - \tfrac{5}{4}x_4 &= -\tfrac{3}{4} \\ \tfrac{17}{8}x_3 - \tfrac{27}{8}x_4 &= -\tfrac{17}{8}. \end{aligned}$$

(v) Re-order to obtain

$$\begin{aligned}
4x_1 + 4x_2 + \quad x_3 + \quad x_4 &= \quad 11\\
-2x_2 - \tfrac{5}{4}x_3 + \tfrac{7}{4}x_4 &= -\tfrac{11}{4}\\
\tfrac{17}{8}x_3 - \tfrac{27}{8}x_4 &= -\tfrac{17}{8}\\
\tfrac{3}{4}x_3 - \tfrac{5}{4}x_4 &= -\tfrac{3}{4}.
\end{aligned}$$

(vi) Use the third equation to eliminate $x_3$ giving

$$\begin{aligned}
4x_1 + 4x_2 + \quad x_3 + \quad x_4 &= \quad 11\\
-2x_2 - \tfrac{5}{4}x_3 + \tfrac{7}{4}x_4 &= -\tfrac{11}{4}\\
\tfrac{17}{8}x_3 - \tfrac{27}{8}x_4 &= -\tfrac{17}{8}\\
-\tfrac{1}{17}x_4 &= \quad 0.
\end{aligned}$$

(vii) Starting from the last equation, use the equations in turn to give

$$x_4 = 0, x_3 = -1, x_2 = 2, x_1 = 1.$$

We see that in both the examples above we have reduced the system of equations $\mathbf{Ax} = \mathbf{b}$ to the form $\mathbf{A'x} = \mathbf{b'}$ where $\mathbf{A'}$ is upper triangular. Written out in matrix form the original system was

$$\begin{bmatrix} 1 & 1 & 1 & -1 \\ 4 & 4 & 1 & 1 \\ 1 & -1 & -1 & 2 \\ 2 & 1 & 2 & -2 \end{bmatrix} \begin{bmatrix} x_1 \\ x_2 \\ x_3 \\ x_4 \end{bmatrix} = \begin{bmatrix} 2 \\ 1 \\ 0 \\ 2 \end{bmatrix}.$$

Gaussian elimination produced the system

$$\begin{bmatrix} \underline{1} & 1 & 1 & -1 \\ 0 & -\underline{1} & 0 & -4 \\ 0 & 0 & -\underline{3} & 5 \\ 0 & 0 & 0 & -\underline{\tfrac{1}{3}} \end{bmatrix} \begin{bmatrix} x_1 \\ x_2 \\ x_3 \\ x_4 \end{bmatrix} = \begin{bmatrix} 2 \\ -2 \\ 3 \\ 0 \end{bmatrix}$$

while with partial pivoting we obtained

$$\begin{bmatrix} \underline{4} & 4 & 1 & 1 \\ 0 & -\underline{2} & -\tfrac{5}{4} & \tfrac{7}{4} \\ 0 & 0 & \underline{\tfrac{17}{8}} & -\tfrac{27}{8} \\ 0 & 0 & 0 & -\underline{\tfrac{1}{17}} \end{bmatrix} \begin{bmatrix} x_1 \\ x_2 \\ x_3 \\ x_4 \end{bmatrix} = \begin{bmatrix} 11 \\ -\tfrac{11}{4} \\ -\tfrac{17}{8} \\ 0 \end{bmatrix}.$$

In both cases we have underlined the pivots. In both cases also, the matrix $\mathbf{A'}$ is obtained from the matrix $\mathbf{A}$ by a sequence of elementary transformations. It hence follows that $\det(\mathbf{A}) = \pm \det(\mathbf{A'})$, the plus sign being appropriate if an even number of equation interchanges has taken place (since these correspond to row interchanges in the matrix) and the minus sign if an odd number has taken place.

But by Theorem 2.17 we have

$$\begin{aligned}\det(\mathbf{A}') &= \text{product of diagonal elements of } \mathbf{A}' \\ &= \text{product of pivots.}\end{aligned}$$

Hence it follows that

$$\det(\mathbf{A}) = \pm \text{ products of pivots}$$

and in general this provides a good method for calculating a determinant.

In the first case above,

$$\begin{aligned}\text{product of pivots} &= (1) \times (-1) \times (-3) \times (-\tfrac{1}{3}) \\ &= -1\end{aligned}$$

and in the second case

$$\begin{aligned}\text{product of pivots} &= (4) \times (-2) \times (\tfrac{17}{8}) \times (-\tfrac{1}{17}) \\ &= 1.\end{aligned}$$

In the first case there were two interchanges of equations and in the second there were three (one at stage (i), one at stage (iii) and one at stage (v)). Hence by both methods we reach the conclusion that

$$\begin{vmatrix} 1 & 1 & 1 & -1 \\ 4 & 4 & 1 & 1 \\ 1 & -1 & -1 & 2 \\ 2 & 1 & 2 & -2 \end{vmatrix} = -1.$$

Either of the above two methods can also be adapted to calculate inverse matrices. For if we wish to calculate the inverse of $\mathbf{A}$ where $\mathbf{A}$ is square of order $n$, this means that we require to find a matrix $\mathbf{X}$ such that $\mathbf{AX} = \mathbf{I}$. If we denote the columns of $\mathbf{X}$ by $\mathbf{X}^{(1)}, \mathbf{X}^{(2)}, \ldots \mathbf{X}^{(n)}$ so that $\mathbf{X} = [\mathbf{X}^{(1)}, \mathbf{X}^{(2)}, \ldots \mathbf{X}^{(n)}]$ then $\mathbf{AX} = \mathbf{I}$ may be written in the form $[\mathbf{AX}^{(1)}, \mathbf{AX}^{(2)}, \ldots \mathbf{AX}^{(i)}, \ldots \mathbf{AX}^{(n)}] = [\mathbf{I}^{(1)}, \mathbf{I}^{(2)}, \ldots \mathbf{I}^{(i)}, \ldots \mathbf{I}^{(n)}]$ where $\mathbf{I}^{(i)}$ is the $i$th column of $\mathbf{I}$. Hence the $i$th column of $\mathbf{X}$ is obtained by solving the system of equations

$$\mathbf{AX}^{(i)} = \mathbf{I}^{(i)} = \begin{bmatrix} 0 \\ 0 \\ \cdot \\ \cdot \\ \cdot \\ 0 \\ 1 \\ 0 \\ 0 \\ \cdot \\ \cdot \\ \cdot \\ 0 \end{bmatrix} \begin{matrix} \\ \\ \\ \\ \\ \\ \leftarrow i\text{th row} \\ \\ \\ \\ \\ \\ \\ \end{matrix}$$

and the complete inverse matrix is obtained by solving $n$ such sets of equations.

Of course the triangular matrix $\mathbf{A}'$ obtained by the elimination process is independent of the column vector $\mathbf{b}$, and hence in the calculation of an inverse we can consider all the right-hand sides at the same time and only have one set of eliminations to perform. This is illustrated in the following example.

*Example*

3.2.3 Use Gaussian elimination with partial pivoting to find the inverse of

$$\mathbf{A} = \begin{bmatrix} 1 & 0 & -4 \\ 2 & 4 & 1 \\ 0 & 3 & 6 \end{bmatrix}.$$

We wish to solve the sets of equations

$$\mathbf{AX}^{(1)} = \begin{bmatrix} 1 \\ 0 \\ 0 \end{bmatrix} \qquad \mathbf{AX}^{(2)} = \begin{bmatrix} 0 \\ 1 \\ 0 \end{bmatrix} \qquad \mathbf{AX}^{(3)} = \begin{bmatrix} 0 \\ 0 \\ 1 \end{bmatrix}$$

giving

$$\mathbf{AX} = \mathbf{I}.$$

We can summarize these by using the notation

$$\begin{aligned} x_1^{(i)} \qquad\qquad - 4x_3^{(i)} &= 1, 0, 0 \\ 2x_1^{(i)} + 4x_2^{(i)} + \ x_3^{(i)} &= 0, 1, 0 \\ 3x_2^{(i)} + 6x_3^{(i)} &= 0, 0, 1. \end{aligned}$$

Carrying out the process of Gaussian elimination with partial pivoting gives in turn the systems of equations:

(i) Re-arrange

$$\begin{aligned} 2x_1^{(i)} + 4x_2^{(i)} + \ x_3^{(i)} &= 0, 1, 0 \\ x_1^{(i)} \qquad\qquad - 4x_3^{(i)} &= 1, 0, 0 \\ 3x_2^{(i)} + 6x_3^{(i)} &= 0, 0, 1. \end{aligned}$$

(ii) Eliminate $x_1^{(i)}$

$$\begin{aligned} 2x_1^{(i)} + 4x_2^{(i)} + \ x_3^{(i)} &= 0, 1, 0 \\ -2x_2^{(i)} - \tfrac{9}{2}x_3^{(i)} &= 1, -\tfrac{1}{2}, 0 \\ 3x_2^{(i)} + 6x_3^{(i)} &= 0, 0, 1. \end{aligned}$$

(iii) Re-arrange

$$\begin{aligned} 2x_1^{(i)} + 4x_2^{(i)} + \ x_3^{(i)} &= 0, 1, 0 \\ 3x_2^{(i)} + 6x_3^{(i)} &= 0, 0, 1 \\ -2x_2^{(i)} - \tfrac{9}{2}x_3^{(i)} &= 1, \tfrac{1}{2}, 0. \end{aligned}$$

(iv) Eliminate $x_2^{(i)}$

$$\begin{aligned} 2x_1^{(i)} + 4x_2^{(i)} + x_3^{(i)} &= 0, 1, 0 \\ 3x_2^{(i)} + 6x_3^{(i)} &= 0, 0, 1 \\ -\tfrac{1}{2}x_3^{(i)} &= 1, -\tfrac{1}{2}, \tfrac{2}{3}. \end{aligned}$$

So for $\mathbf{X}^{(1)}$ we have

$$\begin{aligned} -\tfrac{1}{2}x_3^{(1)} &= 1 && \text{giving } x_3^{(1)} = -2 \\ 3x_2^{(1)} + 6x_3^{(1)} &= 0 && \text{giving } x_2^{(1)} = 4 \\ 2x_1^{(1)} + 4x_2^{(1)} + x_3^{(1)} &= 0 && \text{giving } x_1^{(1)} = -7. \end{aligned}$$

Thus

$$\mathbf{X}^{(1)} = \begin{bmatrix} -7 \\ 4 \\ 2 \end{bmatrix}$$

and similarly we should obtain

$$\mathbf{X}^{(2)} = \begin{bmatrix} 4 \\ -2 \\ 1 \end{bmatrix} \quad \text{and} \quad \mathbf{X}^{(3)} = \begin{bmatrix} -\frac{16}{3} \\ 3 \\ -\frac{4}{3} \end{bmatrix}.$$

Thus

$$\mathbf{A}^{-1} = \mathbf{X} = \begin{bmatrix} -7 & 4 & -\frac{16}{3} \\ 4 & -2 & 3 \\ 2 & 1 & -\frac{4}{3} \end{bmatrix}.$$

We note that in the above example the calculation of the inverse involves the solution of three separate sets of simultaneous equations $\mathbf{Ax} = \mathbf{b}$ with the same $\mathbf{A}$ but different $\mathbf{b}$'s. It hence follows that it could be useful to calculate the inverse $\mathbf{A}^{-1}$ to solve $\mathbf{Ax} = \mathbf{b}$ when we wish to solve the system with many different vectors $\mathbf{b}$. In general for a system of $n$ equations in $n$ unknowns, there will be a saving in calculating the inverse if the solution is required for more than $n$ different vectors $\mathbf{b}$. Otherwise there is less work in merely solving the separate systems.

*Exercise*

3.2.1 Use Gaussian elimination, both with and without partial pivoting for the matrix

$$\mathbf{A} = \begin{bmatrix} 1 & 2 & 0 & 3 \\ -1 & 4 & -1 & -1 \\ 0 & 2 & 0 & -2 \\ 3 & 1 & 1 & 2 \end{bmatrix}$$

to determine

(i) the vector **x** such that

$$\mathbf{Ax} = \begin{bmatrix} 16 \\ -3 \\ -8 \\ 11 \end{bmatrix}$$

(ii) $\mathbf{A}^{-1}$; (iii) det (**A**).

METHOD 3: TRIANGULARIZATION (CHOLESKI–TURING METHOD)

Here we make use of Theorem 2.14, which states that under certain conditions a matrix **A** can be written in the form **A** = **LU** where **L** is a unit lower triangular matrix and **U** is an upper triangular matrix.

Assuming that the matrix **A** of the given set of simultaneous equations **Ax** = **b** satisfies these conditions, we may then write the system of equations in the form **LUx** = **b**. If we introduce the column vector **y** defined by **y** = **Ux** we see that **Ly** = **b** and so the given system is equivalent to the two systems

$$\mathbf{Ly} = \mathbf{b} \tag{3.2.3}$$

and

$$\mathbf{Ux} = \mathbf{y}. \tag{3.2.4}$$

But we have already noted when considering Gaussian elimination that a system of equations with a triangular matrix is easily solved; thus the procedure is to first solve (3.2.3) for **y** and then substitute **y** into (3.2.4) and solve for **x**.

The method of finding matrices **L** and **U** such that **A** = **LU** has already been indicated in Example 2.6.3 on page 73 it consists of writing

$$\mathbf{L} = \begin{bmatrix} 1 & 0 & 0 & 0 & \cdots & 0 \\ l_{21} & 1 & 0 & & & 0 \\ l_{31} & l_{32} & 1 & & & 0 \\ \cdot & & & & & \cdot \\ \cdot & & & & & \cdot \\ \cdot & & & & & \cdot \\ & & & & & 0 \\ l_{n1} & l_{n2} & & & l_{n,n-1} & 1 \end{bmatrix}$$

$$\mathbf{U} = \begin{bmatrix} u_{11} & u_{12} & u_{13} & \cdots & & u_{1n} \\ 0 & u_{22} & u_{23} & & & u_{2n} \\ 0 & 0 & u_{33} & & & u_{3n} \\ \cdot & & & & & \cdot \\ \cdot & & & & & \cdot \\ \cdot & & & & & \cdot \\ 0 & 0 & \cdots\cdots & & 0 & u_{nn} \end{bmatrix}$$

and forming the product **LU**. If **A** and **LU** are then equated element by element in a suitable order, a system of equations is obtained which yields the $l$'s and $u$'s explicitly in terms of the $a$'s and $l$'s and $u$'s already determined. If the condition for decomposition to be valid as given in Theorem 2.14 is not satisfied, then it will turn out that a division by zero will appear in these equations.

As already noted in Chapter 2, use of the triangular decomposition together with Theorem 2.18 enables the determinant of **A** to be readily calculated; we have

$$\begin{aligned}\det(\mathbf{A}) &= \det(\mathbf{LU})\\ &= \det(\mathbf{L}).\det(\mathbf{U}) && \text{by Theorem 2.18}\\ &= 1.(u_{11}u_{22}\ldots u_{nn}) && \text{by Theorem 2.17}\\ &= u_{11}u_{22}\ldots u_{nn}.\end{aligned}$$

The method can also be used to calculate inverses; we have $\mathbf{A} = \mathbf{LU}$ and hence $\mathbf{A}^{-1} = (\mathbf{LU})^{-1} = \mathbf{U}^{-1}\mathbf{L}^{-1}$ by Theorem 2.21. But the inverses of triangular matrices are comparatively easy to calculate. From the result that $\mathbf{A}^{-1} = \text{adj}(\mathbf{A})/\det(\mathbf{A})$ it is fairly easy to show that the inverse of an upper triangular matrix is an upper triangular matrix; and similarly the inverse of a lower triangular matrix is a lower triangular matrix. Thus if

$$\mathbf{U} = \begin{bmatrix} u_{11} & u_{12} & \ldots & u_{1n}\\ 0 & u_{22} & \ldots & u_{2n}\\ 0 & & & \\ \cdot & & & \cdot\\ \cdot & & & \cdot\\ \cdot & & & \cdot\\ 0 & 0 & & u_{nn}\end{bmatrix}$$

we are looking for an upper triangular matrix

$$\mathbf{U}^{-1} = \begin{bmatrix} p_{11} & p_{12} & \ldots & p_{1n}\\ 0 & p_{22} & & p_{2n}\\ 0 & & & \\ \cdot & & & \cdot\\ \cdot & & & \cdot\\ \cdot & & & \cdot\\ 0 & & & p_{nn}\end{bmatrix}$$

such that $\mathbf{UU}^{-1} = \mathbf{I}$. Again the elements of $\mathbf{U}^{-1}$ are obtained by forming the product $\mathbf{UU}^{-1}$, equating the elements to the elements of **I** in an appropriate order and hence obtaining a system of equations which yield the $p$'s explicitly in terms of the $u$'s and previously calculated $p$'s.

*Example*

3.2.4 Use triangular decomposition for the matrix

$$\mathbf{A} = \begin{bmatrix} 1 & 1 & 1 & -1\\ 1 & -1 & -1 & 2\\ 4 & 4 & 1 & 1\\ 2 & 1 & 2 & -2\end{bmatrix}$$

to: (i) solve

$$\mathbf{Ax} = \begin{bmatrix} 2 \\ 0 \\ 11 \\ 2 \end{bmatrix}$$

(ii) obtain det $(\mathbf{A})$

(iii) obtain $\mathbf{A}^{-1}$.

(This is essentially the same problem as was considered in the Gaussian elimination Example 3.2.2 above.)

We first wish to obtain $\mathbf{A}$ in the form $\mathbf{A} = \mathbf{LU}$ where

$$\mathbf{L} = \begin{bmatrix} 1 & 0 & 0 & 0 \\ l_{21} & 1 & 0 & 0 \\ l_{31} & l_{32} & 1 & 0 \\ l_{41} & l_{42} & l_{43} & 1 \end{bmatrix} \quad \text{and} \quad \mathbf{U} = \begin{bmatrix} u_{11} & u_{12} & u_{13} & u_{14} \\ 0 & u_{22} & u_{23} & u_{24} \\ 0 & 0 & u_{33} & u_{34} \\ 0 & 0 & 0 & u_{44} \end{bmatrix}.$$

Forming the product $\mathbf{LU}$ gives

$$\mathbf{LU} = \begin{bmatrix} u_{11} & u_{12} & u_{13} & u_{14} \\ l_{21}u_{11} & l_{21}u_{12} + u_{22} & l_{21}u_{13} + u_{23} & l_{21}u_{14} + u_{24} \\ l_{31}u_{11} & l_{31}u_{12} + l_{32}u_{22} & l_{31}u_{13} + l_{32}u_{23} + u_{33} & l_{31}u_{14} + l_{32}u_{24} + u_{34} \\ l_{41}u_{11} & l_{41}u_{12} + l_{42}u_{22} & l_{41}u_{13} + l_{42}u_{23} + l_{43}u_{33} & l_{41}u_{14} + l_{42}u_{24} + l_{43}u_{34} + u_{44} \end{bmatrix}.$$

Equating the elements of this matrix to the elements of $\mathbf{A}$ row by row gives sixteen equations for the sixteen unknowns:

$$\begin{aligned}
u_{11} &= 1 & l_{31}u_{11} &= 4 \\
u_{12} &= 1 & l_{31}u_{12} + l_{32}u_{22} &= 4 \\
u_{13} &= 1 & l_{31}u_{13} + l_{32}u_{23} + u_{33} &= 1 \\
u_{14} &= -1 & l_{31}u_{14} + l_{32}u_{24} + u_{34} &= 1 \\
l_{21}u_{11} &= 1 & l_{41}u_{11} &= 2 \\
l_{21}u_{12} + u_{22} &= -1 & l_{41}u_{12} + l_{42}u_{22} &= 1 \\
l_{21}u_{13} + u_{23} &= -1 & l_{41}u_{13} + l_{42}u_{23} + l_{43}u_{33} &= 2 \\
l_{21}u_{14} + u_{24} &= 2 & l_{41}u_{14} + l_{42}u_{24} + l_{43}u_{34} + u_{44} &= -2.
\end{aligned}$$

Taken in order these sixteen equations give $u_{11} = 1, u_{12} = 1, u_{13} = 1, u_{14} = -1, l_{21} = 1,$ $u_{22} = -2, u_{23} = -2, u_{24} = 3, l_{31} = 4, l_{32} = 0, u_{33} = -3, u_{34} = 5, l_{41} = 2, l_{42} = \frac{1}{2}, l_{43} = -\frac{1}{3},$ $u_{44} = \frac{1}{6}$. Thus we obtain

$$\mathbf{L} = \begin{bmatrix} 1 & 0 & 0 & 0 \\ 1 & 1 & 0 & 0 \\ 4 & 0 & 1 & 0 \\ 2 & \frac{1}{2} & -\frac{1}{3} & 1 \end{bmatrix} \quad \text{and} \quad \mathbf{U} = \begin{bmatrix} 1 & 1 & 1 & -1 \\ 0 & -2 & -2 & 3 \\ 0 & 0 & -3 & 5 \\ 0 & 0 & 0 & \frac{1}{6} \end{bmatrix}.$$

(i) To solve $\mathbf{Ax} = \mathbf{b}$ we solve the two systems

$$\mathbf{Ly} = \mathbf{b}$$

$$\mathbf{Ux} = \mathbf{y}.$$

In this case $\mathbf{Ly} = \mathbf{b}$ is

$$\begin{bmatrix} 1 & 0 & 0 & 0 \\ 1 & 1 & 0 & 0 \\ 4 & 0 & 1 & 0 \\ 2 & \frac{1}{2} & -\frac{1}{3} & 1 \end{bmatrix} \begin{bmatrix} y_1 \\ y_2 \\ y_3 \\ y_4 \end{bmatrix} = \begin{bmatrix} 2 \\ 0 \\ 11 \\ 2 \end{bmatrix}$$

which when written out gives

$$\begin{aligned} y_1 \qquad\qquad\qquad\qquad &= 2 \\ y_1 + y_2 \qquad\qquad\qquad &= 0 \\ 4y_1 + y_3 \qquad\qquad\quad &= 11 \\ 2y_1 + \tfrac{1}{2}y_2 - \tfrac{1}{3}y_3 + y_4 &= 2. \end{aligned}$$

Using these equations in turn gives $y_1 = 2, y_2 = -2, y_3 = 3, y_4 = 0$ and so $\mathbf{Ux} = \mathbf{y}$ becomes

$$\begin{bmatrix} 1 & 1 & 1 & -1 \\ 0 & -2 & -2 & 3 \\ 0 & 0 & -3 & 5 \\ 0 & 0 & 0 & \frac{1}{6} \end{bmatrix} \begin{bmatrix} x_1 \\ x_2 \\ x_3 \\ x_4 \end{bmatrix} = \begin{bmatrix} 2 \\ -2 \\ 3 \\ 0 \end{bmatrix}$$

which when written out gives

$$\begin{aligned} x_1 + x_2 + x_3 - x_4 &= 2 \\ -2x_2 - 2x_3 + 3x_4 &= -2 \\ -3x_3 + 5x_4 &= 3 \\ \tfrac{1}{6}x_4 &= 0. \end{aligned}$$

Using these in turn, starting from the last, gives $x_4 = 0, x_3 = -1, x_2 = 2, x_1 = 1$ and hence

$$\mathbf{x} = \begin{bmatrix} 1 \\ 2 \\ -1 \\ 0 \end{bmatrix}.$$

(ii) We have

$$\begin{aligned}\det(\mathbf{A}) &= u_{11}u_{22}u_{33}u_{44} \\ &= 1 \times (-2) \times (-3) \times (\tfrac{1}{6}) \\ &= 1.\end{aligned}$$

(iii) Since $\mathbf{A} = \mathbf{LU}$ we have $\mathbf{A}^{-1} = \mathbf{U}^{-1}\mathbf{L}^{-1}$.

Now, if we let

$$\mathbf{U}^{-1} = \begin{bmatrix} p_{11} & p_{12} & p_{13} & p_{14} \\ 0 & p_{22} & p_{23} & p_{24} \\ 0 & 0 & p_{33} & p_{34} \\ 0 & 0 & 0 & p_{44} \end{bmatrix}$$

we require that

$$\begin{bmatrix} u_{11} & u_{12} & u_{13} & u_{14} \\ 0 & u_{22} & u_{23} & u_{24} \\ 0 & 0 & u_{33} & u_{34} \\ 0 & 0 & 0 & u_{44} \end{bmatrix}\begin{bmatrix} p_{11} & p_{12} & p_{13} & p_{14} \\ 0 & p_{22} & p_{23} & p_{24} \\ 0 & 0 & p_{33} & p_{34} \\ 0 & 0 & 0 & p_{44} \end{bmatrix} = \begin{bmatrix} 1 & 0 & 0 & 0 \\ 0 & 1 & 0 & 0 \\ 0 & 0 & 1 & 0 \\ 0 & 0 & 0 & 1 \end{bmatrix}.$$

Multiplying out and equating non-zero elements gives the ten equations

$$u_{11}p_{11} = 1$$

$$u_{22}p_{22} = 1$$

$$u_{33}p_{33} = 1$$

$$u_{44}p_{44} = 1$$

$$u_{11}p_{12} + u_{12}p_{22} = 0$$

$$u_{22}p_{23} + u_{23}p_{33} = 0$$

$$u_{33}p_{34} + u_{34}p_{44} = 0$$

$$u_{11}p_{13} + u_{12}p_{23} + u_{13}p_{33} = 0$$

$$u_{22}p_{24} + u_{23}p_{34} + u_{24}p_{44} = 0$$

$$u_{11}p_{14} + u_{12}p_{24} + u_{13}p_{34} + u_{14}p_{44} = 0.$$

Substituting the values for the $u$'s in this example we obtain the equations

$$p_{11} = 1$$
$$-2p_{22} = 1$$
$$-3p_{33} = 1$$
$$\tfrac{1}{6}p_{44} = 1$$
$$p_{12} + p_{22} = 0$$
$$-2p_{23} - 2p_{33} = 0$$
$$-3p_{34} + 5p_{44} = 0$$
$$p_{13} + p_{23} + p_{33} = 0$$
$$-2p_{24} - 2p_{34} + 3p_{44} = 0$$
$$p_{14} + p_{24} + p_{34} - p_{44} = 0.$$

These may be used in turn to give $p_{11} = 1, p_{22} = -\frac{1}{2}, p_{33} = -\frac{1}{3}, p_{44} = 6, p_{12} = \frac{1}{2}$, $p_{23} = \frac{1}{3}, p_{34} = 10, p_{13} = 0, p_{24} = -1, p_{14} = -3$ and hence

$$\mathbf{U}^{-1} = \begin{bmatrix} 1 & \frac{1}{2} & 0 & -3 \\ 0 & -\frac{1}{2} & \frac{1}{3} & -1 \\ 0 & 0 & -\frac{1}{3} & 10 \\ 0 & 0 & 0 & 6 \end{bmatrix}.$$

In an exactly similar manner we find that

$$\mathbf{L}^{-1} = \begin{bmatrix} 1 & 0 & 0 & 0 \\ -1 & 1 & 0 & 0 \\ -4 & 0 & 1 & 0 \\ -\frac{17}{6} & -\frac{1}{2} & \frac{1}{3} & 1 \end{bmatrix},$$

and hence

$$\mathbf{A}^{-1} = \mathbf{U}^{-1}\mathbf{L}^{-1} = \begin{bmatrix} 1 & \frac{1}{2} & 0 & -3 \\ 0 & -\frac{1}{2} & \frac{1}{3} & -1 \\ 0 & 0 & -\frac{1}{3} & 10 \\ 0 & 0 & 0 & 6 \end{bmatrix} \begin{bmatrix} 1 & 0 & 0 & 0 \\ -1 & 1 & 0 & 0 \\ -4 & 0 & 1 & 0 \\ -\frac{17}{6} & -\frac{1}{2} & \frac{1}{3} & 1 \end{bmatrix}$$

$$= \begin{bmatrix} 9 & 2 & -1 & -3 \\ 2 & 0 & 0 & -1 \\ -27 & -5 & 3 & 10 \\ -17 & -3 & 2 & 6 \end{bmatrix}.$$

*Exercises*

3.2.2 For the matrix

$$\mathbf{A} = \begin{bmatrix} 1 & 1 & 2 & 1 \\ 1 & 1 & -1 & -1 \\ 1 & 2 & 3 & -4 \\ 1 & -1 & -2 & -1 \end{bmatrix}$$

use triangularization to determine

(i) the solution to

$$\mathbf{Ax} = \begin{bmatrix} 1 \\ 0 \\ -10 \\ 1 \end{bmatrix}$$

(ii) the value of det (**A**).

(Hint: **A** cannot be decomposed into the form **LU** as it stands—an interchange of rows must take place.)

3.2.3 Use triangularization to obtain the inverse of

$$\begin{bmatrix} 6 & -2 & -3 \\ 1 & 4 & 12 \\ -1 & 8 & 3 \end{bmatrix}.$$

## 3.3 Indirect Methods of Solution

Indirect methods are iterative methods. This means that we start with an initial approximation or guess, $\mathbf{x}^{(0)}$, to the solution of $\mathbf{Ax} = \mathbf{b}$; then the method provides a rule whereby we may calculate an improved approximation $\mathbf{x}^{(1)}$. From $\mathbf{x}^{(1)}$ we may in turn calculate a still better approximation $\mathbf{x}^{(2)}$. We may proceed in this way as far as we please, calculating at each stage an improved approximation $\mathbf{x}^{(r+1)}$ from the preceding approximation $\mathbf{x}^{(r)}$. If the method is a good one, $\mathbf{x}^{(r)}$ will be a good approximation to the required solution for small values of $r$; in other words, only a few iterations will be needed.

METHOD 4: JACOBI'S METHOD

We first describe the method then afterwards give a justification for its validity.

The given system of equations is

$$\begin{aligned} a_{11}x_1 + a_{12}x_2 + a_{13}x_3 + \ldots + a_{1n}x_n &= b_1 \\ a_{21}x_1 + a_{22}x_2 + a_{23}x_3 + \ldots + a_{2n}x_n &= b_2 \\ &\vdots \\ a_{i1}x_1 + a_{i2}x_2 + a_{i3}x_3 + \ldots + a_{in}x_n &= b_i \\ &\vdots \\ a_{n1}x_1 + a_{n2}x_2 + a_{n3}x_3 + \ldots + a_{nn}x_n &= b_n. \end{aligned} \qquad (3.3.1)$$

If we now rewrite these so that the $i$th equation (for $1 \leqslant i \leqslant n$) contains only $x_i$ on its left-hand side we obtain

$$\begin{aligned}
x_1 &= (b_1 - a_{12}x_2 - a_{13}x_3 - \ldots - a_{1n}x_n)/a_{11} \\
x_2 &= (b_2 - a_{21}x_1 - a_{23}x_3 - \ldots - a_{2n}x_n)/a_{22} \\
&\vdots \\
x_i &= (b_i - a_{i1}x_1 - a_{i2}x_2 - \ldots - a_{i,i-1}x_{i-1} - a_{i,i+1}x_{i+1} - \ldots - a_{in}x_n)/a_{ii} \\
&\vdots \\
x_n &= (b_n - a_{n1}x_1 - a_{n2}x_2 - \ldots - a_{n,n-1}x_{n-1})/a_{nn}.
\end{aligned} \tag{3.3.2}$$

Of course this procedure is only valid if all the $a_{ii}$ are non-zero or if the equations can be suitably re-arranged to make this so.

The Jacobi iterative scheme is then obtained from the system (3.3.2) by substituting the values of $x_i^{(r)}$ at any stage in the iterative process into the right-hand side of (3.3.2) to give the values at the next stage, $x_i^{(r+1)}$. In other words, the iterative scheme is given by the system of equations (3.3.2) with a superscript $r$ on the right-hand side, and a superscript $r + 1$ on the left-hand side:

$$\begin{aligned}
x_1^{(r+1)} &= (b_1 - a_{12}x_2^{(r)} - a_{13}x_3^{(r)} - \ldots - a_{1n}x_n^{(r)})/a_{11} \\
x_2^{(r+1)} &= (b_2 - a_{21}x_1^{(r)} - a_{23}x_3^{(r)} - \ldots - a_{2n}x_n^{(r)})/a_{22} \\
&\vdots \\
x_i^{(r+1)} &= (b_i - a_{i1}x_1^{(r)} - a_{i2}x_2^{(r)} - \ldots - a_{i,i-1}x_{i-1}^{(r)} - a_{i,i+1}x_{i+1}^{(r)} - \ldots - a_{in}x_n^{(r)})/a_{ii} \\
&\vdots \\
x_n^{(r+1)} &= (b_n - a_{n1}x_1^{(r)} - a_{n2}x_2^{(r)} - \ldots - a_{n,n-1}x_{n-1}^{(r)})/a_{nn}.
\end{aligned} \tag{3.3.3}$$

We must now justify the contention that the sequence $\mathbf{x}^{(0)}, \mathbf{x}^{(1)}, \mathbf{x}^{(2)}, \ldots$ generated by the equations (3.3.3) gives, under certain circumstances, a sequence which converges to the solution vector $\mathbf{x}$ which satisfies equations (3.3.1) $\mathbf{Ax} = \mathbf{b}$.

Equations (3.3.3) may be written in matrix form

$$\mathbf{x}^{(r+1)} = \mathbf{v} + \mathbf{B}\mathbf{x}^{(r)}, \tag{3.3.4}$$

where

$$\mathbf{v} = \begin{bmatrix} b_1/a_{11} \\ b_2/a_{22} \\ \vdots \\ b_n/a_{nn} \end{bmatrix}$$

and

$$\mathbf{B} = -\begin{bmatrix} 0 & a_{12}/a_{11} & a_{13}/a_{11} & & \dots & a_{1n}/a_{11} \\ a_{21}/a_{22} & 0 & a_{23}/a_{22} & & \dots & a_{2n}/a_{22} \\ \vdots & & & & & \\ a_{i1}/a_{ii} & \dots & a_{i,i-1}/a_{ii} \quad 0 & a_{i,i+1}/a_{ii} & \dots & a_{in}/a_{ii} \\ \vdots & & & & & \\ a_{n1}/a_{nn} & & \dots\dots\dots & a_{n,n-1}/a_{nn} & & 0 \end{bmatrix}.$$

The original system of equations, $\mathbf{Ax} = \mathbf{b}$, may be written in the form

$$\mathbf{x} = \mathbf{v} + \mathbf{Bx} \tag{3.3.5}$$

since this is just the matrix form of the equations (3.3.2).

It now follows from equation (3.3.4) that we may construct expressions for $\mathbf{x}^{(1)}$, $\mathbf{x}^{(2)}, \dots$ in terms of $\mathbf{x}^{(0)}$:

$$\begin{aligned} \mathbf{x}^{(1)} &= \mathbf{v} + \mathbf{Bx}^{(0)} \\ \mathbf{x}^{(2)} &= \mathbf{v} + \mathbf{Bx}^{(1)} \\ &= \mathbf{v} + \mathbf{B}(\mathbf{v} + \mathbf{Bx}^{(0)}) \\ &= \mathbf{v} + \mathbf{Bv} + \mathbf{B}^2\mathbf{x}^{(0)} \\ \mathbf{x}^{(3)} &= \mathbf{v} + \mathbf{Bx}^{(2)} \\ &= \mathbf{v} + \mathbf{B}(\mathbf{v} + \mathbf{Bv} + \mathbf{B}^2\mathbf{x}^{(0)}) \\ &= \mathbf{v} + \mathbf{Bv} + \mathbf{B}^2\mathbf{v} + \mathbf{B}^3\mathbf{x}^{(0)}. \end{aligned}$$

In general we should obtain

$$\begin{aligned} \mathbf{x}^{(r)} &= \mathbf{v} + \mathbf{Bv} + \mathbf{B}^2\mathbf{v} + \dots + \mathbf{B}^{r-1}\mathbf{v} + \mathbf{B}^r\mathbf{x}^{(0)} \\ &= (\mathbf{I} + \mathbf{B} + \mathbf{B}^2 + \dots + \mathbf{B}^{r-1})\mathbf{v} + \mathbf{B}^r\mathbf{x}^{(0)}. \end{aligned} \tag{3.3.6}$$

Now, if we wish $\mathbf{x}^{(r)}$ to converge to $\mathbf{x}$ (the solution of $\mathbf{Ax} = \mathbf{b}$) as $r \to \infty$ it is obviously necessary that $\mathbf{B}^r\mathbf{x}^{(0)} \to \mathbf{0}$ as $r \to \infty$; otherwise we should obtain in the limit a quantity which depends on the initial guess $\mathbf{x}^{(0)}$. Under the assumption that this happens we now investigate the behaviour as $r \to \infty$ of $(\mathbf{I} + \mathbf{B} + \mathbf{B}^2 + \dots + \mathbf{B}^{r-1})\mathbf{v}$.

We note that $\mathbf{I} + \mathbf{B} + \mathbf{B}^2 + \dots + \mathbf{B}^{r-1}$ is just a matrix geometric progression; we may find an expression for the sum of the $r$ terms with common ratio $\mathbf{B}$ by the same method as is used for geometric progressions of real or complex numbers.

If we let

$$\mathbf{S}_r = \mathbf{I} + \mathbf{B} + \mathbf{B}^2 + \dots + \mathbf{B}^{r-1}$$

then

$$\mathbf{S}_r\mathbf{B} = \mathbf{B} + \mathbf{B}^2 + \dots + \mathbf{B}^{r-1} + \mathbf{B}^r,$$

so

$$\mathbf{S}_r - \mathbf{S}_r\mathbf{B} = \mathbf{I} - \mathbf{B}^r.$$

Hence

$$\mathbf{S}_r(\mathbf{I} - \mathbf{B}) = \mathbf{I} - \mathbf{B}^r,$$

and thus

$$\mathbf{S}_r = (\mathbf{I} - \mathbf{B}^r)(\mathbf{I} - \mathbf{B})^{-1}$$

provided $\mathbf{I} - \mathbf{B}$ is non-singular. Thus we see that as $r \to \infty$ the limit of the sum exists provided $\mathbf{B}^r \to \mathbf{0}$; in this case $\mathbf{S}_r \to \mathbf{S} = (\mathbf{I} - \mathbf{B})^{-1}$.

Thus from equation (3.3.6) we may now deduce that provided $\mathbf{B}^r \to \mathbf{0}$ as $r \to \infty$ then $\mathbf{x}^{(r)}$ will converge to the vector $\mathbf{x}$ given by

$$\mathbf{x} = (\mathbf{I} - \mathbf{B})^{-1}\mathbf{v}.$$

But this means that $(\mathbf{I} - \mathbf{B})\mathbf{x} = \mathbf{v}$ and hence $\mathbf{x} = \mathbf{B}\mathbf{x} + \mathbf{v}$. This is just equation (3.3.5) and hence $\mathbf{x}$, the limit of the sequence $\mathbf{x}^{(r)}$, satisfies the required equation.

To summarize, we have shown that if $\mathbf{B}^r \to \mathbf{0}$ as $r \to \infty$ then the Jacobi sequence (3.3.3) converges to the solution of $\mathbf{A}\mathbf{x} = \mathbf{b}$. Thus $\mathbf{B}^r \to \mathbf{0}$ is a *sufficient* condition for the convergence of the Jacobi sequence. It is also a *necessary* condition, a property which will be discussed at greater length in Chapter 4.

Of course, given a matrix $\mathbf{A}$, it is not practicable to determine directly by forming the matrix $\mathbf{B}$ whether or not $\mathbf{B}^r \to \mathbf{0}$. The important results for practical applications are given by the following theorem which we state without proof meantime; a proof will be provided in Chapter 4 where we shall be able to make use of some results concerning eigenvalues and eigenvectors.

THEOREM 3.3

If $\mathbf{B}$ is any square matrix of order $n$ then any one of the following is a sufficient condition for $\lim_{r\to\infty} \mathbf{B}^r = \mathbf{0}$:

(i) $\sum_{i=1}^{n} |b_{ij}| < 1 \qquad (1 \leqslant j \leqslant n)$

(ii) $\sum_{j=1}^{n} |b_{ij}| < 1 \qquad (1 \leqslant j \leqslant n)$

(iii) $\sum_{i,j=1}^{n} |b_{ij}|^2 < 1.$ ∎

Condition (i) just states that the sum of the moduli of the elements in each column should be less than unity; condition (ii) states the same for rows; condition (iii) states that the sum of the squared moduli of all elements should be less than unity. Again we emphasize the fact that these are *sufficient* conditions—the truth of any one of them guarantees the convergence of Jacobi's method, but it is possible that all of them are false and Jacobi's method still works. In practical terms we have to ensure that the elements of $\mathbf{B}$ are as small in magnitude as possible; since those in the $i$th row all contain $a_{ii}$ in the denominator we seek to re-arrange the equations in such a way that all the $a_{ii}$ (diagonal elements of $\mathbf{A}$) are as large as possible. Not only does this help to guarantee convergence, but the more dominant we can make the $a_{ii}$ the faster we shall obtain convergence.

In practice we must have some criterion for stopping iterations. A common such criterion is to stop when $\mathbf{x}^{(r+1)}$ and $\mathbf{x}^{(r)}$ agree to some specified accuracy; there is of course no guarantee that this implies that $\mathbf{x}^{(r+1)}$ agrees with $\mathbf{x}$, the true answer, to anything even approaching this accuracy, but we usually (rather hopefully) assume this to be the case.

The other matter on which guidance is needed in practice is the choice of $\mathbf{x}^{(0)}$, the initial approximation. If any information is available about the nature of the solution then obviously it should be incorporated into one's choice of $\mathbf{x}^{(0)}$. If no such informatio is available, then a purely arbitrary choice must be made (such as $x_i^{(0)} = 1$ for all $i$); the nr this choice is to the true solution the more rapid will be the convergence. If $|a_{ij}| \ll |a_{ii}|$ $(i \neq j; i, j = 1, \ldots, n)$, then it is expedient to choose

$$x_i^{(0)} = b_i/a_{ii} \quad (i = 1, \ldots, n).$$

*Example*

3.3.1 Use the Jacobi method to obtain the solution of the following system of equations correct to three decimal places:

$$\begin{aligned} 2x - y + 5z &= 15 \\ 2x + y + z &= 7 \\ x + 3y + z &= 10. \end{aligned}$$

We first re-arrange the equations so that we obtain diagonal coefficients as dominant as possible:

$$\begin{aligned} 2x + y + z &= 7 \\ x + 3y + z &= 10 \\ 2x - y + 5z &= 15. \end{aligned}$$

Rewriting in the form of the equations (3.3.2) we have

$$\begin{aligned} x &= (7 - y - z)/2 \\ y &= (10 - x - z)/3 \\ z &= (15 - 2x + y)/5 \end{aligned}$$

and hence we obtain the iterative scheme corresponding to (3.3.3):

$$\begin{aligned} x^{(r+1)} &= (7 - y^{(r)} - z^{(r)})/2 \\ y^{(r+1)} &= (10 - x^{(r)} - z^{(r)})/3 \\ z^{(r+1)} &= (15 - 2x^{(r)} + y^{(r)})/5. \end{aligned}$$

We now have to make a choice for $x^{(0)}$, $y^{(0)}$ and $z^{(0)}$; for lack of any further information we choose $x^{(0)} = y^{(0)} = z^{(0)} = 1$.

Then the above iterative scheme gives

$$x^{(1)} = (7 - y^{(0)} - z^{(0)})/2 = (7 - 1 - 1)/2 = 2{\cdot}5000$$
$$y^{(1)} = (10 - x^{(0)} - z^{(0)})/3 = (10 - 1 - 1)/3 = 2{\cdot}6667$$
$$z^{(1)} = (15 - 2x^{(0)} + y^{(0)})/5 = (15 - 2 + 1)/5 = 2{\cdot}8000.$$

(Since we require the result to three decimal places we carry out intermediate work correct to four.)

Repeating the iteration we obtain

$$x^{(2)} = (7 - y^{(1)} - z^{(1)})/2 = (7 - 2{\cdot}6667 - 2{\cdot}8000)/2 = 0{\cdot}7666$$
$$y^{(2)} = (10 - x^{(1)} - z^{(1)})/3 = (10 - 2{\cdot}5000 - 2{\cdot}8000)/3 = 1{\cdot}5667$$
$$z^{(2)} = (15 - 2x^{(1)} + y^{(1)})/5 = (15 - 2 \times 2{\cdot}5000 + 2{\cdot}6667)/5 = 2{\cdot}5333.$$

Proceeding in this way we obtain

$$x^{(3)} = 1{\cdot}4500 \quad y^{(3)} = 2{\cdot}2334 \quad z^{(3)} = 3{\cdot}0067$$
$$x^{(4)} = 0{\cdot}8800 \quad y^{(4)} = 1{\cdot}8478 \quad z^{(4)} = 2{\cdot}8776$$
$$x^{(5)} = 1{\cdot}1428 \quad y^{(5)} = 2{\cdot}0844 \quad z^{(5)} = 3{\cdot}0176$$
$$x^{(6)} = 0{\cdot}9490 \quad y^{(6)} = 1{\cdot}9465 \quad z^{(6)} = 2{\cdot}9398$$
$$x^{(7)} = 1{\cdot}0469 \quad y^{(7)} = 2{\cdot}0304 \quad z^{(7)} = 3{\cdot}0097$$
$$x^{(8)} = 0{\cdot}9800 \quad y^{(8)} = 1{\cdot}9811 \quad z^{(8)} = 2{\cdot}9873$$
$$x^{(9)} = 1{\cdot}0158 \quad y^{(9)} = 2{\cdot}0109 \quad z^{(9)} = 3{\cdot}0042$$
$$x^{(10)} = 0{\cdot}9925 \quad y^{(10)} = 2{\cdot}0053 \quad z^{(10)} = 2{\cdot}9959$$
$$x^{(11)} = 0{\cdot}9953 \quad y^{(11)} = 2{\cdot}0011 \quad z^{(11)} = 3{\cdot}0041$$
$$x^{(12)} = 0{\cdot}9974 \quad y^{(12)} = 2{\cdot}0002 \quad z^{(12)} = 3{\cdot}0021$$
$$x^{(13)} = 0{\cdot}9989 \quad y^{(13)} = 2{\cdot}0002 \quad z^{(13)} = 3{\cdot}0011$$
$$x^{(14)} = 0{\cdot}9994 \quad y^{(14)} = 2{\cdot}0000 \quad z^{(14)} = 3{\cdot}0005$$
$$x^{(15)} = 0{\cdot}9998 \quad y^{(15)} = 2{\cdot}0000 \quad z^{(15)} = 3{\cdot}0002$$
$$x^{(16)} = 0{\cdot}9999 \quad y^{(16)} = 2{\cdot}0000 \quad z^{(16)} = 3{\cdot}0001$$
$$x^{(17)} = 1{\cdot}0000 \quad y^{(17)} = 2{\cdot}0000 \quad z^{(17)} = 3{\cdot}0000$$
$$x^{(18)} = 1{\cdot}0000 \quad y^{(18)} = 2{\cdot}0000 \quad z^{(18)} = 3{\cdot}0000.$$

So to three decimal places we have the approximate solution $x = 1{\cdot}000$, $y = 2{\cdot}000$, $z = 3{\cdot}000$; in fact the exact solution is $x = 1$, $y = 2$, $z = 3$.

We note that convergence has been slow; this is essentially because the diagonal elements were not particularly dominant. We also note that convergence takes place in the solution vector as a whole, and not in individual elements. Thus, although we started off with the approximation 1 for $x$ (which in fact was the exact solution) it departed from this as far as 2·5 before gradually converging to the approximation of 1·000.

METHOD 5: GAUSS-SEIDEL METHOD

This is essentially a development of Jacobi's method described above. Since the vector $\mathbf{x}^{(r+1)}$ is supposed to be an improvement on the vector $\mathbf{x}^{(r)}$, it intuitively makes sense to use the improved values $x_i^{(r+1)}$ as soon as they become available. Thus in the iterative scheme (3.3.3) $x_1^{(r+1)}$ is calculated by the first equation and so could be used in place of $x_1^{(r)}$ in the second and subsequent equations. Similarly $x_2^{(r+1)}$ is calculated in the second equation and so could be used in place of $x_2^{(r)}$ in the third and subsequent equations. In general $x_i^{(r+1)}$ could replace $x_i^{(r)}$ in the $(i+1)$th and subsequent equations, and then we should obtain the Gauss-Seidel iterative scheme:

$$\begin{aligned}
x_1^{(r+1)} &= (b_1 - a_{12}x_2^{(r)} - a_{13}x_3^{(r)} - \ldots - a_{1n}x_n^{(r)})/a_{11} \\
x_2^{(r+1)} &= (b_2 - a_{21}x_1^{(r+1)} - a_{23}x_3^{(r)} - \ldots - a_{2n}x_n^{(r)})/a_{22} \\
x_3^{(r+1)} &= (b_3 - a_{31}x_1^{(r+1)} - a_{32}x_2^{(r+1)} - a_{34}x_4^{(r)} - \ldots - a_{3n}x_n^{(r)})/a_{33} \\
&\vdots \\
x_i^{(r+1)} &= (b_i - a_{i1}x_1^{(r+1)} - \ldots - a_{i,i-1}x_{i-1}^{(r+1)} - a_{i,i+1}x_{i+1}^{(r)} - \ldots - a_{in}x_n^{(r)})/a_{ii} \\
&\vdots \\
x_n^{(r+1)} &= (b_n - a_{n1}x_1^{(r+1)} - a_{n2}x_2^{(r+1)} - \ldots - a_{n,n-1}x_{n-1}^{(r+1)})/a_{nn}.
\end{aligned} \tag{3.3.7}$$

To investigate the convergence of this scheme we write the system of equations (3.3.7) in matrix form:

$$\mathbf{x}^{(r+1)} = \mathbf{v} + \mathbf{B}_\mathrm{L}\mathbf{x}^{(r+1)} + \mathbf{B}_\mathrm{U}\mathbf{x}^{(r)} \tag{3.3.8}$$

where

$$\mathbf{v} = \begin{bmatrix} b_1/a_{11} \\ b_2/a_{22} \\ \cdot \\ \cdot \\ \cdot \\ b_n/a_{nn} \end{bmatrix},$$

$$\mathbf{B}_\mathrm{L} = -\begin{bmatrix} 0 & 0 & 0 & \ldots & 0 \\ a_{21}/a_{22} & 0 & 0 & & 0 \\ a_{31}/a_{33} & a_{32}/a_{33} & 0 & & \cdot \\ a_{41}/a_{44} & a_{42}/a_{44} & a_{43}/a_{44} & & \cdot \\ \cdot & & & & \cdot \\ \cdot & & & & \cdot \\ \cdot & & & & \cdot \\ a_{n1}/a_{nn} & a_{n2}/a_{nn} & \ldots & a_{n,n-1}/a_{nn} & 0 \end{bmatrix}$$

and

$$\mathbf{B}_\mathrm{U} = -\begin{bmatrix} 0 & a_{12}/a_{11} & a_{13}/a_{11} & \cdots\cdots & & a_{1n}/a_{11} \\ 0 & 0 & a_{23}/a_{22} & \cdots\cdots & & a_{2n}/a_{22} \\ 0 & 0 & 0 & a_{34}/a_{33} & \cdots & a_{3n}/a_{33} \\ \cdot & & & & & \\ \cdot & & & & & \\ \cdot & & & & & \\ 0 & 0 & \cdots\cdots & & 0 & a_{n-1,n}/a_{n-1,n-1} \\ 0 & 0 & \cdots\cdots & & 0 & 0 \end{bmatrix}.$$

$\mathbf{v}$ is the same vector as for Jacobi's method; $\mathbf{B}_\mathrm{L}$ and $\mathbf{B}_\mathrm{U}$ are the lower and upper triangular parts respectively of the Jacobi matrix $\mathbf{B}$ so that $\mathbf{B} = \mathbf{B}_\mathrm{L} + \mathbf{B}_\mathrm{U}$.

Equation (3.3.8) may be rewritten in the form

$$\mathbf{x}^{(r+1)} - \mathbf{B}_\mathrm{L}\mathbf{x}^{(r+1)} = \mathbf{v} + \mathbf{B}_\mathrm{U}\mathbf{x}^{(r)}$$

giving

$$\mathbf{x}^{(r+1)} = (\mathbf{I} - \mathbf{B}_\mathrm{L})^{-1}\mathbf{v} + (\mathbf{I} - \mathbf{B}_\mathrm{L})^{-1}\mathbf{B}_\mathrm{U}\mathbf{x}^{(r)}. \tag{3.3.9}$$

Thus by comparison with the Jacobi system in the form (3.3.4) which was shown to converge to the required solution when $\mathbf{B}^r \to \mathbf{0}$, we have that the system (3.3.9) converges to the required solution provided $\{(\mathbf{I} - \mathbf{B}_\mathrm{L})^{-1}\mathbf{B}_\mathrm{U}\}^r \to \mathbf{0}$.

We shall prove in the next chapter that the conditions of Theorem 3.3 are also sufficient conditions for the above result.

We thus see that provided any of the conditions of Theorem 3.3 are satisfied, both Jacobi and Gauss–Seidel methods are valid; and in general Gauss–Seidel may be shown then to converge at about twice the rate of Jacobi. However, the reader is warned to beware of the fact that there are situations in which the Jacobi method converges and the Gauss–Seidel does not. But all the comments about *practical* criteria for convergence made above for Jacobi's method on page 133 hold also for the Gauss–Seidel method.

*Example*

3.3.2 Use the Gauss–Seidel method to obtain correct to three decimal places the solution of

$$\begin{aligned} 2x - y + 5z &= 15 \\ 2x + y + z &= 7 \\ x + 3y + z &= 10 \end{aligned}$$

(which is the same set as was solved by the Jacobi method above).

As before we re-arrange the equations to give dominant diagonal terms:

$$\begin{aligned} 2x + y + z &= 7 \\ x + 3y + z &= 10 \\ 2x - y + 5z &= 15. \end{aligned}$$

These can be rewritten as

$$x = (7 - y - z)/2$$
$$y = (10 - x - z)/3$$
$$z = (15 - 2x + y)/5$$

which give rise to the Gauss-Seidel iterative scheme:

$$x^{(r+1)} = (7 - y^{(r)} - z^{(r)})/2$$
$$y^{(r+1)} = (10 - x^{(r+1)} - z^{(r)})/3$$
$$z^{(r+1)} = (15 - 2x^{(r+1)} + y^{(r+1)})/5.$$

Again, for lack of any other information, we start off with $x^{(0)} = y^{(0)} = z^{(0)} = 1$. Then the iterative scheme gives

$$x^{(1)} = (7 - y^{(0)} - z^{(0)})/2 = (7 - 1 - 1)/2 = 2{\cdot}5000$$
$$y^{(1)} = (10 - x^{(1)} - z^{(0)})/3 = (10 - 2{\cdot}5000 - 1)/3 = 2{\cdot}1667$$
$$z^{(1)} = (15 - 2x^{(1)} + y^{(1)})/5 = (15 - 2 \times 2{\cdot}5000 + 2{\cdot}1667)/5 = 2{\cdot}4333.$$

Repeating the iteration gives

$$x^{(2)} = (7 - y^{(1)} - z^{(1)})/2 = (7 - 2{\cdot}1667 - 2{\cdot}4333)/2 = 1{\cdot}2000$$
$$y^{(2)} = (10 - x^{(2)} - z^{(1)})/3 = (10 - 1{\cdot}2000 - 2{\cdot}4333)/3 = 2{\cdot}1222$$
$$z^{(2)} = (15 - 2x^{(2)} + y^{(2)})/5 = (15 - 2 \times 1{\cdot}2000 + 2{\cdot}1222)/5 = 2{\cdot}9444.$$

Proceeding in this way we obtain

$$x^{(3)} = 0{\cdot}9667 \qquad y^{(3)} = 2{\cdot}0296 \qquad z^{(3)} = 3{\cdot}0192$$
$$x^{(4)} = 0{\cdot}9756 \qquad y^{(4)} = 2{\cdot}0017 \qquad z^{(4)} = 3{\cdot}0101$$
$$x^{(5)} = 0{\cdot}9941 \qquad y^{(5)} = 1{\cdot}9986 \qquad z^{(5)} = 3{\cdot}0021$$
$$x^{(6)} = 0{\cdot}9997 \qquad y^{(6)} = 1{\cdot}9994 \qquad z^{(6)} = 3{\cdot}0000$$
$$x^{(7)} = 1{\cdot}0003 \qquad y^{(7)} = 1{\cdot}9999 \qquad z^{(7)} = 2{\cdot}9999$$
$$x^{(8)} = 1{\cdot}0001 \qquad y^{(8)} = 2{\cdot}0000 \qquad z^{(8)} = 3{\cdot}0000$$
$$x^{(9)} = 1{\cdot}0000 \qquad y^{(9)} = 2{\cdot}0000 \qquad z^{(9)} = 3{\cdot}0000$$
$$x^{(10)} = 1{\cdot}0000 \qquad y^{(10)} = 2{\cdot}0000 \qquad z^{(10)} = 3{\cdot}0000.$$

The above example, by converging in ten iterations for the Gauss–Seidel method while taking eighteen for the Jacobi method illustrates the improvement generally to be obtained by using the Gauss–Seidel method.

*Exercises*

3.3.1 For each of the following systems of equations show that both the Jacobi and Gauss–Seidel methods converge:

(i)
$$\begin{aligned} x + 2y + z &= 0 \\ 3x + y - z &= 0 \\ x - y + 4z &= 3 \end{aligned}$$

(ii)
$$\begin{aligned} w + x + y + 2z &= 3 \\ 4w + x - y - z &= 6 \\ -w + 6x + 2y - z &= 7 \\ w + x + 7y + z &= 3. \end{aligned}$$

3.3.2 Use both the Jacobi and Gauss–Seidel methods to solve the system (ii) in (3.3.1) above correct to three significant figures.

3.3.3 For the system of equations $\mathbf{Ax} = \mathbf{b}$ show that the general iterative scheme $\mathbf{Mx}^{(r+1)} = \mathbf{Nx}^{(r)} + \mathbf{b}$ where $\mathbf{M} - \mathbf{N} = \mathbf{A}$ gives a sequence which converges to the required solution vector if $(\mathbf{M}^{-1}\mathbf{N})^r \to \mathbf{0}$ as $r \to \infty$.

## 3.4 Validity of Solutions

We have now seen several methods by which we may obtain a numerical solution to a system of equations of the form $\mathbf{Ax} = \mathbf{b}$. We now wish to enquire how accurate are the numbers which a method yields as a supposed solution. There are two main sources of possible trouble—ill-conditioning and instability. The first arises when a problem is such that small errors in the initial data lead to large errors in the final answer, and the second when rounding errors accumulate in such a way as to produce disastrously bad results. All that we shall do here is to illustrate by examples some of the pitfalls of which the user of numerical methods must beware. The reader who is interested in more detail is advised to consult a more specialized text on numerical analysis.†

*Example 1* ‡

Consider the problem of solving $\mathbf{Ax} = \mathbf{b}$ when

$$\mathbf{A} = \begin{bmatrix} \frac{1}{2} & \frac{1}{3} & \frac{1}{4} & \frac{1}{5} \\ \frac{1}{3} & \frac{1}{4} & \frac{1}{5} & \frac{1}{6} \\ \frac{1}{4} & \frac{1}{5} & \frac{1}{6} & \frac{1}{7} \\ \frac{1}{5} & \frac{1}{6} & \frac{1}{7} & \frac{1}{8} \end{bmatrix}.$$

We are interested in seeing what effect small changes in $\mathbf{b}$ have on the solution vector $\mathbf{x}$; this is appropriate to the situation in which $\mathbf{b}$ might be the result of experimental measurements. At any rate let us assume that $\mathbf{b}$ is in error from some true vector $\mathbf{b}_0$

† The numerical solution of systems of linear equations is considered at various levels of mathematical sophistication in the following references listed in the bibliography: (4), (8), (11), (15), (19), (20), (22).

‡ This example is due to Fox (8). The matrix $\mathbf{A}$ is an example of the well-known Hilbert matrix.

by the amount $\boldsymbol{\epsilon}$ so that $\mathbf{b} = \mathbf{b}_0 + \boldsymbol{\epsilon}$. Then the true solution $\mathbf{x}_0$ is given by $\mathbf{A}\mathbf{x}_0 = \mathbf{b}_0$, that is $\mathbf{x}_0 = \mathbf{A}^{-1}\mathbf{b}_0$; the solution $\mathbf{x}$ obtained from using the erroneous $\mathbf{b}$ is given by $\mathbf{A}\mathbf{x} = \mathbf{b}_0 + \boldsymbol{\epsilon}$, that is $\mathbf{x} = \mathbf{A}^{-1}\mathbf{b}_0 + \mathbf{A}^{-1}\boldsymbol{\epsilon} = \mathbf{x}_0 + \mathbf{A}^{-1}\boldsymbol{\epsilon}$. Thus the error in the final answer is given by $\mathbf{A}^{-1}\boldsymbol{\epsilon}$ and whether or not the problem is ill-conditioned depends on whether this is large or small compared with $\mathbf{x}_0 = \mathbf{A}^{-1}\mathbf{b}_0$.

To show what can happen we shall consider some different values for $\mathbf{b}$. To bring out the effect of errors in $\mathbf{b}$ we calculate $\mathbf{A}^{-1}$ exactly; this may be done using the result that $\mathbf{A}^{-1} = \operatorname{adj}(\mathbf{A})/\det(\mathbf{A})$ and we obtain

$$\mathbf{A}^{-1} = \begin{bmatrix} 200 & -1200 & 2100 & -1120 \\ -1200 & 8100 & -15120 & 8400 \\ 2100 & -15120 & 29400 & -16800 \\ -1120 & 8400 & -16800 & 9800 \end{bmatrix}.$$

Suppose that $\mathbf{b}$ is subject to the error $\boldsymbol{\epsilon}$ such that $|\epsilon_i| \leqslant 10^{-4}$. The worst possible error $\mathbf{A}^{-1}\boldsymbol{\epsilon}$ in the result arises when

$$\boldsymbol{\epsilon} = 10^{-4} \times \begin{bmatrix} 1 \\ -1 \\ 1 \\ -1 \end{bmatrix};$$

the alternating signs here arise because of the alternating signs in each row of $\mathbf{A}^{-1}$ and in this case we obtain, correct to two decimal places,

$$\mathbf{A}^{-1}\boldsymbol{\epsilon} = \begin{bmatrix} 0{\cdot}46 \\ -3{\cdot}28 \\ 6{\cdot}34 \\ -3{\cdot}61 \end{bmatrix}.$$

Whether this is significant or not depends on the value of $\mathbf{A}^{-1}\mathbf{b}_0$. If

$$\mathbf{b}_0 = \begin{bmatrix} 1{\cdot}0000 \\ -1{\cdot}0000 \\ 1{\cdot}0000 \\ -1{\cdot}0000 \end{bmatrix} \quad \text{then} \quad \mathbf{A}^{-1}\mathbf{b}_0 = \begin{bmatrix} 4620 \\ -32820 \\ 63420 \\ -36120 \end{bmatrix}$$

and the errors are negligible ($\sim 10^{-2}\%$).

If

$$\mathbf{b}_0 = \begin{bmatrix} 1{\cdot}0000 \\ 1{\cdot}0000 \\ 1{\cdot}0000 \\ 1{\cdot}0000 \end{bmatrix} \quad \text{then} \quad \mathbf{A}^{-1}\mathbf{b}_0 = \begin{bmatrix} -20 \\ 180 \\ -420 \\ 280 \end{bmatrix}$$

and the errors are rather more significant ($\sim 2\%$).

Finally, if

$$\mathbf{b}_0 = \begin{bmatrix} 1\cdot0000 \\ 0\cdot7227 \\ 0\cdot5697 \\ 0\cdot4714 \end{bmatrix} \quad \text{then} \quad \mathbf{A}^{-1}\mathbf{b}_0 = \begin{bmatrix} -1\cdot16 \\ -0\cdot23 \\ 2\cdot44 \\ -0\cdot56 \end{bmatrix}$$

and here we see that the errors completely swamp the solution in such a way that not a single figure in the computed answer can be regarded as significant.

It is obvious, therefore, that we must be continually on the look out for such ill-conditioned situations.

*Example 2*

A favourite, but rather naïve, test for accuracy of a solution is to substitute an approximate solution $\mathbf{x}'$ into the original equation $\mathbf{Ax} = \mathbf{b}$ and see how closely it is satisfied. $\boldsymbol{\epsilon} = \mathbf{Ax}' - \mathbf{b}$ is called the *residual vector*, and if this is small there is the temptation to say that $\mathbf{x}'$ is a good approximation to the true $\mathbf{x}$. That this can lead to catastrophic results is shown by consideration of the problem $\mathbf{Ax} = \mathbf{b}$ with

$$\mathbf{A} = \begin{bmatrix} 10 & 7 & 8 & 7 \\ 7 & 5 & 6 & 5 \\ 8 & 6 & 10 & 9 \\ 7 & 5 & 9 & 10 \end{bmatrix} \quad \text{and} \quad \mathbf{b} = \begin{bmatrix} 32 \\ 23 \\ 33 \\ 31 \end{bmatrix}.$$

If we choose

$$\mathbf{x}' = \begin{bmatrix} 6\cdot0 \\ -7\cdot2 \\ 2\cdot9 \\ -0\cdot1 \end{bmatrix} \quad \text{then} \quad \mathbf{Ax}' = \begin{bmatrix} 32\cdot1 \\ 22\cdot9 \\ 32\cdot9 \\ 31\cdot1 \end{bmatrix} \quad \text{and} \quad \boldsymbol{\epsilon} = \begin{bmatrix} 0\cdot1 \\ -0\cdot1 \\ -0\cdot1 \\ 0\cdot1 \end{bmatrix},$$

an apparent error of around 1%.

If next we choose

$$\mathbf{x}' = \begin{bmatrix} 1\cdot50 \\ 0\cdot18 \\ 1\cdot19 \\ 0\cdot89 \end{bmatrix} \quad \text{then} \quad \mathbf{Ax}' = \begin{bmatrix} 32\cdot01 \\ 22\cdot99 \\ 32\cdot99 \\ 31\cdot01 \end{bmatrix} \quad \text{and} \quad \boldsymbol{\epsilon} = \begin{bmatrix} 0\cdot01 \\ -0\cdot01 \\ -0\cdot01 \\ 0\cdot01 \end{bmatrix}$$

and $\mathbf{Ax}'$ agrees with $\mathbf{b}$ to around 0·1%. But any hope that $\mathbf{x}'$ approximates the true solution to anything like this accuracy is dashed when we note that the true solution is in fact

$$\begin{bmatrix} 1 \\ 1 \\ 1 \\ 1 \end{bmatrix}.$$

The above two examples indicate how necessary it is to be careful concerning the validity of numerical results obtained from the procedures outlined in Section 3.3.

# 4. Eigenvalues and Eigenvectors

## 4.1 Basic Properties of Eigenvalues and Eigenvectors

It is difficult to proceed very far with the use of matrix methods in many situations without encountering eigenvalues and eigenvectors. Already, in the appropriate chapters, we have mentioned that further investigation of quadratic forms and the convergence of Jacobi and Gauss–Seidel iterative schemes depended on results to be presented in this chapter. Eigenvalue theory is important, too, in topics as apparently diverse as systems of differential equations, mechanical vibrations and quantum theory.

In the present chapter we present many of the basic results of eigenvalue theory and indicate some numerical methods for the calculation of eigenvalues and eigenvectors.

DEFINITION 4.1

If $\mathbf{A}$ is a square matrix of order $n$ and if $\mathbf{x}$ is a non-zero vector such that $\mathbf{Ax} = \lambda\mathbf{x}$, where $\lambda$ is some number, then $\mathbf{x}$ is said to be an eigenvector of $\mathbf{A}$ with corresponding eigenvalue $\lambda$. □

The above definition contains the fundamental property of eigenvalues and eigenvectors—(pre)multiplying an eigenvector by the appropriate matrix just produces a multiple of the eigenvector, the multiplying factor being the eigenvalue.

At the moment we have no reason to assume the existence of eigenvalues and eigenvectors for all (or any) square matrices. We can, however, show from the definition that if $\mathbf{A}$ possesses an eigenvector then it is not unique. For, if $\mathbf{Ax} = \lambda\mathbf{x}$, then multiplying both sides by any constant $k$ gives $k\mathbf{Ax} = k\lambda\mathbf{x}$ or $\mathbf{A}(k\mathbf{x}) = \lambda(k\mathbf{x})$, implying that if $\mathbf{x}$ is an eigenvector then so also is $k\mathbf{x}$ where $k$ is any non-zero constant. This situation is reminiscent of that occurring in the study of vectors in ordinary three-dimensional space; multiplying a vector by a constant leaves its direction unaltered but changes its magnitude. We can extend this idea to vectors with $n$ components which we shall call vectors in $n$-dimensional space. In three dimensions the length of a vector with components $x_1, x_2, x_3$ is given by $\sqrt{(x_1^2 + x_2^2 + x_3^2)}$; in $n$ dimensions we *define* the length of a vector $\mathbf{x}$ with components $x_1, x_2, x_3, \ldots, x_n$ by $\sqrt{(x_1^2 + x_2^2 + \ldots + x_n^2)}$

and denote it by $\|\mathbf{x}\|$†. Two vectors which differ only by a constant factor will then be said to have the same direction, but different lengths. Also a unit vector $\mathbf{u}$ is one such that $\|\mathbf{u}\| = 1$. We may then remove the non-uniqueness concerning eigenvectors mentioned above by insisting that they be of unit magnitude or length. Thus, if $\mathbf{x}$ is any eigenvector, one of unit length is given by $\mathbf{u} = (1/\|\mathbf{x}\|)\mathbf{x}$. Such an eigenvector is called a *normalized* eigenvector. *Henceforth, unless stated to the contrary, we shall assume that all eigenvectors are normalized.*

We shall now show that the eigenvalues of a square matrix of order $n$ satisfy an $n$th degree polynomial equation. Thus in general there will be $n$ eigenvalues (not necessarily all distinct); these need not be real, but if the matrix elements are real they will either be real or else occur in complex conjugate pairs.

THEOREM 4.1

If $\mathbf{A}$ is a square matrix of order $n$, any eigenvalue $\lambda$ satisfies the $n$th degree polynomial equation $\det(\mathbf{A} - \lambda\mathbf{I}) = 0$; this equation is known as the *characteristic equation* of $\mathbf{A}$.

*Proof*

We seek $\lambda$ and (non-zero) $\mathbf{x}$ such that $\mathbf{Ax} = \lambda\mathbf{x}$ or $(\mathbf{A} - \lambda\mathbf{I})\mathbf{x} = \mathbf{0}$. Now, this is a system of $n$ homogeneous equations in the $n$ unknowns $x_1, x_2, \ldots, x_n$ and hence, by Theorem 3.2, if we require a non-zero solution $\mathbf{x}$ then $\mathbf{A} - \lambda\mathbf{I}$ must be singular or, in other words, $\det(\mathbf{A} - \lambda\mathbf{I}) = 0$. Also, Theorem 3.2 guarantees that if $\det(\mathbf{A} - \lambda\mathbf{I}) = 0$ then such a non-zero solution exists. Since

$$\det(\mathbf{A} - \lambda\mathbf{I}) = \begin{vmatrix} a_{11} - \lambda & a_{12} & a_{13} & \cdots & a_{1n} \\ a_{21} & a_{22} - \lambda & a_{23} & \cdots & a_{2n} \\ a_{31} & a_{32} & a_{33} - \lambda & \cdots & a_{3n} \\ \vdots & & & & \\ a_{n1} & a_{n2} & \cdots & a_{n,n-1} & a_{nn} - \lambda \end{vmatrix}$$

it follows, by considering the expansion of the determinant according to the rules of Chapter 1, that we obtain an expression for $\det(\mathbf{A} - \lambda\mathbf{I})$ which contains the product $(a_{11} - \lambda)(a_{22} - \lambda) \ldots (a_{nn} - \lambda)$; this is the term containing the highest power of $\lambda$ and hence the equation $\det(\mathbf{A} - \lambda\mathbf{I}) = 0$ is a polynomial equation of the $n$th degree in $\lambda$. This proves the required result. ∎

We thus see that we have $n$ eigenvalues (not necessarily all distinct) and $n$ corresponding eigenvectors. We denote the eigenvalues by $\lambda_1, \lambda_2, \ldots, \lambda_n$ and the corresponding eigenvectors by $\mathbf{x}^{(1)}, \mathbf{x}^{(2)}, \ldots, \mathbf{x}^{(n)}$ so that we have

$$\mathbf{Ax}^{(r)} = \lambda_r \mathbf{x}^{(r)} \qquad (1 \leqslant r \leqslant n).$$

† Strictly speaking this definition is valid only for vectors with real components; to cater for the complex case we have the definition $\|\mathbf{x}\| = \sqrt{(|x_1|^2 + |x_2|^2 + \ldots + |x_n|^2)}$. Notice also that

$$\begin{aligned} \|\mathbf{x}\| &= \sqrt{(\bar{\mathbf{x}}^{\mathrm{T}}\mathbf{x})} && \text{in general} \\ &= \sqrt{(\mathbf{x}^{\mathrm{T}}\mathbf{x})} && \text{for real } \mathbf{x}. \end{aligned}$$

In principle, Theorem 4.1 provides a method for the calculation of eigenvalues and eigenvectors. Expanding $\det(\mathbf{A} - \lambda\mathbf{I})$ leads to an explicit polynomial equation, the roots of which give all the eigenvalues $\lambda_1, \lambda_2, \ldots, \lambda_n$. Substituting each $\lambda_r$ into the equation $(\mathbf{A} - \lambda\mathbf{I})\mathbf{x} = \mathbf{0}$ gives the $n$ sets of equations $(\mathbf{A} - \lambda_r\mathbf{I})\mathbf{x}^{(r)} = \mathbf{0}$ which when solved give the $\mathbf{x}^{(r)}$. In practice, this method cannot be recommended for matrices of order greater than about 3 or 4 because of the excessive amount of work involved in expanding the determinant. Nowadays it would only be used for low-order matrices when no computing equipment is available. Nevertheless, that situation occurs sufficiently often to warrant presenting a worked example making use of the method.

*Examples*

4.1.1 Find the characteristic equation for the matrix

$$\mathbf{A} = \begin{bmatrix} 2 & 1 & -1 \\ 0 & -2 & -2 \\ 1 & 1 & 0 \end{bmatrix};$$

hence find the eigenvalues and corresponding eigenvectors of **A**. Here

$$\mathbf{A} - \lambda\mathbf{I} = \begin{bmatrix} 2-\lambda & 1 & -1 \\ 0 & -2-\lambda & -2 \\ 1 & 1 & -\lambda \end{bmatrix}$$

so

$$\det(\mathbf{A} - \lambda\mathbf{I}) = \begin{vmatrix} 2-\lambda & 1 & -1 \\ 0 & -2-\lambda & -2 \\ 1 & 1 & -\lambda \end{vmatrix}$$

$$= (2-\lambda)\{(2+\lambda)\lambda + 2\} - 1\{0(-\lambda) + 2\} - 1\{0(1) + (2+\lambda)\}$$

$$= -\lambda^3 + \lambda \qquad \text{after simplification.}$$

Thus the characteristic equation, $\det(\mathbf{A} - \lambda\mathbf{I}) = 0$, is just $\lambda^3 - \lambda = 0$ and hence the eigenvalues, the roots of this equation, are given by

$$\lambda_1 = 0, \qquad \lambda_2 = 1, \qquad \lambda_3 = -1.$$

The eigenvectors corresponding to these eigenvalues may now be calculated in turn.

(1) $\lambda_1 = 0$.

We have to solve $(\mathbf{A} - \lambda_1\mathbf{I})\mathbf{x} = \mathbf{0}$ for $\mathbf{x}$ to obtain $\mathbf{x}^{(1)}$. But

$$\mathbf{A} - \lambda_1\mathbf{I} = \begin{bmatrix} 2 & 1 & -1 \\ 0 & -2 & -2 \\ 1 & 1 & 0 \end{bmatrix}$$

and so the system of equations to be solved is just

$$\begin{aligned} 2x_1 + x_2 - x_3 &= 0 \\ -2x_2 - 2x_3 &= 0 \\ x_1 + x_2 &= 0. \end{aligned}$$

These yield immediately $x_1 = -x_2 = x_3$ and hence the solution of the system is given by $x_1 = k, x_2 = -k, x_3 = k$ for any $k$. Thus an eigenvector is given by

$$\begin{bmatrix} k \\ -k \\ k \end{bmatrix}$$

and it will be normalized if we choose $k$ such that $k^2 + k^2 + k^2 = 1$, i.e. $k = 1/\sqrt{3}$. Thus the normalized eigenvector corresponding to $\lambda_1$ is given by

$$\mathbf{x}^{(1)} = \frac{1}{\sqrt{3}} \begin{bmatrix} 1 \\ -1 \\ 1 \end{bmatrix}.$$

(2) $\lambda_2 = 1$.

We have to solve $(\mathbf{A} - \lambda_2\mathbf{I})\mathbf{x} = \mathbf{0}$ which reduces to

$$\begin{bmatrix} 1 & 1 & -1 \\ 0 & -3 & -2 \\ 1 & 1 & -1 \end{bmatrix} \begin{bmatrix} x_1 \\ x_2 \\ x_3 \end{bmatrix} = \begin{bmatrix} 0 \\ 0 \\ 0 \end{bmatrix}$$

giving

$$\begin{aligned} x_1 + x_2 - x_3 &= 0 \\ -3x_2 - 2x_3 &= 0 \\ x_1 + x_2 - x_3 &= 0. \end{aligned}$$

These equations yield $x_1 = \frac{5}{3}x_3, x_2 = -\frac{2}{3}x_3$, and hence an eigenvector is given by

$$\begin{bmatrix} \frac{5}{3}k \\ -\frac{2}{3}k \\ k \end{bmatrix};$$

it will be normalized if we choose $k$ such that

$$\tfrac{25}{9}k^2 + \tfrac{4}{9}k^2 + k^2 = 1, \qquad \text{i.e. } k = 3/\sqrt{(38)}.$$

Thus the normalized eigenvector corresponding to $\lambda_2$ is given by

$$\mathbf{x}^{(2)} = \frac{1}{\sqrt{(38)}} \begin{bmatrix} 5 \\ -2 \\ 3 \end{bmatrix}.$$

(3) $\lambda_3 = -1$.

In exactly the same manner as in (2) above we find the normalized eigenvector

$$\mathbf{x}^{(3)} = \frac{1}{\sqrt{6}} \begin{bmatrix} 1 \\ -2 \\ 1 \end{bmatrix}.$$

So, finally, we have the eigenvalues and corresponding normalized eigenvectors given by

$$\lambda_1 = 0, \mathbf{x}^{(1)} = \frac{1}{\sqrt{3}}\begin{bmatrix} 1 \\ -1 \\ 1 \end{bmatrix}; \qquad \lambda_2 = 1, \mathbf{x}^{(2)} = \frac{1}{\sqrt{(38)}}\begin{bmatrix} 5 \\ -2 \\ 3 \end{bmatrix};$$

$$\lambda_3 = -1, \mathbf{x}^{(3)} = \frac{1}{\sqrt{6}}\begin{bmatrix} 1 \\ -2 \\ 1 \end{bmatrix}.$$

4.1.2 Show that if **A** is a square matrix of order $n$ and has rank $r$ then the number of eigenvalues equal to zero is at least $(n - r)$.

The characteristic equation is $\det(\mathbf{A} - \lambda\mathbf{I}) = 0$ which, when written out in full, is

$$\begin{vmatrix} a_{11} - \lambda & a_{12} & a_{13} & \dots & a_{1n} \\ a_{21} & a_{22} - \lambda & a_{23} & & \cdot \\ a_{31} & a_{32} & a_{33} - \lambda & & \cdot \\ \cdot & & & & \cdot \\ \cdot & & & & \cdot \\ \cdot & & & & \cdot \\ a_{n_1} & & & & a_{nn} - \lambda \end{vmatrix} = 0$$

giving a polynomial equation of the form

$$\lambda^n + c_1\lambda^{n-1} + c_2\lambda^{n-2} + \dots + c_{n-1}\lambda + c_n = 0.$$

Now, from the expansion properties of the determinant it follows that $c_n = (-1)^n \det(A$ that $c_{n-1}$ contains subdeterminants of $\det(\mathbf{A})$ of order $(n - 1)$, $c_{n-2}$ contains subdeterminants of order $(n - 2)$ and in general $c_i$ contains subdeterminants of order $i$. Thus the definition of rank implies that if **A** is of rank $r$ then $c_{r+1} = c_{r+2} = \dots = c_n = 0$, since all determinants appearing in these coefficients are zero. Thus the characteristic equation reduces to

$$\lambda^n + c_1\lambda^{n-1} + \dots + c_r\lambda^{n-r} = 0$$

or

$$\lambda^{n-r}(\lambda^r + c_1\lambda^{r-1} + \dots + c_r) = 0$$

implying that $(n - r)$ roots are equal to zero, and hence $(n - r)$ eigenvalues are zero; we have not shown that $c_r \neq 0$ (indeed it is not necessarily true) and hence there may be more than $(n - r)$ zero roots.

*Exercises*

4.1.1 Find the characteristic equation for the matrix

$$\mathbf{A} = \begin{bmatrix} 10 & -2 & 4 \\ -20 & 4 & -10 \\ -30 & 6 & -13 \end{bmatrix}.$$

Solve the characteristic equation to obtain the eigenvalues and hence obtain the corresponding normalized eigenvectors.

4.1.2 Show that

$$\mathbf{A} = \begin{bmatrix} 78 & -60 & 15 \\ 150 & -117 & 30 \\ 200 & -160 & 43 \end{bmatrix} \quad \text{and} \quad \mathbf{B} = \begin{bmatrix} -1 & 2 & -3 \\ 0 & 1 & -1 \\ 1 & -1 & 2 \end{bmatrix}$$

both have two coincident eigenvalues. Show, however, that **A** possesses two distinct eigenvectors corresponding to the two coincident eigenvalues whereas **B** possesses only one eigenvector corresponding to the two coincident eigenvalues.

4.1.3 If the eigenvalues of **A** are $\lambda_1, \lambda_2, \ldots, \lambda_n$ with corresponding eigenvectors $\mathbf{x}^{(1)}$, $\mathbf{x}^{(2)}, \ldots, \mathbf{x}^{(n)}$ show that if $p$ is a positive integer then the eigenvalues of $\mathbf{A}^p$ are $\lambda_1^p$, $\lambda_2^p, \ldots, \lambda_n^p$ with corresponding eigenvectors $\mathbf{x}^{(1)}, \mathbf{x}^{(2)}, \ldots, \mathbf{x}^{(n)}$. Deduce that if $F(\mathbf{A})$ is a matrix polynomial then the eigenvalues of $F(\mathbf{A})$ are $F(\lambda_1), F(\lambda_2), \ldots, F(\lambda_n)$.

4.1.4 Show that if **T** is any non-singular matrix of the same order as **A** then **A** and $\mathbf{T}^{-1}\mathbf{AT}$ have the same characteristic equation.

## 4.2 Further Properties of Eigenvalues and Eigenvectors

In the previous section we have seen that any square matrix of order $n$ possesses $n$ eigenvalues (not necessarily all distinct). Corresponding to these eigenvalues are (normalized) eigenvectors; in the worked examples and exercises considered above, distinct eigenvalues led to distinct eigenvectors whereas a pair of coincident eigenvalues in one case had two distinct eigenvectors associated with them and in another had only one.

In this section we wish to develop some results concerning eigenvalues and eigenvectors and their relationships. The first property which we shall prove is the generalization of what we have seen above—that the eigenvectors corresponding to distinct eigenvalues are themselves distinct.

THEOREM 4.2

If **A** is a square matrix then the eigenvectors corresponding to distinct eigenvalues (if any) are themselves distinct.

*Proof*

Suppose the contrary is true so that we have the same eigenvector **x** corresponding to the distinct eigenvalues $\lambda_r$ and $\lambda_s$. We shall show that this supposition is self-contradictory and hence false.

The supposition gives $\mathbf{Ax} = \lambda_r\mathbf{x}$ and $\mathbf{Ax} = \lambda_s\mathbf{x}$ which on subtraction yield $(\lambda_r - \lambda_s)\mathbf{x} = \mathbf{0}$. But since $\lambda_r$ and $\lambda_s$ are distinct, $\lambda_r - \lambda_s \neq 0$ and thus we have $\mathbf{x} = \mathbf{0}$. But this cannot be, since **x** is an eigenvector and thus by definition non-zero. Hence the supposition must be false and so the theorem is proved. ■

An important type of matrix in engineering applications is the real symmetric matrix (i.e. a matrix $\mathbf{A}$ such that $\bar{\mathbf{A}} = \mathbf{A}$ and $\mathbf{A}^{\mathrm{T}} = \mathbf{A}$ – all elements are real and there is symmetry about the leading diagonal). In quantum mechanical applications we are often interested in Hermitian matrices (i.e. matrices such that $\bar{\mathbf{A}}^{\mathrm{T}} = \mathbf{A}$). Hermitian matrices include real symmetric matrices as a subset, and hence any property we prove for Hermitian matrices automatically implies its truth for real symmetric matrices.

THEOREM 4.3

The eigenvalues of a Hermitian matrix are all real.

*Proof*

Suppose $\mathbf{A}$ is Hermitian and has an eigenvalue $\lambda$ with corresponding eigenvector $\mathbf{x}$. Then

$$\mathbf{A}\mathbf{x} = \lambda\mathbf{x} \tag{4.2.1}$$

and multiplying both sides of this equation by $\bar{\mathbf{x}}^{\mathrm{T}}$ gives

$$\bar{\mathbf{x}}^{\mathrm{T}}\mathbf{A}\mathbf{x} = \lambda\bar{\mathbf{x}}^{\mathrm{T}}\mathbf{x}. \tag{4.2.2}$$

Taking the simultaneous transpose and complex conjugate of equation (4.2.1) gives

$$(\bar{\mathbf{A}}\,\bar{\mathbf{x}})^{\mathrm{T}} = \bar{\lambda}\,\bar{\mathbf{x}}^{\mathrm{T}}$$

or

$$\bar{\mathbf{x}}^{\mathrm{T}}\bar{\mathbf{A}}^{\mathrm{T}} = \bar{\lambda}\,\bar{\mathbf{x}}^{\mathrm{T}}$$

so that

$$\bar{\mathbf{x}}^{\mathrm{T}}\mathbf{A} = \bar{\lambda}\,\bar{\mathbf{x}}^{\mathrm{T}} \tag{4.2.3}$$

since $\mathbf{A}$ is Hermitian.

Postmultiplying both sides of equation (4.2.3) by $\mathbf{x}$ now gives

$$\bar{\mathbf{x}}^{\mathrm{T}}\mathbf{A}\mathbf{x} = \bar{\lambda}\,\bar{\mathbf{x}}^{\mathrm{T}}\mathbf{x} \tag{4.2.4}$$

and if we now compare equations (4.2.2) and (4.2.4) we see that we must have $\lambda = \bar{\lambda}$; this implies that $\lambda$ is real, as required. ■

THEOREM 4.4

The eigenvectors of a real symmetric matrix can be chosen to be real.

*Proof*

Let $\mathbf{A}$ be a real symmetric matrix with an eigenvector $\mathbf{x}$ and corresponding eigenvalue $\lambda$. Theorem 4.3 tells us that $\lambda$ must be real. Then the corresponding eigenvector $\mathbf{x}$ is obtained by solving the system of linear equations $(\mathbf{A} - \lambda\mathbf{I})\mathbf{x} = \mathbf{0}$. Since the coefficients are real, the solution must be real (provided the arbitrary multiplicative constant associated with the solution of a homogeneous system is chosen to be real). ■

THEOREM 4.5

If $\mathbf{A}$ is a Hermitian matrix then the eigenvectors $\mathbf{x}^{(r)}$ and $\mathbf{x}^{(s)}$ corresponding to distinct eigenvalues $\lambda_r$ and $\lambda_s$ are orthogonal in the sense that $\bar{\mathbf{x}}^{(r)\mathrm{T}}\mathbf{x}^{(s)} = 0$.

*Proof*

We have

$$\mathbf{A}\mathbf{x}^{(r)} = \lambda_r \mathbf{x}^{(r)} \tag{4.2.5}$$

and

$$\mathbf{A}\mathbf{x}^{(s)} = \lambda_s \mathbf{x}^{(s)} \tag{4.2.6}$$

where, by Theorem 4.3, $\lambda_r$ and $\lambda_s$ are real.

Multiplying equation (4.2.6) on the left by $\bar{\mathbf{x}}^{(r)\mathrm{T}}$ gives

$$\bar{\mathbf{x}}^{(r)\mathrm{T}}\mathbf{A}\mathbf{x}^{(s)} = \lambda_s \bar{\mathbf{x}}^{(r)\mathrm{T}}\mathbf{x}^{(s)}. \tag{4.2.7}$$

Taking the transpose and complex conjugate of equation (4.2.5) gives

$$\bar{\mathbf{x}}^{(r)\mathrm{T}}\mathbf{A} = \lambda_r \bar{\mathbf{x}}^{(r)\mathrm{T}} \tag{4.2.8}$$

on using the facts that $\bar{\mathbf{A}}^{\mathrm{T}} = \mathbf{A}$ and $\bar{\lambda}_r = \lambda_r$.

If we now multiply equation (4.2.8) on the right by $\mathbf{x}^{(s)}$ we obtain

$$\bar{\mathbf{x}}^{(r)\mathrm{T}}\mathbf{A}\mathbf{x}^{(s)} = \lambda_r \bar{\mathbf{x}}^{(r)\mathrm{T}}\mathbf{x}^{(s)}. \tag{4.2.9}$$

Comparing equations (4.2.7) and (4.2.9) we see that

$$\lambda_s \bar{\mathbf{x}}^{(r)\mathrm{T}}\mathbf{x}^{(s)} = \lambda_r \bar{\mathbf{x}}^{(r)\mathrm{T}}\mathbf{x}^{(s)}$$

which, since $\lambda_s \neq \lambda_r$, implies that $\bar{\mathbf{x}}^{(r)\mathrm{T}}\mathbf{x}^{(s)} = 0$. ■

THEOREM 4.6

The eigenvectors $\mathbf{x}^{(1)}, \mathbf{x}^{(2)}, \ldots, \mathbf{x}^{(r)}$ corresponding to the distinct eigenvalues $\lambda_1, \lambda_2, \ldots, \lambda_r$ of a square matrix $\mathbf{A}$ of order $n$ are linearly independent in the sense that if $c_1\mathbf{x}^{(1)} + c_2\mathbf{x}^{(2)} + \ldots + c_r\mathbf{x}^{(r)} = \mathbf{0}$ then $c_1 = c_2 = \ldots = c_r = 0$.

*Proof*

For clarity of exposition we consider the case of three distinct eigenvalues; the method generalizes in an obvious manner.

We have, then, that

$$\mathbf{A}\mathbf{x}^{(1)} = \lambda_1\mathbf{x}^{(1)}, \qquad \mathbf{A}\mathbf{x}^{(2)} = \lambda_2\mathbf{x}^{(2)}, \qquad \mathbf{A}\mathbf{x}^{(3)} = \lambda_3\mathbf{x}^{(3)}$$

(where $\lambda_1, \lambda_2, \lambda_3$ are all different) and we wish to show that if

$$c_1\mathbf{x}^{(1)} + c_2\mathbf{x}^{(2)} + c_3\mathbf{x}^{(3)} = \mathbf{0} \tag{4.2.10}$$

then $c_1 = c_2 = c_3 = 0$.

Premultiplying equation (4.2.10) by $\mathbf{A}$ and $\mathbf{A}^2$ gives in turn the equations

$$c_1\lambda_1\mathbf{x}^{(1)} + c_2\lambda_2\mathbf{x}^{(2)} + c_3\lambda_3\mathbf{x}^{(3)} = \mathbf{0}$$

$$c_1\lambda_1^2\mathbf{x}^{(1)} + c_2\lambda_2^2\mathbf{x}^{(2)} + c_3\lambda_3^2\mathbf{x}^{(3)} = \mathbf{0}$$

which, combined with equation (4.2.10) itself may be written in the partitioned matrix form

$$\begin{bmatrix} 1 & 1 & 1 \\ \lambda_1 & \lambda_2 & \lambda_3 \\ \lambda_1^2 & \lambda_2^2 & \lambda_3^2 \end{bmatrix} \begin{bmatrix} c_1\mathbf{x}^{(1)} \\ c_2\mathbf{x}^{(2)} \\ c_3\mathbf{x}^{(3)} \end{bmatrix} = \begin{bmatrix} 0 \\ 0 \\ 0 \end{bmatrix}$$

or

$$\mathbf{B} \begin{bmatrix} c_1\mathbf{x}^{(1)} \\ c_2\mathbf{x}^{(2)} \\ c_3\mathbf{x}^{(3)} \end{bmatrix} = \begin{bmatrix} 0 \\ 0 \\ 0 \end{bmatrix} \quad \text{where } \mathbf{B} = \begin{bmatrix} 1 & 1 & 1 \\ \lambda_1 & \lambda_2 & \lambda_3 \\ \lambda_1^2 & \lambda_2^2 & \lambda_3^2 \end{bmatrix}.$$

This equation has solution given by

$$\begin{bmatrix} c_1\mathbf{x}^{(1)} \\ c_2\mathbf{x}^{(2)} \\ c_3\mathbf{x}^{(3)} \end{bmatrix} = \mathbf{B}^{-1} \begin{bmatrix} 0 \\ 0 \\ 0 \end{bmatrix} = \begin{bmatrix} 0 \\ 0 \\ 0 \end{bmatrix} \tag{4.2.11}$$

provided $\mathbf{B}$ is non-singular.

But $\det(\mathbf{B}) = (\lambda_1 - \lambda_2)(\lambda_2 - \lambda_3)(\lambda_3 - \lambda_1) \neq 0$ since the eigenvalues are all distinct. Thus $\mathbf{B}$ is non-singular, the solution (4.2.11) is valid and we may deduce that $c_1\mathbf{x}^{(1)} = \mathbf{0}$, $c_2\mathbf{x}^{(2)} = \mathbf{0}$ and $c_3\mathbf{x}^{(3)} = \mathbf{0}$. Since the eigenvectors are non-null (by definition) it follows that $c_1 = c_2 = c_3 = 0$ as required. ∎

From the above theorem we may deduce that if **A** is of order $n$ and has $n$ distinct eigenvalues then *any* vector of $n$ components may be expressed as a linear combination of the eigenvectors of **A**; this is the content of the following theorem.

THEOREM 4.7

If **A** is a square matrix of order $n$ with distinct eigenvalues $\lambda_1, \lambda_2, \ldots, \lambda_n$ and corresponding eigenvectors $\mathbf{x}^{(1)}, \mathbf{x}^{(2)}, \ldots, \mathbf{x}^{(n)}$ then *any* vector $\mathbf{x}$ with $n$ components can be written in the form

$$\mathbf{x} = k_1\mathbf{x}^{(1)} + k_2\mathbf{x}^{(2)} + \ldots + k_n\mathbf{x}^{(n)}.$$

*Proof*

Let us denote by **M** the square matrix of order $n$ whose columns are the eigenvectors of **A** so that

$$\mathbf{M} = [\mathbf{x}^{(1)} \quad \mathbf{x}^{(2)} \quad \ldots \quad \mathbf{x}^{(n)}].$$

We show that **M** is non-singular. By Theorem 4.6 the system of homogeneous equations $\mathbf{Mc} = \mathbf{0}$ has only the solution $\mathbf{c} = \mathbf{0}$, and thus by Theorem 3.2 **M** is non-singular.

Thus, given any vector $\mathbf{x}$ we can define a vector $\mathbf{k}$ by $\mathbf{k} = \mathbf{M}^{-1}\mathbf{x}$ which is equivalent to

$$\mathbf{x} = \mathbf{M}\mathbf{k} = [\mathbf{x}^{(1)}\ \mathbf{x}^{(2)}\ \ldots\ \mathbf{x}^{(n)}] \begin{bmatrix} k_1 \\ k_2 \\ \cdot \\ \cdot \\ \cdot \\ k_n \end{bmatrix}$$

$$= k_1\mathbf{x}^{(1)} + k_2\mathbf{x}^{(2)} + \ldots + k_n\mathbf{x}^{(n)};$$

this is the required result. ∎

The result of the above theorem may be extended to Hermitian matrices even though they may have coincident eigenvalues. The appropriate property is given in the following theorem.

THEOREM 4.8

If $\mathbf{A}$ is a Hermitian matrix of order $n$, and if it possesses $m$ coincident eigenvalues then $m$ distinct mutually orthogonal eigenvectors can be found corresponding to these eigenvalues.

*Proof*

We postpone this until later; the method to be adopted makes use of some results still to be proved (but not depending on this theorem!). ∎

We now come to some theorems which give information about the eigenvalues; we shall see how for certain special matrices the eigenvalues may be obtained by inspection and we shall also prove two relationships between the eigenvalues for general matrices.

THEOREM 4.9

If $\mathbf{A}$ is a triangular matrix then its eigenvalues are its diagonal elements.

*Proof*

We provide the proof for the case when $\mathbf{A}$ is lower triangular; an exactly similar proof holds for upper triangular $\mathbf{A}$.

Let

$$\mathbf{A} = \begin{bmatrix} a_{11} & 0 & 0 & 0 & \ldots & 0 \\ a_{21} & a_{22} & 0 & & & 0 \\ a_{31} & a_{32} & a_{33} & & & 0 \\ \cdot & & & & & \cdot \\ \cdot & & & & & \cdot \\ \cdot & & & & & \cdot \\ a_{n1} & a_{n2} & a_{n3} & & & a_{nn} \end{bmatrix}.$$

We may, in this case, readily determine the eigenvalues from the characteristic equation. For

$$\mathbf{A} - \lambda\mathbf{I} = \begin{bmatrix} a_{11} - \lambda & 0 & 0 & \dots & 0 \\ a_{21} & a_{22} - \lambda & 0 & & 0 \\ a_{31} & a_{32} & a_{33} - \lambda & & 0 \\ \cdot & & & & \cdot \\ \cdot & & & & \cdot \\ \cdot & & & & \cdot \\ a_{n1} & a_{n2} & a_{n3} & & a_{nn} - \lambda \end{bmatrix}$$

and thus, by Theorem 2.17

$$\det(\mathbf{A} - \lambda\mathbf{I}) = (a_{11} - \lambda)(a_{22} - \lambda)(a_{33} - \lambda) \dots (a_{nn} - \lambda).$$

It follows that the characteristic equation is just

$$(a_{11} - \lambda)(a_{22} - \lambda)(a_{33} - \lambda) \dots (a_{nn} - \lambda) = 0$$

with roots $a_{11}, a_{22}, a_{33}, \dots, a_{nn}$, which proves the required result. ■

*Corollary*

If $\mathbf{A}$ is a diagonal matrix its eigenvalues are just its diagonal elements $a_{11}, a_{22}, \dots, a_{nn}$; also the eigenvector corresponding to $a_{ii}$ is just the vector whose $i$th component is 1 and all other components are 0.

*Proof*

The first part of the corollary is just a special case of the theorem, since a diagonal matrix is just a special case of a triangular one.

The second part of the corollary may immediately be verified by noting that

$$\begin{bmatrix} a_{11} & & & & \\ & a_{22} & & 0 & \\ & & a_{33} & & \\ & 0 & & \ddots & \\ & & & & a_{nn} \end{bmatrix} \begin{bmatrix} 0 \\ 0 \\ \vdots \\ 0 \\ 1 \\ 0 \\ \vdots \\ 0 \end{bmatrix} = a_{ii} \begin{bmatrix} 0 \\ 0 \\ \vdots \\ 0 \\ 1 \\ 0 \\ \vdots \\ 0 \end{bmatrix}$$

where the 1 in the column vector appears in the $i$th position. ■

DEFINITION 4.2

The sum of the diagonal elements of a square matrix $\mathbf{A}$ is called the trace of $\mathbf{A}$ and is denoted by $\mathrm{Tr}(\mathbf{A})$. Thus if $\mathbf{A}$ is of order $n$,

$$\mathrm{Tr}(\mathbf{A}) = \sum_{i=1}^{n} a_{ii}.$$

□

THEOREM 4.10

If **A** is a square matrix of order $n$ with eigenvalues $\lambda_1, \lambda_2, \ldots, \lambda_n$ then the sum and product of all the eigenvalues are given by

(i) $\sum_{i=1}^{n} \lambda_i = \mathrm{Tr}(\mathbf{A})$

(ii) $\prod_{i=1}^{n} \lambda_i = \det(\mathbf{A})$.

*Proof*

$\lambda_1, \lambda_2, \ldots, \lambda_n$ are the roots of the characteristic equation $\det(\mathbf{A} - \lambda\mathbf{I}) = 0$ which when written out in full is just

$$\begin{vmatrix} a_{11} - \lambda & a_{12} & a_{13} & \cdots & a_{1n} \\ a_{21} & a_{22} - \lambda & a_{23} & & a_{2n} \\ a_{31} & a_{32} & a_{33} - \lambda & & a_{3n} \\ \vdots & & & & \vdots \\ a_{n1} & \cdots & \cdots & \cdots & a_{nn} - \lambda \end{vmatrix} = 0. \qquad (4.2.12)$$

Expanding the determinant will give a polynomial equation of the form

$$(-\lambda)^n + c_1(-\lambda)^{n-1} + c_2(-\lambda)^{n-2} + \ldots + c_{n-1}(-\lambda) + c_n = 0 \qquad (4.2.13)$$

where the $c_i$ are functions of the matrix elements $a_{ij}$.

Since the roots of this equation are just $\lambda_1, \lambda_2, \ldots \lambda_n$ it follows that it must have the factorized form

$$(-1)^n(\lambda - \lambda_1)(\lambda - \lambda_2) \ldots (\lambda - \lambda_n) = 0$$

or, equivalently,

$$(-\lambda + \lambda_1)(-\lambda + \lambda_2) \ldots (-\lambda + \lambda_n) = 0. \qquad (4.2.14)$$

Comparing the coefficients of the powers of $\lambda^{n-1}$ and $\lambda^0$ in equations (4.2.13) and (4.2.14) gives us respectively

$$\lambda_1 + \lambda_2 + \lambda_3 + \ldots + \lambda_n = c_1 \qquad (4.2.15)$$

$$\lambda_1\lambda_2\lambda_3 \ldots \lambda_n = c_n. \qquad (4.2.16)$$

However, from the expansion properties (Definition 1.4) of the determinant in equation (4.2.12) we see that terms in $\lambda^{n-1}$ can only arise from taking the product of $(n - 1)$ diagonal elements $(-\lambda)$ together with the remaining diagonal element $(a_{ii})$. This can be done in the $n$ different ways in which $a_{ii}$ may be selected and hence in the expansion of the determinant the $n$ terms in $(-\lambda)^{n-1}$ give rise to a coefficient of $(-\lambda)^{n-1}$ equal to $a_{11} + a_{22} + \ldots + a_{nn}$. Thus $c_1 = a_{11} + a_{22} + \ldots + a_{nn}$ and so, from equation (4.2.15) we have

$$\lambda_1 + \lambda_2 + \lambda_3 + \ldots + \lambda_n = a_{11} + a_{22} + a_{33} + \ldots + a_{nn}$$

or, in other words

$$\sum_{i=1}^{n} \lambda_i = \mathrm{Tr}(\mathbf{A}).$$

Also, we see from equation (4.2.13) that $c_n$ is just the value of the polynomial when $\lambda = 0$; this means that $c_n = \det(\mathbf{A} - \lambda\mathbf{I})$ with $\lambda$ set equal to 0, and hence $c_n = \det(\mathbf{A})$. Equation (4.2.16) now gives

$$\lambda_1\lambda_2\lambda_3 \ldots \lambda_n = \det(\mathbf{A}) \qquad \text{or} \qquad \prod_{i=1}^{n} \lambda_i = \det(\mathbf{A}),$$

as required. ■

We now come to a group of results which relate the eigenvalues and eigenvectors of a given matrix to those of another matrix obtained from the original matrix by some transformation. They have fundamental importance in both the theoretical basis behind certain methods for the calculation of eigenvalues and in the application of eigenvalue theory in various fields, a little of which we shall see in Section 4.5.

DEFINITION 4.3

If $\mathbf{A}$ is a square matrix of order $n$, $\mathbf{T}$ a non-singular square matrix of order $n$ and $\mathbf{B} = \mathbf{T}^{-1}\mathbf{A}\mathbf{T}$ then $\mathbf{B}$ and $\mathbf{A}$ are said to be *similar* matrices, or to be related by a *similarity transformation.* □

THEOREM 4.11

If $\mathbf{A}$ and $\mathbf{B}$ are similar matrices then they have the same eigenvalues; also, if $\mathbf{B} = \mathbf{T}^{-1}\mathbf{A}\mathbf{T}$ and if $\mathbf{x}$ is an eigenvector of $\mathbf{A}$ then $\mathbf{T}^{-1}\mathbf{x}$ is an eigenvector of $\mathbf{B}$.

*Proof*

Suppose that the eigenvalue of $\mathbf{A}$ corresponding to the eigenvector $\mathbf{x}$ is $\lambda$. Then we have

$$\begin{aligned}
\mathbf{B}(\mathbf{T}^{-1}\mathbf{x}) &= (\mathbf{T}^{-1}\mathbf{A}\mathbf{T})\mathbf{T}^{-1}\mathbf{x} \\
&= \mathbf{T}^{-1}\mathbf{A}\mathbf{T}\mathbf{T}^{-1}\mathbf{x} \\
&= \mathbf{T}^{-1}\mathbf{A}\mathbf{x} \\
&= \mathbf{T}^{-1}\lambda\mathbf{x} \\
&= \lambda(\mathbf{T}^{-1}\mathbf{x}).
\end{aligned}$$

Thus $\mathbf{T}^{-1}\mathbf{x}$ is an eigenvector of $\mathbf{B}$ with eigenvalue $\lambda$, and hence the result is proved. ■

Theorem 4.11 will be used in situations where, for instance, the matrix $\mathbf{B}$ will have much simpler properties than the matrix $\mathbf{A}$. Thus, if $\mathbf{B}$ were diagonal or triangular then we know (Theorem 4.9) that its eigenvalues are just its diagonal elements and hence if we could find such a transformation we could thus determine the eigenvalues. Alternatively, knowing the eigenvalues we could possibly make use of the diagonal form to simplify certain problems. The following theorem shows that in the case of a matrix with distinct eigenvalues such a transformation to a diagonal matrix is possible.

THEOREM 4.12

If $\mathbf{A}$ is a square matrix of order $n$ with distinct eigenvalues then there exists a matrix $\mathbf{T}$ such that $\mathbf{T}^{-1}\mathbf{AT}$ is diagonal with the eigenvalues of $\mathbf{A}$ as diagonal elements.

*Proof*

We provide a proof by explicitly constructing a matrix $\mathbf{T}$ with the required properties.

Suppose that $\mathbf{A}$ has eigenvectors $\mathbf{x}^{(i)}$ with eigenvalues $\lambda_i$ so that $\mathbf{Ax}^{(i)} = \lambda_i \mathbf{x}^{(i)}$ for $1 \leqslant i \leqslant n$.

Let us choose $\mathbf{T} = [\mathbf{x}^{(1)}\ \mathbf{x}^{(2)} \ldots \mathbf{x}^{(n)}]$ so that the $i$th column of $\mathbf{T}$ is just the $i$th eigenvector of $\mathbf{A}$.

We can show immediately that $\mathbf{T}$ is non-singular. For the columns of $\mathbf{T}$ are linearly independent (Theorem 4.6) and in the proof of Theorem 4.7 we have already shown that such a matrix is non-singular.

Now, since $\mathbf{T}^{-1}\mathbf{T} = \mathbf{I}$ and also

$$\begin{aligned} \mathbf{T}^{-1}\mathbf{T} &= \mathbf{T}^{-1}[\mathbf{x}^{(1)} \quad \mathbf{x}^{(2)} \quad \ldots \quad \mathbf{x}^{(n)}] \\ &= [\mathbf{T}^{-1}\mathbf{x}^{(1)} \quad \mathbf{T}^{-1}\mathbf{x}^{(2)} \quad \ldots \quad \mathbf{T}^{-1}\mathbf{x}^{(n)}] \end{aligned}$$

it follows that

$$\mathbf{T}^{-1}\mathbf{x}^{(i)} = \mathbf{I}^{(i)} \tag{4.2.17}$$

where $\mathbf{I}^{(i)}$ is the $i$th column of the unit matrix $\mathbf{I}$. Then

$$\begin{aligned} \mathbf{T}^{-1}\mathbf{AT} &= \mathbf{T}^{-1}\mathbf{A}\,[\mathbf{x}^{(1)} \quad \mathbf{x}^{(2)} \quad \ldots \quad \mathbf{x}^{(n)}] \\ &= \mathbf{T}^{-1}[\mathbf{Ax}^{(1)} \quad \mathbf{Ax}^{(2)} \quad \ldots \quad \mathbf{Ax}^{(n)}] \\ &= \mathbf{T}^{-1}[\lambda_1\mathbf{x}^{(1)} \quad \lambda_2\mathbf{x}^{(2)} \quad \ldots \quad \lambda_n\mathbf{x}^{(n)}] \\ &= [\lambda_1\mathbf{T}^{-1}\mathbf{x}^{(1)} \quad \lambda_2\mathbf{T}^{-1}\mathbf{x}^{(2)} \quad \ldots \quad \lambda_n\mathbf{T}^{-1}\mathbf{x}^{(n)}] \\ &= [\lambda_1\mathbf{I}^{(1)} \quad \lambda_2\mathbf{I}^{(2)} \quad \ldots \quad \lambda_n\mathbf{I}^{(n)}] \\ &\qquad \text{by equation (4.2.17)} \\ &= \begin{bmatrix} \lambda_1 & 0 & \ldots & 0 \\ 0 & \lambda_2 & \ldots & 0 \\ 0 & 0 & & 0 \\ 0 & 0 & & 0 \\ \cdot & \cdot & & \cdot \\ \cdot & \cdot & & \cdot \\ \cdot & \cdot & & \cdot \\ 0 & 0 & & \lambda_n \end{bmatrix} \end{aligned}$$

which proves the required result. ■

For a general matrix (with the possibility of coincident eigenvalues) it is not possible to prove as strong a result as that contained in the above theorem. We can, however, prove that transformation to a triangular matrix may be accomplished, and this will have consequences for the Hermitian case; the appropriate results are contained in the following two theorems.

THEOREM 4.13

For an arbitrary square matrix there exists a similarity transformation such that $\mathbf{T}^{-1}\mathbf{AT}$ is triangular; moreover the transformation matrix is unitary (i.e. $\mathbf{T}^{-1} = \mathbf{T}^{+}$).

*Proof*†

We use the method of induction. Assuming the result to be true for all matrices of order $N-1$ we shall prove it true for all matrices of order $N$. We shall then prove it true for all matrices of order 2 and hence it must be true for all matrices of all orders.

Suppose then, that $\mathbf{A}$ is of order $N$ and that $\mathbf{x}^{(1)}$ is an eigenvector of $\mathbf{A}$ corresponding to eigenvalue $\lambda_1$.

We first show that we can construct an orthonormal set $\mathbf{y}^{(1)}(=\mathbf{x}^{(1)}), \mathbf{y}^{(2)}, \mathbf{y}^{(3)} \ldots \mathbf{y}^{(N)}$ in the sense that

$$\mathbf{y}^{(i)+}\mathbf{y}^{(j)} = \begin{cases} 1 & \text{if } i = j \\ 0 & \text{if } i \neq j. \end{cases}$$

This construction may be effected explicitly by the so-called Gram–Schmidt orthogonalization procedure. We note that out of the $N+1$ vectors $\mathbf{x}^{(1)}, \mathbf{I}^{(1)}, \mathbf{I}^{(2)}, \ldots, \mathbf{I}^{(N)}$ (where $\mathbf{I}^{(k)}$ is the $k$th column of the unit matrix of order $N$) we can pick $\mathbf{x}^{(1)}$ and $(N-1)$ of the others to form an independent set (this is taken as intuitively obvious, but if desired the reader can construct an explicit proof for himself). Let us label this independent set as

$$\mathbf{e}^{(1)}\ (=\mathbf{x}^{(1)}), \mathbf{e}^{(2)}, \mathbf{e}^{(3)}, \ldots \mathbf{e}^{(N)}.$$

If we now define the set of vectors $\mathbf{y}^{(i)}$ $(1 \leqslant i \leqslant N)$ by

$$\mathbf{y}^{(1)} = \mathbf{e}^{(1)}\ (=\mathbf{x}^{(1)})$$

$$\mathbf{y}^{(i)} = \left(\mathbf{e}^{(i)} - \sum_{k=1}^{i-1} (\mathbf{e}^{(i)+}\mathbf{y}^{(k)})\mathbf{y}^{(k)}\right) \Bigg/ \left\| \mathbf{e}^{(i)} - \sum_{k=1}^{i-1} (\mathbf{e}^{(i)+}\mathbf{y}^{(k)})\mathbf{y}^{(k)} \right\| \qquad (2 \leqslant i \leqslant N)$$

then it is easy to verify that the set $\mathbf{y}^{(i)}$ has the required orthonormal properties.

We next define the matrix $\mathbf{T}_1$ by

$$\mathbf{T}_1 = [\mathbf{y}^{(1)} \quad \mathbf{y}^{(2)} \quad \mathbf{y}^{(3)} \quad \ldots \quad \mathbf{y}^{(N)}].$$

We see immediately that the orthonormality of the $\mathbf{y}^{(i)}$ implies the unitarity of $\mathbf{T}_1$; for we have

$$\mathbf{T}_1^{+}\mathbf{T}_1 = \begin{bmatrix} \mathbf{y}^{(1)+} \\ \mathbf{y}^{(2)+} \\ \mathbf{y}^{(3)+} \\ \cdot \\ \cdot \\ \cdot \\ \mathbf{y}^{(N)+} \end{bmatrix} [\mathbf{y}^{(1)} \quad \mathbf{y}^{(2)} \quad \mathbf{y}^{(3)} \quad \ldots \quad \mathbf{y}^{(N)}]$$

† This proof may be omitted on a first reading.

$$= \begin{bmatrix} \mathbf{y}^{(1)+}\mathbf{y}^{(1)} & \mathbf{y}^{(1)+}\mathbf{y}^{(2)} & \dots & \mathbf{y}^{(1)+}\mathbf{y}^{(N)} \\ \mathbf{y}^{(2)+}\mathbf{y}^{(1)} & \mathbf{y}^{(2)+}\mathbf{y}^{(2)} & \dots & \mathbf{y}^{(2)+}\mathbf{y}^{(N)} \\ \cdot & & & \\ \cdot & & & \\ \cdot & & & \\ \mathbf{y}^{(N)+}\mathbf{y}^{(1)} & \mathbf{y}^{(N)+}\mathbf{y}^{(2)} & \dots & \mathbf{y}^{(N)+}\mathbf{y}^{(N)} \end{bmatrix}$$

$$= \begin{bmatrix} 1 & 0 & \dots & 0 \\ 0 & 1 & \dots & 0 \\ \cdot & & & \\ \cdot & & & \\ \cdot & & & \\ 0 & 0 & \dots & 1 \end{bmatrix} \quad \text{by the orthonormality property of the } \mathbf{y}^{(i)}$$

$$= \mathbf{I}.$$

Then

$$\mathbf{T}_1^{-1}\mathbf{A}\mathbf{T}_1 = \mathbf{T}_1^{+}\mathbf{A}\mathbf{T}_1$$

$$= \begin{bmatrix} \mathbf{y}^{(1)+} \\ \mathbf{y}^{(2)+} \\ \cdot \\ \cdot \\ \cdot \\ \mathbf{y}^{(N)+} \end{bmatrix} \mathbf{A}[\mathbf{y}^{(1)} \quad \mathbf{y}^{(2)} \quad \dots \quad \mathbf{y}^{(N)}]$$

$$= \begin{bmatrix} \mathbf{y}^{(1)+} \\ \mathbf{y}^{(2)+} \\ \cdot \\ \cdot \\ \cdot \\ \mathbf{y}^{(N)+} \end{bmatrix} [\lambda_1\mathbf{y}^{(1)} \quad \mathbf{A}\mathbf{y}^{(2)} \quad \dots \quad \mathbf{A}\mathbf{y}^{(N)}]$$

since $\mathbf{y}^{(1)} = \mathbf{x}^{(1)}$ is an eigenvector of $\mathbf{A}$ with eigenvalue $\lambda_1$

$$= \begin{bmatrix} \lambda_1 & \mathbf{y}^{(1)+}\mathbf{A}\mathbf{y}^{(2)} & \mathbf{y}^{(1)+}\mathbf{A}\mathbf{y}^{(3)} & \dots & \mathbf{y}^{(1)+}\mathbf{A}\mathbf{y}^{(N)} \\ 0 & \mathbf{y}^{(1)+}\mathbf{A}\mathbf{y}^{(2)} & & & \mathbf{y}^{(2)+}\mathbf{A}\mathbf{y}^{(N)} \\ \cdot & \cdot & & & \\ \cdot & \cdot & & & \\ \cdot & \cdot & & & \\ 0 & \mathbf{y}^{(N)+}\mathbf{A}\mathbf{y}^{(2)} & \dots & \dots & \mathbf{y}^{(N)+}\mathbf{A}\mathbf{y}^{(N)} \end{bmatrix}$$

again making use of the orthonormality of the $\mathbf{y}^{(i)}$.

We may write this last result in the form

$$\mathbf{T}_1^{-1}\mathbf{A}\mathbf{T}_1 = \begin{bmatrix} \lambda_1 & \mathbf{w} \\ \mathbf{0} & \mathbf{B} \end{bmatrix} \tag{4.2.18}$$

where $\mathbf{0}$ is a null column vector and $\mathbf{w}$ is a row vector, both with $N-1$ components; $\mathbf{B}$ is a square matrix of order $N-1$. However, by our assumption that the theorem holds true for square matrices of order $N-1$ it follows that there exists a unitary matrix $\mathbf{U}$ such that $\mathbf{U}^{-1}\mathbf{BU}=\mathbf{Z}$ where $\mathbf{Z}$ is upper triangular. If we now define $\mathbf{T}_2$ by

$$\mathbf{T}_2=\begin{bmatrix}1 & \mathbf{0}^T\\ \mathbf{0} & \mathbf{U}\end{bmatrix}$$

(which is unitary since $\mathbf{U}$ is unitary) and $\mathbf{T}$ by $\mathbf{T}=\mathbf{T}_1\mathbf{T}_2$ (which is unitary since both $\mathbf{T}_1$ and $\mathbf{T}_2$ are) it follows that

$$\begin{aligned}\mathbf{T}^{-1}\mathbf{AT}&=(\mathbf{T}_1\mathbf{T}_2)^{-1}\mathbf{A}(\mathbf{T}_1\mathbf{T}_2)\\ &=\mathbf{T}_2^{-1}(\mathbf{T}_1^{-1}\mathbf{AT}_1)\mathbf{T}_2\\ &=\begin{bmatrix}1 & \mathbf{0}^T\\ \mathbf{0} & \mathbf{U}^{-1}\end{bmatrix}\begin{bmatrix}\lambda_1 & \mathbf{w}\\ \mathbf{0} & \mathbf{B}\end{bmatrix}\begin{bmatrix}1 & \mathbf{0}^T\\ \mathbf{0} & \mathbf{U}\end{bmatrix}\end{aligned}$$

by equation (4.2.18) and the definition of $\mathbf{T}_2$ above,

$$\begin{aligned}&=\begin{bmatrix}1 & \mathbf{0}^T\\ \mathbf{0} & \mathbf{U}^{-1}\end{bmatrix}\begin{bmatrix}\lambda_1 & \mathbf{wU}\\ \mathbf{0} & \mathbf{BU}\end{bmatrix}\\ &=\begin{bmatrix}\lambda_1 & \mathbf{wU}\\ \mathbf{0} & \mathbf{U}^{-1}\mathbf{BU}\end{bmatrix}\\ &=\begin{bmatrix}\lambda_1 & \mathbf{wU}\\ \mathbf{0} & \mathbf{Z}\end{bmatrix}.\end{aligned}$$

But $\mathbf{Z}$ is upper triangular; thus $\mathbf{T}^{-1}\mathbf{AT}$ is upper triangular and hence we have proved the first part of the inductive proof (that if all matrices of order $N-1$ can be reduced to upper triangular form then so can all matrices of order $N$).

But in fact the above method shows directly that matrices of order 2 may be reduced to upper triangular form, for in this case equation (4.2.18) has $\mathbf{0}$, $\mathbf{B}$ and $\mathbf{w}$ all as single element matrices and thus the reduction to upper triangular form has been accomplished in this case by $\mathbf{T}_1$.

Thus both parts of the inductive proof have been completed and we have proved the required result. ∎

As a direct deduction from this theorem we can prove that any Hermitian matrix is similar to a diagonal matrix. This we do in the following theorem.

THEOREM 4.14

If $\mathbf{A}$ is a Hermitian matrix then there exists a unitary matrix $\mathbf{T}$ such that $\mathbf{T}^{-1}\mathbf{AT}$ is diagonal; the diagonal elements are the eigenvalues of $\mathbf{A}$ and the columns of $\mathbf{T}$ are the eigenvectors of $\mathbf{A}$.

*Proof*

By Theorem 4.13 we know that there exists a unitary matrix $\mathbf{T}$ such that $\mathbf{B} = \mathbf{T}^+\mathbf{A}\mathbf{T}$ with $\mathbf{B}$ upper triangular. Since $\mathbf{A}$ is Hermitian we may show that $\mathbf{B}$ is also Hermitian; for

$$\begin{aligned}\mathbf{B}^+ &= (\mathbf{T}^+\mathbf{A}\mathbf{T})^+ \\ &= \mathbf{T}^+\mathbf{A}^+(\mathbf{T}^+)^+\end{aligned}$$

since by Exercise 2.9.6 the Hermitian conjugate of a product is the product of the Hermitian conjugates taken in reverse order. Thus

$$\begin{aligned}\mathbf{B}^+ &= \mathbf{T}^+\mathbf{A}\mathbf{T} \qquad \text{since} \quad \mathbf{A}^+ = \mathbf{A} \\ &= \mathbf{B}.\end{aligned}$$

Thus $\mathbf{B}$ is Hermitian. But $\mathbf{B}$ is upper triangular, so that $B_{ij} = 0$ if $i > j$. Thus we now have for $i < j$

$$\begin{aligned}B_{ij} &= (\overline{B_{ji}}) \qquad \text{(since } \mathbf{B} \text{ is Hermitian)} \\ &= \overline{0} \\ &= 0.\end{aligned}$$

Thus $\mathbf{B}$ is diagonal, and since it is Hermitian the diagonal elements must be real.

So we have $\mathbf{B} = \mathbf{T}^+\mathbf{A}\mathbf{T} = \mathbf{T}^{-1}\mathbf{A}\mathbf{T}$ with $\mathbf{B}$ diagonal and $\mathbf{T}$ unitary. Hence $\mathbf{T}\mathbf{B} = \mathbf{A}\mathbf{T}$ and on writing

$$\mathbf{T} = [\mathbf{t}^{(1)} \quad \mathbf{t}^{(2)} \quad \ldots \quad \mathbf{t}^{(n)}]$$

where $\mathbf{t}^{(i)}$ is the $i$th column of $\mathbf{T}$ we obtain

$$\mathbf{A}[\mathbf{t}^{(1)} \quad \mathbf{t}^{(2)} \quad \ldots \quad \mathbf{t}^{(n)}] = [\mathbf{t}^{(1)}\ \mathbf{t}^{(2)} \ \ldots \ \mathbf{t}^{(n)}] \begin{bmatrix} b_1 & 0 & 0 & \ldots & 0 \\ 0 & b_2 & 0 & & \\ 0 & 0 & b_3 & & \\ \cdot & \cdot & \cdot & & \\ \cdot & \cdot & \cdot & & \\ \cdot & \cdot & \cdot & & \\ 0 & 0 & 0 & & b_n \end{bmatrix}$$

which, on multiplying out the partitioned forms gives

$$[\mathbf{A}\mathbf{t}^{(1)} \quad \mathbf{A}\mathbf{t}^{(2)} \quad \ldots \quad \mathbf{A}\mathbf{t}^{(n)}] = [b_1\mathbf{t}^{(1)} \quad b_2\mathbf{t}^{(2)} \quad \ldots \quad b_n\mathbf{t}^{(n)}].$$

Thus $\mathbf{A}\mathbf{t}^{(i)} = b_i\mathbf{t}^{(i)}$ implying that $\mathbf{t}^{(i)}$ is an eigenvector of $\mathbf{A}$ belonging to eigenvalue $b_i$; that is, the columns of $\mathbf{T}$ are eigenvectors of $\mathbf{A}$, and the diagonal elements of $\mathbf{B}$ are the eigenvalues of $\mathbf{A}$, as required. ∎

We are now in a position to provide the proof of Theorem 4.8, as promised previously.

*Proof of Theorem 4.8*

With the notation of Theorem 4.14 we have

$$\mathbf{T}^+\mathbf{T} = \mathbf{I}$$

giving

$$\begin{bmatrix} \mathbf{t}^{(1)+} \\ \mathbf{t}^{(2)+} \\ \cdot \\ \cdot \\ \cdot \\ \mathbf{t}^{(n)+} \end{bmatrix} [\mathbf{t}^{(1)} \quad \mathbf{t}^{(2)} \quad \dots \quad \mathbf{t}^{(n)}] = \begin{bmatrix} 1 & 0 & 0 & \dots & 0 \\ 0 & 1 & 0 & \dots & 0 \\ 0 & 0 & 1 & \dots & 0 \\ \cdot & & & & \\ \cdot & & & & \\ \cdot & & & & \\ 0 & \dots & & \dots & 1 \end{bmatrix}$$

and hence

$$\mathbf{t}^{(i)+}\,\mathbf{t}^{(j)} = \begin{cases} 1 & \text{if } i = j \\ 0 & \text{if } i \neq j. \end{cases}$$

But the $\mathbf{t}^{(i)}$ are just the eigenvectors of the Hermitian matrix **A**; thus **A** possesses an orthogonal set of eigenvectors and hence a linearly independent set, irrespective of whether or not the eigenvalues are distinct. This completes the proof of Theorem 4.8.

We also note that Theorem 4.14 gives a special result when **A** is a real symmetric matrix (which is just a special case of a Hermitian matrix). For then the eigenvectors may be chosen to be real (Theorem 4.4) and it follows that **T** must also be real; hence the unitary property of **T**, $\mathbf{T}^{-1} = \mathbf{T}^{+}$ just becomes $\mathbf{T}^{-1} = \mathbf{T}^{\mathrm{T}}$ so that in fact **T** is orthogonal. Thus we have proved the following theorem.

THEOREM 4.15

If **A** is real and symmetric then there exists an orthogonal matrix **T** such that $\mathbf{T}^{-1}\mathbf{AT}$ is diagonal; the diagonal elements are the eigenvalues of **A** and the columns of **T** are the eigenvectors of **A**. ■

Theorems 4.12, 4.13, 4.14 and 4.15 have provided results concerning the transformation of all but the general matrix with coincident eigenvalues. The proof of the general result for this case is beyond the scope of this book, but for completeness we state the result in the following theorem.

THEOREM 4.16

For an arbitrary square matrix **A** there exists a similarity transformation such that

$$\mathbf{T}^{-1}\mathbf{AT} = \begin{bmatrix} \mathbf{\Lambda}_1 & \mathbf{0} & \mathbf{0} & \dots & \mathbf{0} \\ \mathbf{0} & \mathbf{\Lambda}_2 & \mathbf{0} & & \\ \mathbf{0} & \mathbf{0} & \mathbf{\Lambda}_3 & & \\ \cdot & & & \cdot & \\ \cdot & & & & \\ \cdot & & & & \\ \mathbf{0} & & & & \mathbf{\Lambda}_m \end{bmatrix}$$

(i.e. to 'block diagonal form') where the matrices $\mathbf{\Lambda}_1, \mathbf{\Lambda}_2, \ldots \mathbf{\Lambda}_m$ are such that

$$\mathbf{\Lambda}_i = \begin{bmatrix} \lambda_j & 1 & 0 & 0 & 0 & \cdots & 0 \\ 0 & \lambda_j & 1 & 0 & & & \\ 0 & 0 & \lambda_j & 1 & & & \\ \cdot & & & & \cdot & & \\ \cdot & & & & & \cdot & \\ \cdot & & & & & \cdot & 1 \\ 0 & & & & & \cdot & \lambda_j \end{bmatrix};$$

the $\lambda_j$ are the eigenvalues of $\mathbf{A}$ and a given eigenvalue may appear as the diagonal elements of more than one $\mathbf{\Lambda}_i$.

This block diagonal form is often called the Jordan canonical form of $\mathbf{A}$; the determinants $\det(\mathbf{\Lambda}_i - \lambda\mathbf{I})$ are called the elementary divisors of $\mathbf{A}$. It may further be shown that if the elementary divisors are linear in $\lambda$ then the eigenvectors form an independent set and any vector $\mathbf{x}$ of dimension $n$ can be expressed as a linear combination of them.

*Examples*

4.2.1 Show that the eigenvalues of a unitary matrix have unit modulus.
Let $\mathbf{A}$ be unitary so that $\mathbf{A}^{-1} = \mathbf{A}^+$ and suppose it has an eigenvalue $\lambda$ with corresponding eigenvector $\mathbf{x}$, so that $\mathbf{Ax} = \lambda\mathbf{x}$.
Taking the Hermitian conjugate of this gives

$$\mathbf{x}^+\mathbf{A}^+ = \bar{\lambda}\mathbf{x}^+.$$

Multiplying these two results together (remember that they are conformable for multiplication) gives

$$\mathbf{x}^+\mathbf{A}^+\mathbf{Ax} = \bar{\lambda}\lambda\mathbf{x}^+\mathbf{x}$$

or

$$\mathbf{x}^+\mathbf{x} = |\lambda|^2\mathbf{x}^+\mathbf{x}$$

since $\mathbf{A}^+\mathbf{A} = \mathbf{I}$. Thus $|\lambda|^2 = 1$ which proves the required result.

4.2.2 Illustrate the results of Theorems 4.3, 4.5, 4.10 and 4.14 for the matrix

$$\mathbf{A} = \begin{bmatrix} 1 & 1+\mathrm{i} \\ 1-\mathrm{i} & 1 \end{bmatrix}.$$

Now, $\mathbf{A}$ is Hermitian, so we wish to check that the results concerning Hermitian matrices fit this case.
We first wish to find the eigenvalues and eigenvectors of $\mathbf{A}$; the only method we know yet for this is to use the characteristic equation

$$\det(\mathbf{A} - \lambda\mathbf{I}) = 0.$$

This gives

$$\begin{vmatrix} 1-\lambda & 1+\mathrm{i} \\ 1-\mathrm{i} & 1-\lambda \end{vmatrix} = 0$$

or

$$(1-\lambda)^2-(1-\mathrm{i})(1+\mathrm{i})=0,$$

which reduces to

$$\lambda^2-2\lambda-1=0$$

with roots

$$\lambda_1=1+\sqrt{2} \qquad \text{and} \qquad \lambda_2=1-\sqrt{2}.$$

The eigenvectors are obtained from solving $\mathbf{A}\mathbf{x}^{(i)}=\lambda_i\mathbf{x}^{(i)}$. For $\lambda_1=1+\sqrt{2}$ this gives

$$\begin{bmatrix} 1 & 1+\mathrm{i} \\ 1-\mathrm{i} & 1 \end{bmatrix}\begin{bmatrix} x_1 \\ x_2 \end{bmatrix}=(1+\sqrt{2})\begin{bmatrix} x_1 \\ x_2 \end{bmatrix}.$$

Hence

$$x_1+(1+\mathrm{i})x_2=(1+\sqrt{2})x_1$$

giving

$$(1+\mathrm{i})x_2=x_1\sqrt{2}$$

and hence an eigenvector

$$\begin{bmatrix} 1+\mathrm{i} \\ \sqrt{2} \end{bmatrix}$$

which when normalized gives

$$\mathbf{x}^{(1)}=\tfrac{1}{2}\begin{bmatrix} 1+\mathrm{i} \\ \sqrt{2} \end{bmatrix}.$$

Similarly, for $\lambda_2=1-\sqrt{2}$ we should obtain

$$\mathbf{x}^{(2)}=\tfrac{1}{2}\begin{bmatrix} 1+\mathrm{i} \\ -\sqrt{2} \end{bmatrix}.$$

We can now illustrate the results of the various theorems:

*Theorem 4.3*

The eigenvalues $(1\pm\sqrt{2})$ are real (verifying that the eigenvalues of a Hermitian matrix are real).

*Theorem 4.5*

We wish to show that the eigenvectors are orthogonal in the sense that $\mathbf{x}^{(1)+}\mathbf{x}^{(2)}=0$. But

$$\mathbf{x}^{(1)}=\tfrac{1}{2}\begin{bmatrix} 1+\mathrm{i} \\ \sqrt{2} \end{bmatrix} \qquad \text{gives } \mathbf{x}^{(1)+}=\tfrac{1}{2}[1-\mathrm{i} \quad \sqrt{2}]$$

and so

$$\mathbf{x}^{(1)+}\mathbf{x}^{(2)} = \tfrac{1}{2}[1-\mathrm{i} \quad \sqrt{2}]\,\tfrac{1}{2}\begin{bmatrix}1+\mathrm{i}\\ \sqrt{2}\end{bmatrix}$$
$$= \tfrac{1}{4}\{(1-\mathrm{i})(1+\mathrm{i})-2\}$$
$$= 0, \qquad \text{as required.}$$

*Theorem 4.10*

We wish to show that $\lambda_1 + \lambda_2 = \mathrm{Tr}(\mathbf{A})$ and $\lambda_1\lambda_2 = \det(\mathbf{A})$. But $\lambda_1 + \lambda_2 = 2$, $\lambda_1\lambda_2 = -1$,

$$\mathrm{Tr}(\mathbf{A}) = 1 + 1 = 2, \qquad \det(\mathbf{A}) = \begin{vmatrix}1 & 1+\mathrm{i}\\ 1-\mathrm{i} & 1\end{vmatrix}$$
$$= 1-(1-\mathrm{i})(1+\mathrm{i})$$
$$= 1-2$$
$$= -1$$

and hence the result is verified.

*Theorem 4.14*

We require to show that with $\mathbf{T} = [\mathbf{x}^{(1)} \quad \mathbf{x}^{(2)}]$ we have

$$\mathbf{T}^{+}\mathbf{A}\mathbf{T} = \begin{bmatrix}\lambda_1 & 0\\ 0 & \lambda_2\end{bmatrix}.$$

But

$$\mathbf{T} = \tfrac{1}{2}\begin{bmatrix}1+\mathrm{i} & 1+\mathrm{i}\\ \sqrt{2} & -\sqrt{2}\end{bmatrix}$$

so

$$\mathbf{T}^{+} = \tfrac{1}{2}\begin{bmatrix}1-\mathrm{i} & \sqrt{2}\\ 1-\mathrm{i} & -\sqrt{2}\end{bmatrix}$$

and hence

$$\mathbf{T}^{+}\mathbf{A}\mathbf{T} = \tfrac{1}{4}\begin{bmatrix}1-\mathrm{i} & \sqrt{2}\\ 1-\mathrm{i} & -\sqrt{2}\end{bmatrix}\begin{bmatrix}1 & 1+\mathrm{i}\\ 1-\mathrm{i} & 1\end{bmatrix}\begin{bmatrix}1+\mathrm{i} & 1+\mathrm{i}\\ \sqrt{2} & -\sqrt{2}\end{bmatrix}$$

which, on carrying out the matrix multiplications yields

$$\mathbf{T}^{+}\mathbf{A}\mathbf{T} = \begin{bmatrix}1+\sqrt{2} & 0\\ 0 & 1-\sqrt{2}\end{bmatrix} \qquad \text{as required.}$$

4.2.3 Show that $\mathbf{A}$ has a zero eigenvalue if and only if it is singular.

We first suppose that $\mathbf{A}$ is singular; then $\det(\mathbf{A}) = 0$ and since, by Theorem 4.10 $\det(\mathbf{A}) = \Pi_{i=1}^{n}\lambda_i$ it follows that $\Pi_{i=1}^{n}\lambda_i = 0$ and hence one of the eigenvalues is zero.

Similarly, if one of the eigenvalues is zero then $\Pi_{i=1}^{n}\lambda_i = 0$ so that $\det(\mathbf{A}) = 0$ and hence the matrix is singular.

4.2.4 A generalized eigenvalue problem is that of finding a non-zero vector $\mathbf{x}$ such that $\mathbf{Ax} = \lambda\mathbf{Bx}$ where $\mathbf{A}$ and $\mathbf{B}$ are given square matrices of the same order; $\mathbf{x}$ is then the generalized eigenvector and $\lambda$ the generalized eigenvalue (the ordinary case is included by taking $\mathbf{B} = \mathbf{I}$). Find the analogue of the characteristic equation for the generalized problem, and show that if $\mathbf{B}$ is non-singular the problem can be reduced to the ordinary eigenvalue problem. Show also that if $\mathbf{A}$ and $\mathbf{B}$ are symmetric the problem can be reduced to the usual eigenvalue problem for a symmetric matrix.

The problem $\mathbf{Ax} = \lambda\mathbf{Bx}$ is equivalent to finding $\lambda$ such that the homogeneous system $(\mathbf{A} - \lambda\mathbf{B})\mathbf{x} = \mathbf{0}$ has a non-null solution $\mathbf{x}$. This we know to be the case if and only if $\det(\mathbf{A} - \lambda\mathbf{B}) = \mathbf{0}$ and this is the required equation.

If $\mathbf{B}^{-1}$ exists, the problem is equivalent to $\mathbf{B}^{-1}\mathbf{Ax} = \lambda\mathbf{x}$ which is the ordinary eigenvalue problem for the matrix $\mathbf{B}^{-1}\mathbf{A}$.

If both $\mathbf{A}$ and $\mathbf{B}$ are symmetric, although it is true that $\mathbf{B}^{-1}$ is also symmetric it is not necessarily true that $\mathbf{B}^{-1}\mathbf{A}$ is symmetric (see Example 2.4.7) so that the ordinary eigenvalue problem $\mathbf{B}^{-1}\mathbf{Ax} = \lambda\mathbf{x}$ does not in general have a symmetric matrix. But if $\mathbf{B}$ is symmetric and non-singular, the results of Section 2.6 show that it can be decomposed into the form $\mathbf{B} = \mathbf{LL}^{\mathrm{T}}$ where $\mathbf{L}$ is a non-singular lower triangular matrix. Then $\mathbf{Ax} = \lambda\mathbf{Bx}$ becomes

$$\mathbf{Ax} = \lambda\mathbf{LL}^{\mathrm{T}}\mathbf{x}$$

which, when writing $\mathbf{y} = \mathbf{L}^{\mathrm{T}}\mathbf{x}$ gives

$$\mathbf{L}^{-1}\mathbf{AL}^{-1\mathrm{T}}\mathbf{y} = \lambda\mathbf{y}.$$

But $\mathbf{L}^{-1}\mathbf{AL}^{-1\mathrm{T}}$ *is* symmetric; for

$$(\mathbf{L}^{-1}\mathbf{AL}^{-1\mathrm{T}})^{\mathrm{T}} = \mathbf{L}^{-1}\mathbf{A}^{\mathrm{T}}\mathbf{L}^{-1\mathrm{T}}$$
$$= \mathbf{L}^{-1}\mathbf{AL}^{-1\mathrm{T}} \qquad \text{since } \mathbf{A} \text{ is symmetric.}$$

Thus the given problem has been transformed to the problem $\mathbf{L}^{-1}\mathbf{AL}^{-1\mathrm{T}}\mathbf{y} = \lambda\mathbf{y}$ which is an ordinary eigenvalue problem with the symmetric matrix $\mathbf{L}^{-1}\mathbf{AL}^{-1\mathrm{T}}$.

*Exercises*

4.2.1 Show that if $\lambda$ is an eigenvalue of an orthogonal matrix then so is $\lambda^{-1}$.

4.2.2 Show that if $\mathbf{A}$ is similar to $\mathbf{B}$ and $\mathbf{B}$ is similar to $\mathbf{C}$ then $\mathbf{A}$ is similar to $\mathbf{C}$.

4.2.3 If

$$\mathbf{S} = \begin{bmatrix} 0 & 1 & 1 \\ 1 & 0 & 1 \\ 1 & 1 & 0 \end{bmatrix} \quad \text{and} \quad \mathbf{A} = \begin{bmatrix} b+c & c-a & b-a \\ c-b & c+a & a-b \\ b-c & a-c & a+b \end{bmatrix}$$

show that $\mathbf{SAS}^{-1}$ is diagonal; hence obtain the eigenvalues and eigenvectors of $\mathbf{A}$, checking that the sum of the eigenvalues is consistent with Theorem 4.10.

4.2.4 Use the matrices

$$\mathbf{A} = \begin{bmatrix} 2 & 1 & -1 \\ 0 & -2 & -2 \\ 1 & 1 & 0 \end{bmatrix} \quad \text{and} \quad \mathbf{B} = \begin{bmatrix} 5 & 2 & 2 \\ 2 & 2 & 1 \\ 2 & 1 & 2 \end{bmatrix}$$

to illustrate as many of the results of this section as possible.

## 4.3 The Cayley-Hamilton Theorem and Matrix Functions

In several applications of matrix algebra it becomes necessary to calculate matrix polynomials, and indeed to generalize such polynomials to infinite power series of matrices. In this section we deal with the definition of such series and, very briefly, with questions of convergence. First, however, we introduce the Cayley-Hamilton theorem which will prove invaluable in the practical computation of both matrix polynomials and power series.

THEOREM 4.17

Any square matrix $\mathbf{A}$ satisfies its own characteristic equation in the sense that if

$$\det(\mathbf{A} - \lambda\mathbf{I}) = c_n\lambda^n + c_{n-1}\lambda^{n-1} + \ldots + c_1\lambda + c_0$$

so that the characteristic equation is

$$c_n\lambda^n + c_{n-1}\lambda^{n-1} + \ldots + c_1\lambda + c_0 = 0$$

then $\mathbf{A}$ satisfies the equation

$$c_n\mathbf{A}^n + c_{n-1}\mathbf{A}^{n-1} + \ldots + c_1\mathbf{A} + c_0\mathbf{I} = \mathbf{0}.$$

*Proof*

Consider the adjoint of $\mathbf{A} - \lambda\mathbf{I}$, defined in Definition 2.18. Since the adjoint is the transposed matrix of cofactors, it follows that each element of $\text{adj}(\mathbf{A} - \lambda\mathbf{I})$ must be a polynomial of degree $(n - 1)$ or less in $\lambda$. Thus we may write

$$\text{adj}(\mathbf{A} - \lambda\mathbf{I}) = \mathbf{A}_{n-1}\lambda^{n-1} + \mathbf{A}_{n-2}\lambda^{n-2} + \ldots + \mathbf{A}_1\lambda + \mathbf{A}_0 \tag{4.3.1}$$

where the matrices $\mathbf{A}_i$ are all independent of $\lambda$. However, we know from the property of the inverse (Theorem 2.19) that

$$\{\text{adj}(\mathbf{A} - \lambda\mathbf{I})\}\,(\mathbf{A} - \lambda\mathbf{I}) = \det(\mathbf{A} - \lambda\mathbf{I})\mathbf{I}$$

and hence, using equation (4.3.1) and the form for $\det(\mathbf{A} - \lambda\mathbf{I})$ introduced in the statement of the theorem, we have

$$(\mathbf{A}_{n-1}\lambda^{n-1} + \mathbf{A}_{n-2}\lambda^{n-2} + \ldots + \mathbf{A}_1\lambda + \mathbf{A}_0)(\mathbf{A} - \lambda\mathbf{I})$$
$$= (c_n\lambda^n + c_{n-1}\lambda^{n-1} + \ldots + c_1\lambda + c_0)\mathbf{I}. \tag{4.3.2}$$

Equation (4.3.2) is an identity in $\lambda$, true for all values of $\lambda$; it follows that the coefficients of the same power of $\lambda$ on both sides must be equal. Hence we have

$$\begin{aligned}
c_n\mathbf{I} &= -\mathbf{A}_{n-1}\\
c_{n-1}\mathbf{I} &= \mathbf{A}_{n-1}\mathbf{A} - \mathbf{A}_{n-2}\\
c_{n-2}\mathbf{I} &= \mathbf{A}_{n-2}\mathbf{A} - \mathbf{A}_{n-3}\\
&\vdots\\
c_1\mathbf{I} &= \mathbf{A}_1\mathbf{A} - \mathbf{A}_0\\
c_0\mathbf{I} &= \mathbf{A}_0\mathbf{A}.
\end{aligned} \tag{4.3.3}$$

If we multiply the first equation of (4.3.3) by $\mathbf{A}^n$, the second by $\mathbf{A}^{n-1}$, the third by $\mathbf{A}^{n-2}$ etc, and then add the resulting equations, we obtain

$$c_n\mathbf{A}^n + c_{n-1}\mathbf{A}^{n-1} + c_{n-2}\mathbf{A}^{n-2} + \ldots + c_1\mathbf{A} + c_0\mathbf{I} = \mathbf{0}$$

since the terms on the right-hand side cancel in pairs. Thus the required result is proved. ■

*Example*

4.3.1 Verify the Cayley–Hamilton theorem for the matrix

$$\mathbf{A} = \begin{bmatrix} 2 & -1 \\ 1 & 0 \end{bmatrix}$$ and use it to calculate $\mathbf{A}^4$ and $\mathbf{A}^{-1}$.

The characteristic equation is given by $\det(\mathbf{A} - \lambda\mathbf{I}) = 0$, yielding

$$\begin{bmatrix} 2-\lambda & -1 \\ 1 & -\lambda \end{bmatrix} = 0$$

and hence

$$\lambda^2 - 2\lambda + 1 = 0.$$

We must verify that $\mathbf{A}$ satisfies the equation

$$\mathbf{A}^2 - 2\mathbf{A} + \mathbf{I} = \mathbf{0}.$$

But

$$\mathbf{A} = \begin{bmatrix} 2 & -1 \\ 1 & 0 \end{bmatrix}$$

so

$$\begin{aligned}
\mathbf{A}^2 &= \begin{bmatrix} 2 & -1 \\ 1 & 0 \end{bmatrix} \begin{bmatrix} 2 & -1 \\ 1 & 0 \end{bmatrix}\\
&= \begin{bmatrix} 3 & -2 \\ 2 & -1 \end{bmatrix}
\end{aligned}$$

and hence

$$\mathbf{A}^2 - 2\mathbf{A} + \mathbf{I} = \begin{bmatrix} 3 & -2 \\ 2 & -1 \end{bmatrix} - 2\begin{bmatrix} 2 & -1 \\ 1 & 0 \end{bmatrix} + \begin{bmatrix} 1 & 0 \\ 0 & 1 \end{bmatrix}$$

$$= \begin{bmatrix} 3 - 4 + 1 & -2 + 2 \\ 2 - 2 & -1 + 1 \end{bmatrix}$$

$$= \begin{bmatrix} 0 & 0 \\ 0 & 0 \end{bmatrix}$$

$$= \mathbf{0}, \qquad \text{as required.}$$

Also, since $\mathbf{A}^2 = 2\mathbf{A} - \mathbf{I}$ we notice that

$$\begin{aligned} \mathbf{A}^4 &= (2\mathbf{A} - \mathbf{I})(2\mathbf{A} - \mathbf{I}) \\ &= 4\mathbf{A}^2 - 4\mathbf{A} + \mathbf{I} \\ &= 4(2\mathbf{A} - \mathbf{I}) - 4\mathbf{A} + \mathbf{I} \\ &= 4\mathbf{A} - 3\mathbf{I} \\ &= 4\begin{bmatrix} 2 & -1 \\ 1 & 0 \end{bmatrix} - 3\begin{bmatrix} 1 & 0 \\ 0 & 1 \end{bmatrix} \\ &= \begin{bmatrix} 5 & -4 \\ 4 & -3 \end{bmatrix}. \end{aligned}$$

To obtain $\mathbf{A}^{-1}$ we notice that multiplying throughout the equation

$$\mathbf{A}^2 - 2\mathbf{A} + \mathbf{I} = \mathbf{0}$$

by $\mathbf{A}^{-1}$ gives

$$\mathbf{A} - 2\mathbf{I} + \mathbf{A}^{-1} = \mathbf{0}$$

and hence

$$\begin{aligned} \mathbf{A}^{-1} &= -\mathbf{A} + 2\mathbf{I} \\ &= -\begin{bmatrix} 2 & -1 \\ 1 & 0 \end{bmatrix} + 2\begin{bmatrix} 1 & 0 \\ 0 & 1 \end{bmatrix} \\ &= \begin{bmatrix} 0 & 1 \\ -1 & 2 \end{bmatrix}. \end{aligned}$$

The above is a simple example of one of the most common uses of the Cayley-Hamilton theorem; we expressed both $\mathbf{A}^4$ and $\mathbf{A}^{-1}$ in terms of $\mathbf{A}$ and used these expressions for computational purposes. These results can be generalized as follows.

If $\mathbf{A}$ is a square matrix of order $n$ then any positive power of $\mathbf{A}$ (or negative power if $\mathbf{A}$ is non-singular) can be expressed as a polynomial of degree $n - 1$ in $\mathbf{A}$.

We now illustrate this idea with a further example using the Cayley-Hamilton theorem to calculate matrix powers.

*Example*

4.3.2 Calculate $\mathbf{A}^{99}$ where

$$\text{(i) } \mathbf{A} = \begin{bmatrix} 1 & 2 \\ -1 & 4 \end{bmatrix}, \qquad \text{(ii) } \mathbf{A} = \begin{bmatrix} 2 & -1 \\ 1 & 0 \end{bmatrix}.$$

(i) We readily verify that

$$\det(\mathbf{A} - \lambda\mathbf{I}) = \lambda^2 - 5\lambda + 6$$

giving the characteristic equation

$$\lambda^2 - 5\lambda + 6 = 0$$

with roots $\lambda = 3, 2$ (and hence the eigenvalues of $\mathbf{A}$ are 3 and 2).

Now, by using the Cayley–Hamilton theorem we know that

$$\mathbf{A}^{99} = c_1\mathbf{A} + c_0\mathbf{I}$$

where $c_1$ and $c_0$ have still to be determined.

But if $\mathbf{x}^{(1)}$ and $\mathbf{x}^{(2)}$ are eigenvectors corresponding to $\lambda_1 = 3$ and $\lambda_1 = 2$ respectively, we obtain

$$\mathbf{A}^{99}\mathbf{x}^{(1)} = c_1\mathbf{A}\mathbf{x}^{(1)} + c_0\mathbf{I}\mathbf{x}^{(1)}$$

giving

$$\lambda_1^{99}\mathbf{x}^{(1)} = c_1\lambda_1\mathbf{x}^{(1)} + c_0\mathbf{x}^{(1)}$$

and hence, since $\mathbf{x}^{(1)} \neq \mathbf{0}$,

$$\lambda_1^{99} = c_1\lambda_1 + c_0.$$

Similarly we also obtain

$$\lambda_2^{99} = c_1\lambda_2 + c_0$$

and thus we have the two equations

$$3^{99} = 3c_1 + c_0$$
$$2^{99} = 2c_1 + c_0$$

which solve to give

$$c_1 = 3^{99} - 2^{99} \simeq 3^{99} = 1{\cdot}7 \times 10^{47}$$
$$c_0 = 3^{99} - 3^{100} + 3 \times 2 \simeq -2 \times 3^{99} = -3{\cdot}4 \times 10^{47}$$

and hence

$$\mathbf{A}^{99} = 1{\cdot}7 \times 10^{47} \begin{bmatrix} 1 & 2 \\ -1 & 4 \end{bmatrix} - 3{\cdot}4 \times 10^{47} \begin{bmatrix} 1 & 0 \\ 0 & 1 \end{bmatrix}$$

$$= 1{\cdot}7 \times 10^{47} \begin{bmatrix} -1 & 2 \\ -1 & 2 \end{bmatrix}.$$

(ii) Here the characteristic equation can be readily shown to be $(\lambda - 1)^2 = 0$ with coincident roots $\lambda_1 = \lambda_2 = 1$. Thus the above method is no longer valid, since it would not lead to two distinct equations for $c_1$ and $c_0$. However, it can be modified as follows.

The Cayley–Hamilton theorem still gives us that

$$\mathbf{A}^{99} = c_1\mathbf{A} + c_0\mathbf{I}.$$

Multiplying on the right by an eigenvector with eigenvalue $\lambda$ gives, as before

$$\lambda^{99} = c_1\lambda + c_0 \tag{4.3.4}$$

and setting $\lambda = 1$ gives

$$1 = c_1 + c_0.$$

Because the characteristic equation has a double root it follows that we may obtain a second equation for $c_1$ and $c_0$ by differentiating equation (4.3.4) with respect to $\lambda$ and then setting $\lambda = 1$ (for a full justification of this procedure see example 4.3.5). Thus we obtain $99\lambda^{98} = c_1$ with $\lambda = 1$; hence $c_1 = 99$ and $c_0 = -98$.

So we obtain

$$\mathbf{A}^{99} = 99\begin{bmatrix} 1 & 2 \\ -1 & 4 \end{bmatrix} - 98\begin{bmatrix} 1 & 0 \\ 0 & 1 \end{bmatrix}$$

$$= \begin{bmatrix} 1 & 198 \\ -99 & 298 \end{bmatrix}.$$

The method, carried out in detail for two square matrices of order 2 in the above example can be generalized in the following manner.

Suppose $\mathbf{A}$ is a square matrix of order $n$ with eigenvalues $\lambda_1, \lambda_2 \ldots, \lambda_n$ and it is desired to calculate $F(\mathbf{A}) = \alpha_m\mathbf{A}^m + \alpha_{m-1}\mathbf{A}^{m-1} + \ldots + \alpha_1\mathbf{A} + \alpha_0\mathbf{I}$ for $m > n$. Then by the Cayley–Hamilton theorem we have $F(\mathbf{A}) = c_{n-1}\mathbf{A}^{n-1} + c_{n-2}\mathbf{A}^{n-2} + \ldots + c_1\mathbf{A} + c_0\mathbf{I}$. If the eigenvalues $\lambda_1, \lambda_2, \ldots, \lambda_n$ are all distinct then the constants $c_{n-1}$, $c_{n-2}, \ldots, c_1, c_0$ can be obtained from the $n$ linear equations

$$F(\lambda_1) = c_{n-1}\lambda_1^{n-1} + c_{n-2}\lambda_1^{n-2} + \ldots + c_1\lambda_1 + c_0$$
$$F(\lambda_2) = c_{n-1}\lambda_2^{n-1} + c_{n-2}\lambda_2^{n-2} + \ldots + c_1\lambda_2 + c_0$$
$$\vdots$$
$$F(\lambda_n) = c_{n-1}\lambda_n^{n-1} + c_{n-2}\lambda_n^{n-2} + \ldots + c_1\lambda_n + c_0.$$

If a subset $\lambda_{p_1}, \lambda_{p_2}, \ldots, \lambda_{pk}$ are coincident ($= \lambda_p$, say) then the equations $F(\lambda) = \Sigma_{i=0}^{n-1} c_i\lambda^i$ with $\lambda = \lambda_{p_1}, \lambda_{p_2}, \ldots$ (which would all be identical) have to be replaced by the equations

$$F(\lambda) = \sum_{i=0}^{n-1} c_i\lambda^i$$

$$\frac{\mathrm{d}F}{\mathrm{d}\lambda} = \frac{\mathrm{d}}{\mathrm{d}\lambda}\sum_{i=0}^{n-1} c_i\lambda^i$$

$$\vdots$$

$$\frac{\mathrm{d}^{k-1}F}{\mathrm{d}\lambda^{k-1}} = \frac{\mathrm{d}^{k-1}}{\mathrm{d}\lambda^{k-1}}\sum_{i=0}^{n-1} c_i\lambda^i$$

in each of which $\lambda$ is set equal to $\lambda_p$.

We note that the method described above depends on knowing the eigenvalues, but neither the corresponding eigenvectors nor the characteristic equation. Thus, provided we have some method for determining the eigenvalues, no direct use needs to be made of the characteristic equation.

As mentioned before, we need also to consider the question of infinite series of matrices; we then need to investigate questions of convergence. This may be readily defined in terms of convergence of usual algebraic sequences.

DEFINITION 4.4

The sequence of matrices $\mathbf{A}^{(1)}, \mathbf{A}^{(2)}, \mathbf{A}^{(3)}, \ldots, \mathbf{A}^{(k)} \ldots$ is said to converge to the matrix $\mathbf{A}$ if every sequence of elements $A_{ij}^{(k)}$ converges to $A_{ij}$. □

DEFINITION 4.5

The infinite series $\mathbf{A}^{(1)} + \mathbf{A}^{(2)} + \mathbf{A}^{(3)} + \ldots + \mathbf{A}^{(k)} + \ldots$ is said to be convergent with sum $\mathbf{S}$ if the sequence of partial sums $\mathbf{S}^{(n)} = \mathbf{A}^{(1)} + \mathbf{A}^{(2)} + \ldots + \mathbf{A}^{(n)}$ converges to $\mathbf{S}$. □

We next state the central theorem concerning the convergence of a matrix power series; its proof lies beyond the scope of this book.

THEOREM 4.18

Let

$$F(z) = c_0 + c_1 z + c_2 z^2 + \ldots + c_k z^k + \ldots$$

be a power series with radius of convergence $R$ (i.e. it is absolutely convergent for $|z| < R$). Then if $\mathbf{A}$ is a square matrix of order $n$ with eigenvalues $\lambda_i$ $(1 \leqslant i \leqslant n)$ the matrix series

$$F(\mathbf{A}) = c_0\mathbf{I} + c_1\mathbf{A} + c_2\mathbf{A}^2 + \ldots + c_k\mathbf{A}^k + \ldots$$

will be convergent provided $|\lambda_i| < R$ for all $i$. ■

One of the most important matrix power series is the exponential series defined by

$$\exp(\mathbf{A}) = \mathbf{I} + \mathbf{A} + \frac{1}{2!}\mathbf{A}^2 + \ldots + \frac{1}{k!}\mathbf{A}^k + \ldots.$$

This arises particularly in the study of linear differential equations. Since the power series

$$\exp(z) = \sum_{k=0}^{\infty} \frac{1}{k!} z^k$$

is known to be absolutely convergent for all $z$, it follows from Theorem 4.18 that $\exp(\mathbf{A})$ is convergent for all matrices $\mathbf{A}$ (since the radius of convergence is infinite, all eigenvalues have the required property since they are finite).

Convergent matrix series may now be calculated using exactly the same method as that described above for polynomials. We illustrate with some examples.

*Examples*

4.3.3 Given that the eigenvalues of

$$\mathbf{A} = \begin{bmatrix} -5 & -12 & -16 \\ 6 & 10 & 8 \\ -2 & -2 & 1 \end{bmatrix} \qquad \text{are } 1, 2, 3$$

find (i) $\mathbf{A}^{10} - 9\mathbf{A}^8$ and (ii) $\exp(\mathbf{A})$.

(i) We know, from the Cayley–Hamilton theorem, that

$$\mathbf{A}^{10} - 9\mathbf{A}^8 = c_2\mathbf{A}^2 + c_1\mathbf{A} + c_0\mathbf{I}$$

and hence, if $\lambda$ is an eigenvalue,

$$\lambda^{10} - 9\lambda^8 = c_2\lambda^2 + c_1\lambda + c_0.$$

Setting $\lambda = 1, 2, 3$ in turn gives

$$\begin{aligned} c_2 + \phantom{2}c_1 + c_0 &= -8 \\ 4c_2 + 2c_1 + c_0 &= -1280 \\ 9c_2 + 3c_2 + c_0 &= 0 \end{aligned}$$

which, when solved, yield $c_2 = 1276$, $c_1 = -5100$, $c_0 = 3816$.

Also

$$\mathbf{A}^2 = \begin{bmatrix} -5 & -12 & -16 \\ 6 & 10 & 8 \\ -2 & -2 & 1 \end{bmatrix} \begin{bmatrix} -5 & -12 & -16 \\ 6 & 10 & 8 \\ -2 & -2 & 1 \end{bmatrix}$$

$$= \begin{bmatrix} -15 & -28 & -32 \\ 14 & 12 & -8 \\ -4 & 2 & 17 \end{bmatrix}$$

and so

$$\mathbf{A}^{10} - 9\mathbf{A}^8 = 1276 \begin{bmatrix} -15 & -28 & -32 \\ 14 & 12 & -8 \\ -4 & 2 & 17 \end{bmatrix} -5100 \begin{bmatrix} -5 & -12 & -16 \\ 6 & 10 & 8 \\ -2 & -2 & 1 \end{bmatrix}$$

$$+ 3816 \begin{bmatrix} 1 & 0 & 0 \\ 0 & 1 & 0 \\ 0 & 0 & 1 \end{bmatrix}$$

$$= \begin{bmatrix} 10176 & 25472 & 40768 \\ -12736 & -31872 & -51008 \\ 5096 & 12752 & 20408 \end{bmatrix}.$$

(ii) Again we have from the Cayley–Hamilton theorem

$$\exp(\mathbf{A}) = c_2\mathbf{A}^2 + c_1\mathbf{A} + c_0\mathbf{I}.$$

Thus, if $\lambda$ is an eigenvalue we have

$$\exp(\lambda) = c_2\lambda^2 + c_1\lambda + c_0$$

giving, in this case,

$$\begin{aligned} c_2 + c_1 + c_0 &= e = 2{\cdot}718 \\ 4c_2 + 2c_1 + c_0 &= e^2 = 7{\cdot}389 \\ 9c_2 + 3c_1 + c_0 &= e^3 = 20{\cdot}086 \end{aligned}$$

which, when solved, give $c_2 = 4{\cdot}013$, $c_1 = -7{\cdot}369$, $c_0 = 6{\cdot}074$. Thus

$$\exp(\mathbf{A}) = 4{\cdot}013 \begin{bmatrix} -15 & -28 & -32 \\ 14 & 12 & -8 \\ -4 & 2 & 17 \end{bmatrix} - 7{\cdot}369 \begin{bmatrix} -5 & -12 & -16 \\ 6 & 10 & 8 \\ -2 & -2 & 1 \end{bmatrix}$$

$$+ 6{\cdot}074 \begin{bmatrix} 1 & 0 & 0 \\ 0 & 1 & 0 \\ 0 & 0 & 1 \end{bmatrix}.$$

$$= \begin{bmatrix} -17{\cdot}276 & -23{\cdot}936 & -10{\cdot}512 \\ 11{\cdot}968 & -19{\cdot}460 & -91{\cdot}056 \\ -1{\cdot}314 & 22{\cdot}764 & 66{\cdot}926 \end{bmatrix}.$$

4.3.4 Find $\exp(\mathbf{A})$ when

$$\mathbf{A} = \begin{bmatrix} 1 & 0 & 0 \\ 7 & 1 & 0 \\ -5 & 8 & 1 \end{bmatrix}.$$

Here $\mathbf{A}$ is triangular; thus (Theorem 4.9) its eigenvalues are its diagonal elements and so are given by $\lambda_1 = \lambda_2 = \lambda_3 = 1$.

By the Cayley–Hamilton theorem we then have

$$\exp(\mathbf{A}) = c_2\mathbf{A}^2 + c_1\mathbf{A} + c_0\mathbf{I},$$

and if $\lambda$ is an eigenvalue then

$$e^\lambda = c_2\lambda^2 + c_1\lambda + c_0.$$

Since $\lambda$ is a triply-repeated eigenvalue, we will also have the first and second derivatives of this relationship:

$$\begin{aligned} e^\lambda &= 2c_2\lambda + c_1 \\ e^\lambda &= 2c_2. \end{aligned}$$

Setting $\lambda = 1$ we obtain

$$\begin{aligned} c_2 + c_1 + c_0 &= e \\ 2c_2 + c_1 &= e \\ 2c_2 &= e \end{aligned}$$

which solve to give $c_2 = e/2, c_1 = 0, c_0 = e/2$. Thus

$$\exp(\mathbf{A}) = \tfrac{1}{2}e(\mathbf{A}^2 + \mathbf{I})$$

$$= \frac{e}{2}\begin{bmatrix} 1 & 0 & 0 \\ 14 & 1 & 0 \\ 46 & 16 & 1 \end{bmatrix} + \frac{e}{2}\begin{bmatrix} 1 & 0 & 0 \\ 0 & 1 & 0 \\ 0 & 0 & 1 \end{bmatrix}$$

$$= e\begin{bmatrix} 1 & 0 & 0 \\ 7 & 1 & 0 \\ 23 & 8 & 1 \end{bmatrix}.$$

4.3.5 If $\det(\mathbf{A} - \lambda\mathbf{I}) = d(\lambda)$ and $f(x)$ is any polynomial in $x$, show that $f(\mathbf{A}) = r(\mathbf{A})$ where $r(x)$ is the remainder on division of $f(x)$ by $d(x)$. Show also that $f(\lambda_i) = r(\lambda_i)$ where $\lambda_i$ is any eigenvalue of $\mathbf{A}$ and deduce that if $\lambda_i$ is an eigenvalue of multiplicity $k$ then

$$f(\lambda_i) = r(\lambda_i),\ f'(\lambda_i) = r'(\lambda_i), \ldots \left(\frac{d^{k-1}f}{dx^{k-1}}\right)_{x=\lambda_i} = \left(\frac{d^{k-1}r}{dx^{k-1}}\right)_{x=\lambda_i}.$$

We write $f(x) = q(x)d(x) + r(x)$ in the sense of polynomial division (so that since $d(x)$ is of degree $n$, if $f(x)$ is of degree $m > n$ then $q(x)$ will be of degree $m - n$ and $r(x)$ of degree $n - 1$).

Thus we must have also $f(\mathbf{A}) = q(\mathbf{A})d(\mathbf{A}) + r(\mathbf{A})$.

But by the Cayley–Hamilton theorem we have $d(\mathbf{A}) = \mathbf{0}$ and hence $f(\mathbf{A}) = r(\mathbf{A})$, which was the first result to be proved.

Since $\lambda_i$ is a root of the characteristic equation we must have $d(\lambda_i) = 0$. But

$$f(x) = q(x)d(x) + r(x)$$

so

$$f(\lambda_i) = q(\lambda_i)d(\lambda_i) + r(\lambda_i)$$
$$= r(\lambda_i).$$

Also if $\lambda_i$ is a double root of the characteristic equation then

$$d(\lambda_i) = d'(\lambda_i) = 0.$$

But

$$f'(x) = q'(x)d(x) + q(x)d'(x) + r'(x)$$

so

$$f'(\lambda_i) = q'(\lambda_i)d(\lambda_i) + q(\lambda_i)d'(\lambda_i) + r'(\lambda_i)$$
$$= r'(\lambda_i).$$

Similarly if $\lambda_i$ is a triple root then $d(\lambda_i) = d'(\lambda_i) = d''(\lambda_i) = 0$ and so in addition to the above results we make use of

$$f''(x) = q''(x)d(x) + 2q'(x)d'(x) + q(x)d''(x) + r''(x)$$

to obtain $f''(\lambda_i) = r''(\lambda_i)$.

Similarly if $\lambda_i$ is a repeated root with multiplicity $k$ then

$$d(\lambda_i) = d'(\lambda_i) = d''(\lambda_i) = \ldots = d^{(k-1)}(\lambda_i) = 0$$

leading to

$$\begin{aligned} f(\lambda_i) &= d(\lambda_i) \\ f'(\lambda_i) &= d'(\lambda_i) \\ &\vdots \\ f^{(k-1)}(\lambda_i) &= d^{(k-1)}(\lambda_i). \end{aligned}$$

This is the required result, and provides a justification of the procedure described in the text for dealing with the situation of multiple eigenvalues.

*Exercises*

4.3.1 Verify that the matrix

$$\mathbf{A} = \begin{bmatrix} -5 & -12 & -16 \\ 6 & 10 & 8 \\ -2 & -2 & 1 \end{bmatrix}$$

of Example 4.3.3 satisfies the Cayley–Hamilton theorem, and hence find $\mathbf{A}^{-1}$.

4.3.2 Explain how matrix power series can be used to define sin(**A**) and cos(**A**) for a suitable matrix **A**.

4.3.3 For the matrix

$$\mathbf{A} = \begin{bmatrix} 1 & 2 & 4 \\ 0 & 2 & 5 \\ 0 & 0 & -1 \end{bmatrix}$$

evaluate (i) $\mathbf{A}^6$, (ii) $\mathbf{A}^{10} + 8\mathbf{A}^8$, (iii) exp(**A**), (iv) sin(**A**).

4.3.4 For the matrix

$$\mathbf{A} = \begin{bmatrix} 1 & -2 & 3 \\ 0 & -1 & 1 \\ -1 & 1 & -2 \end{bmatrix}$$

evaluate (i) $\mathbf{A}^{100}$, (ii) exp(**A**).

4.3.5 For

$$\mathbf{A} = \begin{bmatrix} 1 & 1 \\ 1 & 1 \end{bmatrix} \quad \text{and} \quad \mathbf{B} = \begin{bmatrix} 1 & 2 \\ 2 & 1 \end{bmatrix}$$

evaluate exp(**A**) exp(**B**), exp(**A** + **B**) and exp(**B**) exp(**A**). Under what conditions would you expect the result exp(**A** + **B**) = exp(**A**) exp(**B**) = exp(**B**) exp(**A**)
to be true?

4.3.6 Justify the result that $\{\exp(\mathbf{A})\}^{-1} = \exp(-\mathbf{A})$. Deduce that $\exp(\mathbf{A})$ is always non-singular.

The method described above for calculating matrix power series and polynomials involves the solution of a system of simultaneous equations for the coefficients of the powers of the matrix. An alternative procedure in which the coefficients are obtained explicitly is given by the following theorem.

THEOREM 4.19 (Sylvester's theorem)

If $\mathbf{A}$ is a square matrix of order $n$ with distinct eigenvalues $\lambda_1, \lambda_2, \ldots, \lambda_n$ and if

$$f(z) = \sum_{k=0}^{\infty} c_k z^k$$

then

$$f(\mathbf{A}) = \sum_{i=1}^{n} f(\lambda_i) \frac{(\mathbf{A}-\lambda_1\mathbf{I})(\mathbf{A}-\lambda_2\mathbf{I})\ldots(\mathbf{A}-\lambda_{i-1}\mathbf{I})(\mathbf{A}-\lambda_{i+1}\mathbf{I})\ldots(\mathbf{A}-\lambda_n\mathbf{I})}{(\lambda_i-\lambda_1)(\lambda_i-\lambda_2)\ldots(\lambda_i-\lambda_{i-1})(\lambda_i-\lambda_{i+1})\ldots(\lambda_i-\lambda_n)}$$

$$\left(= \sum_{i=1}^{n} f(\lambda_i) \prod_{\substack{j=1 \\ j\neq i}}^{n} \frac{(\mathbf{A}-\lambda_j\mathbf{I})}{(\lambda_i-\lambda_j)}\right).$$

*Proof*

We provide a proof for the case when $n = 3$; the generalization to arbitrary $n$ is trivial but the details are cumbersome and add nothing to the understanding of the problem. Thus, given a square matrix $\mathbf{A}$ of order 3 with distinct eigenvalues $\lambda_1, \lambda_2, \lambda_3$ and $f(z) = \sum_{k=0}^{\infty} c_k k^k$ we wish to show that

$$f(\mathbf{A}) = f(\lambda_1)\frac{(\mathbf{A}-\lambda_2\mathbf{I})(\mathbf{A}-\lambda_3\mathbf{I})}{(\lambda_1-\lambda_2)(\lambda_1-\lambda_3)} + f(\lambda_2)\frac{(\mathbf{A}-\lambda_1\mathbf{I})(\mathbf{A}-\lambda_3\mathbf{I})}{(\lambda_2-\lambda_1)(\lambda_2-\lambda_3)}$$

$$+ f(\lambda_3)\frac{(\mathbf{A}-\lambda_1\mathbf{I})(\mathbf{A}-\lambda_2\mathbf{I})}{(\lambda_3-\lambda_1)(\lambda_3-\lambda_2)}. \tag{4.3.5}$$

Now, we know by Theorem 4.12 that there exists a matrix $\mathbf{T}$ such that

$$\mathbf{T}^{-1}\mathbf{A}\mathbf{T} = \mathbf{D} \qquad \text{where } \mathbf{D} = \begin{bmatrix} \lambda_1 & 0 & 0 \\ 0 & \lambda_2 & 0 \\ 0 & 0 & \lambda_3 \end{bmatrix}.$$

Then

$$\begin{aligned} \mathbf{T}^{-1}\mathbf{A}^k\mathbf{T} &= \mathbf{T}^{-1}\mathbf{A}\mathbf{A}\ldots\mathbf{A}\mathbf{T} \\ &= \mathbf{T}^{-1}\mathbf{A}(\mathbf{T}\mathbf{T}^{-1})\mathbf{A}(\mathbf{T}\mathbf{T}^{-1})\ldots(\mathbf{T}\mathbf{T}^{-1})\mathbf{A}\mathbf{T} \\ &= (\mathbf{T}^{-1}\mathbf{A}\mathbf{T})(\mathbf{T}^{-1}\mathbf{A}\mathbf{T})\ldots(\mathbf{T}^{-1}\mathbf{A}\mathbf{T}) \\ &= (\mathbf{T}^{-1}\mathbf{A}\mathbf{T})^k \\ &= \mathbf{D}^k \\ &= \begin{bmatrix} \lambda_1^k & 0 & 0 \\ 0 & \lambda_2^k & 0 \\ 0 & 0 & \lambda_3^k \end{bmatrix}. \end{aligned}$$

Thus it follows that

$$\mathbf{T}^{-1}f(\mathbf{A})\mathbf{T} = f(\mathbf{T}^{-1}\mathbf{A}\mathbf{T})$$
$$= f(\mathbf{D})$$
$$= \begin{bmatrix} f(\lambda_1) & 0 & 0 \\ 0 & f(\lambda_2) & 0 \\ 0 & 0 & f(\lambda_3) \end{bmatrix}.$$

If we now denote the right-hand side of equation (4.3.5) by **B**, the required result will be proved if we can show that $\mathbf{T}^{-1}\mathbf{B}\mathbf{T} = f(\mathbf{D})$. But

$$\mathbf{T}^{-1}\mathbf{B}\mathbf{T} = \frac{f(\lambda_1)}{(\lambda_1 - \lambda_2)(\lambda_1 - \lambda_3)}\mathbf{T}^{-1}(\mathbf{A} - \lambda_2\mathbf{I})(\mathbf{A} - \lambda_3\mathbf{I})\mathbf{T}$$

$$+ \frac{f(\lambda_2)}{(\lambda_2 - \lambda_1)(\lambda_2 - \lambda_3)}\mathbf{T}^{-1}(\mathbf{A} - \lambda_1\mathbf{I})(\mathbf{A} - \lambda_3\mathbf{I})\mathbf{T}$$

$$+ \frac{f(\lambda_3)}{(\lambda_3 - \lambda_1)(\lambda_3 - \lambda_2)}\mathbf{T}^{-1}(\mathbf{A} - \lambda_1\mathbf{I})(\mathbf{A} - \lambda_2\mathbf{I})\mathbf{T}$$

$$= \frac{f(\lambda_1)}{(\lambda_1 - \lambda_2)(\lambda_1 - \lambda_3)}(\mathbf{D} - \lambda_2\mathbf{I})(\mathbf{D} - \lambda_3\mathbf{I})$$

$$+ \frac{f(\lambda_2)}{(\lambda_2 - \lambda_1)(\lambda_2 - \lambda_3)}(\mathbf{D} - \lambda_1\mathbf{I})(\mathbf{D} - \lambda_3\mathbf{I})$$

$$+ \frac{f(\lambda_3)}{(\lambda_3 - \lambda_1)(\lambda_3 - \lambda_2)}(\mathbf{D} - \lambda_1\mathbf{I})(\mathbf{D} - \lambda_2\mathbf{I})$$

$$= \frac{f(\lambda_1)}{(\lambda_1 - \lambda_2)(\lambda_1 - \lambda_3)}\begin{bmatrix} \lambda_1 - \lambda_2 & 0 & 0 \\ 0 & 0 & 0 \\ 0 & 0 & \lambda_3 - \lambda_2 \end{bmatrix}\begin{bmatrix} \lambda_1 - \lambda_3 & 0 & 0 \\ 0 & \lambda_2 - \lambda_3 & 0 \\ 0 & 0 & 0 \end{bmatrix}$$

$$+ \frac{f(\lambda_2)}{(\lambda_2 - \lambda_1)(\lambda_2 - \lambda_3)}\begin{bmatrix} 0 & 0 & 0 \\ 0 & \lambda_2 - \lambda_1 & 0 \\ 0 & 0 & \lambda_3 - \lambda_1 \end{bmatrix}\begin{bmatrix} \lambda_1 - \lambda_3 & 0 & 0 \\ 0 & \lambda_2 - \lambda_3 & 0 \\ 0 & 0 & 0 \end{bmatrix}$$

$$+ \frac{f(\lambda_3)}{(\lambda_3 - \lambda_1)(\lambda_3 - \lambda_2)}\begin{bmatrix} 0 & 0 & 0 \\ 0 & \lambda_2 - \lambda_1 & 0 \\ 0 & 0 & \lambda_3 - \lambda_1 \end{bmatrix}\begin{bmatrix} \lambda_1 - \lambda_2 & 0 & 0 \\ 0 & 0 & 0 \\ 0 & 0 & \lambda_3 - \lambda_2 \end{bmatrix}$$

$$= \frac{f(\lambda_1)}{(\lambda_1 - \lambda_2)(\lambda_1 - \lambda_3)}\begin{bmatrix} (\lambda_1 - \lambda_2)(\lambda_1 - \lambda_3) & 0 & 0 \\ 0 & 0 & 0 \\ 0 & 0 & 0 \end{bmatrix}$$

$$+\frac{f(\lambda_2)}{(\lambda_2-\lambda_1)(\lambda_2-\lambda_3)}\begin{bmatrix}0 & 0 & 0\\ 0 & (\lambda_2-\lambda_1)(\lambda_2-\lambda_3) & 0\\ 0 & 0 & 0\end{bmatrix}$$

$$+\frac{f(\lambda_3)}{(\lambda_3-\lambda_1)(\lambda_3-\lambda_2)}\begin{bmatrix}0 & 0 & 0\\ 0 & 0 & 0\\ 0 & 0 & (\lambda_3-\lambda_1)(\lambda_3-\lambda_2)\end{bmatrix}$$

$$=\begin{bmatrix}f(\lambda_1) & 0 & 0\\ 0 & f(\lambda_2) & 0\\ 0 & 0 & f(\lambda_3)\end{bmatrix}.$$

Thus the required result is proved (since $\mathbf{T}^{-1}f(\mathbf{A})\mathbf{T} = \mathbf{T}^{-1}\mathbf{B}\mathbf{T}$ it follows that $f(\mathbf{A}) = \mathbf{B}$) and the proof is complete. ■

*Example*

4.3.5 Use Sylvester's theorem for the matrix

$$\mathbf{A}=\begin{bmatrix}-5 & -12 & -16\\ 6 & 10 & 8\\ -2 & -2 & 1\end{bmatrix}$$

of Example 4.3.3 to determine $\exp(\mathbf{A})$; the eigenvalues of $\mathbf{A}$ are 1,2,3.

Since the eigenvalues are distinct we can use Sylvester's theorem. It gives

$$\exp(\mathbf{A}) = e^{\lambda_1}\frac{(\mathbf{A}-\lambda_2\mathbf{I})(\mathbf{A}-\lambda_3\mathbf{I})}{(\lambda_1-\lambda_2)(\lambda_1-\lambda_3)} + e^{\lambda_2}\frac{(\mathbf{A}-\lambda_1\mathbf{I})(\mathbf{A}-\lambda_3\mathbf{I})}{(\lambda_2-\lambda_1)(\lambda_2-\lambda_3)}$$
$$+ e^{\lambda_3}\frac{(\mathbf{A}-\lambda_1\mathbf{I})(\mathbf{A}-\lambda_2\mathbf{I})}{(\lambda_3-\lambda_1)(\lambda_3-\lambda_2)}.$$

Setting $\lambda_1 = 1, \lambda_2 = 2, \lambda_3 = 3$ this gives

$$\exp(\mathbf{A}) = \tfrac{1}{2}e(\mathbf{A}-2\mathbf{I})(\mathbf{A}-3\mathbf{I}) - e^2(\mathbf{A}-\mathbf{I})(\mathbf{A}-3\mathbf{I}) + \tfrac{1}{2}e^3(\mathbf{A}-\mathbf{I})(\mathbf{A}-2\mathbf{I})$$
$$= \tfrac{1}{2}e(\mathbf{A}^2 - 5\mathbf{A} + 6\mathbf{I}) - e^2(\mathbf{A}^2 - 4\mathbf{A} + 3\mathbf{I}) + \tfrac{1}{2}e^3(\mathbf{A}^2 - 3\mathbf{A} + 2\mathbf{I})$$
$$= (\tfrac{1}{2}e - e^2 + \tfrac{1}{2}e^3)\mathbf{A}^2 + (-\tfrac{5}{2}e + 4e^2 - \tfrac{3}{2}e^3)\mathbf{A} + (3e - 3e^2 + e^3)\mathbf{I}.$$

Substituting

$e = 2{\cdot}718$

$e^2 = 7{\cdot}389$

$e^3 = 20{\cdot}086$

leads to $\exp(\mathbf{A}) = 4{\cdot}013\mathbf{A}^2 - 7{\cdot}369\mathbf{A} + 6{\cdot}074\mathbf{I}$, which agrees with the result previously obtained in Example 4.3.3.

*Exercises*

4.3.6 Use Sylvester's theorem to calculate $\exp(\mathbf{A})$, $\ln(\mathbf{A})$ and $\mathbf{A}^{1/2}$ when

$$\mathbf{A} = \begin{bmatrix} \frac{3}{2} & 1 \\ 0 & \frac{1}{2} \end{bmatrix}.$$

Verify first that $\ln(\mathbf{A})$ and $\mathbf{A}^{1/2}$ are expressible as power series. (Hint:

$$\ln(1+z) = \sum_{k=1}^{\infty} \frac{(-1)^{k-1}}{k} z^k \qquad \text{for } |z| < 1$$

$$(1+z)^n = \sum_{k=0}^{\infty} \binom{n}{k} z^k \qquad \text{for } |z| < 1.)$$

4.3.7 If $\mathbf{A}$ is a square matrix with distinct eigenvalues and $t$ is an arbitrary parameter, show that

$$\det(\exp(\mathbf{A}t)) = e^{(\mathrm{Tr}\,\mathbf{A})t}.$$

4.3.8 If

$$\mathbf{A} = \begin{bmatrix} \pi/2 & 3\pi/2 \\ \pi & \pi \end{bmatrix} \qquad \text{find } \sin(\mathbf{A}).$$

## 4.4 Quadratic Forms

We have already introduced in Section 2.11 the idea of a quadratic form $Q = \mathbf{x}^{\mathrm{T}}\mathbf{A}\mathbf{x}$, (where $\mathbf{x}$ is a column vector and $\mathbf{A}$ a real symmetric matrix). In the problems for which quadratic forms are useful, it often happens that $\mathbf{x}$ is a vector of coordinates used to describe the system; coordinates other than $\mathbf{x}$ may be more useful in the sense that equations describing the system are more readily solved in terms of these new coordinates. The simplest relationship between the coordinate vector $\mathbf{x}$ and a new vector $\mathbf{y}$ is a linear one which, when expressed in matrix form, is $\mathbf{x} = \mathbf{T}\mathbf{y}$ with $\mathbf{T}$ non-singular, so that we also have $\mathbf{y} = \mathbf{T}^{-1}\mathbf{x}$. Expressed in terms of the new coordinates the quadratic form is $Q = \mathbf{y}^{\mathrm{T}}\mathbf{T}^{\mathrm{T}}\mathbf{A}\mathbf{T}\mathbf{y}$. However, by Theorem 4.15 we know that since $\mathbf{A}$ is real and symmetric if we choose $\mathbf{T}$ to be the matrix of eigenvectors of $\mathbf{A}$ then $\mathbf{T}^{\mathrm{T}}\mathbf{A}\mathbf{T} = \mathbf{D}$ where $\mathbf{D}$ is diagonal with the eigenvalues of $\mathbf{A}$ as diagonal elements. Denoting these eigenvalues by $\lambda_1, \lambda_2, \ldots, \lambda_n$ we obtain

$$Q = [y_1 \quad y_2 \quad \ldots \quad y_n] \begin{bmatrix} \lambda_1 & 0 & \ldots & 0 \\ 0 & \lambda_2 & & \\ \vdots & & \ddots & \\ 0 & & & \lambda_n \end{bmatrix} \begin{bmatrix} y_1 \\ y_2 \\ \vdots \\ y_n \end{bmatrix}$$

$$= \lambda_1 y_1^2 + \lambda_2 y_2^2 + \ldots + \lambda_n y_n^2$$

$$= \sum_{i=1}^{n} \lambda_i y_i^2.$$

We have noted before that a positive definite quadratic form $Q = \mathbf{x}^{\mathrm{T}}\mathbf{A}\mathbf{x}$ is one which is positive for all vectors $\mathbf{x}$; then the matrix $\mathbf{A}$ is also said to be positive definite. From the above expression for $Q$ we may immediately deduce the following theorem.

THEOREM 4.20

The quadratic form $Q = \mathbf{x}^{\mathrm{T}}\mathbf{A}\mathbf{x}$ (and the matrix $\mathbf{A}$) is positive definite if and only if $\mathbf{A}$ has positive eigenvalues.

If $\mathbf{A}$ is positive definite, we have from above that we may write $Q = \mathbf{x}^{\mathrm{T}}\mathbf{A}\mathbf{x}$ in the form

$$\begin{aligned} Q &= \mathbf{y}^{\mathrm{T}}\mathbf{D}\mathbf{y} \\ &= \sum_{i=1}^{n} \lambda_i y_i^2 \qquad \text{where } \lambda_i > 0. \end{aligned}$$

By making the further transformation

$$\mathbf{z} = \mathbf{M}\mathbf{y}$$

where

$$\mathbf{M} = \begin{bmatrix} \sqrt{\lambda_1} & 0 & \dots & 0 \\ 0 & \sqrt{\lambda_2} & & \\ \cdot & & \cdot & \\ \cdot & & & \cdot \\ \cdot & & & \cdot \\ 0 & & & \sqrt{\lambda_n} \end{bmatrix}$$

it follows that we may write $Q$ in the form

$$Q = \mathbf{z}^{\mathrm{T}}\mathbf{z} = \sum_{i=1}^{n} z_i^2.$$

This is called the *normal* form of $Q$. However, we note that the transformation is now no longer orthogonal since $\mathbf{M}$ is not orthogonal.

We are now in a position to prove a special property of positive definite matrices, as contained in the following theorem.

THEOREM 4.21

$\mathbf{A}$ is positive definite and symmetric if and only if $\mathbf{A} = \mathbf{P}^{\mathrm{T}}\mathbf{P}$ where $\mathbf{P}$ is non-singular.

*Proof*

If $\mathbf{A}$ is positive definite and symmetric then we have shown above that there exists a (non-singular) transformation matrix $\mathbf{Q}$ such that if $\mathbf{x} = \mathbf{Q}\mathbf{z}$ then

$$\mathbf{x}^{\mathrm{T}}\mathbf{A}\mathbf{x} = \mathbf{z}^{\mathrm{T}}\mathbf{z}.$$

Thus

$$\mathbf{z}^{\mathrm{T}}\mathbf{Q}^{\mathrm{T}}\mathbf{A}\mathbf{Q}\mathbf{z} = \mathbf{z}^{\mathrm{T}}\mathbf{z}$$

and since this is true for all $\mathbf{z}$ we must have

$$\mathbf{Q}^{\mathrm{T}}\mathbf{A}\mathbf{Q} = \mathbf{I}$$

giving $\mathbf{A} = \mathbf{P}^{\mathrm{T}}\mathbf{P}$ with $\mathbf{P} = \mathbf{Q}^{-1}$.

On the other hand, if $\mathbf{A} = \mathbf{P}^{\mathrm{T}}\mathbf{P}$ then the transformation $\mathbf{z} = \mathbf{P}\mathbf{x}$ gives $\mathbf{x}^{\mathrm{T}}\mathbf{A}\mathbf{x} = \mathbf{z}^{\mathrm{T}}\mathbf{z}$ and thus $\mathbf{A}$ is positive definite, as required. ■

In the application of the ideas of the present section to dynamics, it is found that much simplification occurs when the problem is expressed in terms of coordinates which diagonalize the quadratic form. Indeed in some problems we may wish to consider the simultaneous diagonalization of two quadratic forms. That this can in fact be accomplished under certain circumstances is proved in the following theorem.

THEOREM 4.22

Let $\mathbf{x}^{\mathrm{T}}\mathbf{A}\mathbf{x}$ and $\mathbf{x}^{\mathrm{T}}\mathbf{B}\mathbf{x}$ be quadratic forms such that $\mathbf{x}^{\mathrm{T}}\mathbf{B}\mathbf{x}$ is positive definite. Then a non-singular transformation $\mathbf{x} = \mathbf{T}\mathbf{z}$ can be found such that

$$\mathbf{x}^{\mathrm{T}}\mathbf{A}\mathbf{x} = \mu_1 z_1^2 + \mu_2 z_2^2 + \ldots + \mu_n z_n^2$$

and

$$\mathbf{x}^{\mathrm{T}}\mathbf{B}\mathbf{x} = z_1^2 + z_2^2 + \ldots + z_n^2.$$

Also the $\mu_i$ are the roots of $\det(\mathbf{A} - \mu\mathbf{B}) = 0$.

*Proof*

Since $\mathbf{B}$ is positive definite we know that there exists a non-singular transformation $\mathbf{x} = \mathbf{L}\mathbf{y}$ such that

$$\mathbf{x}^{\mathrm{T}}\mathbf{B}\mathbf{x} = \mathbf{y}^{\mathrm{T}}\mathbf{y};$$

then the form $\mathbf{x}^{\mathrm{T}}\mathbf{A}\mathbf{x}$ transforms to

$$\begin{aligned}\mathbf{x}^{\mathrm{T}}\mathbf{A}\mathbf{x} &= \mathbf{y}^{\mathrm{T}}\mathbf{L}^{\mathrm{T}}\mathbf{A}\mathbf{L}\mathbf{y}\\ &= \mathbf{y}^{\mathrm{T}}\mathbf{C}\mathbf{y} \qquad \text{with } \mathbf{C} = \mathbf{L}^{\mathrm{T}}\mathbf{A}\mathbf{L}.\end{aligned}$$

Suppose the eigenvalues of $\mathbf{C}$ (which is real and symmetric) are $\mu_1, \mu_2, \ldots, \mu_n$; then there exists an orthogonal transformation $\mathbf{y} = \mathbf{M}\mathbf{z}$ such that

$$\begin{aligned}\mathbf{y}^{\mathrm{T}}\mathbf{C}\mathbf{y} &= \mathbf{z}^{\mathrm{T}}\mathbf{D}\mathbf{z} \qquad \text{where } \mathbf{D} = \begin{bmatrix} \mu_1 & 0 & \ldots & 0 \\ 0 & \mu_2 & & \\ \cdot & & \cdot & \\ \cdot & & & \cdot \\ \cdot & & & \cdot \\ 0 & & & \mu_n \end{bmatrix}\\ &= \sum_{i=1}^{n} \mu_i z_i^2.\end{aligned}$$

Also

$$\mathbf{x}^T\mathbf{B}\mathbf{x} = \mathbf{y}^T\mathbf{y}$$
$$= \mathbf{z}^T\mathbf{M}^T\mathbf{M}\mathbf{z}$$
$$= \mathbf{z}^T\mathbf{z} \qquad \text{since } \mathbf{M} \text{ is orthogonal.}$$

The required overall transformation is then given by $\mathbf{x} = \mathbf{L}\mathbf{y} = \mathbf{L}\mathbf{M}\mathbf{z}$; so $\mathbf{T} = \mathbf{L}\mathbf{M}$ is the required transformation matrix.

The $\mu_i$ are the eigenvalues of $\mathbf{C}$ so they satisfy the equation $\det(\mathbf{C} - \mu\mathbf{I}) = 0$. But since $\mathbf{C} = \mathbf{L}^T\mathbf{A}\mathbf{L}$ and $\mathbf{L}^T\mathbf{B}\mathbf{L} = \mathbf{I}$ this equation becomes

$$\det\{\mathbf{L}^T(\mathbf{A} - \mu\mathbf{B})\mathbf{L}\} = 0$$

and since $\mathbf{L}$ is non-singular we may deduce that

$$\det(\mathbf{A} - \mu\mathbf{B}) = 0.$$

■

As already mentioned this theorem is of widespread importance in dynamics. In practical applications, it is often not necessary to know the actual transformation matrix $\mathbf{T}$, but merely to be guaranteed that it exists to give the simultaneous diagonalization of the two quadratic forms.

*Examples*

4.4.1 Show that the surface

$$x^2 + y^2 + z^2 - xy + xz - yz = 1$$

represents an ellipsoid, and find the lengths of the axes.

In terms of a suitable coordinate system an ellipsoid has equation

$$\frac{X^2}{a^2} + \frac{Y^2}{b^2} + \frac{Z^2}{c^2} = 1;$$

$a, b, c$ are then the lengths of the semi-axes.

Thus, here we wish to transform the quadratic form $x^2 + y^2 + z^2 - xy + xz - yz$ to a diagonal form; the given equation will represent an ellipsoid if the form is positive definite, and the lengths of the axes can be found directly from the coefficients.

The quadratic form here is

$$x^2 + y^2 + z^2 - xy + xz - yz = [x \quad y \quad z] \begin{bmatrix} 1 & -\frac{1}{2} & \frac{1}{2} \\ -\frac{1}{2} & 1 & -\frac{1}{2} \\ \frac{1}{2} & -\frac{1}{2} & 1 \end{bmatrix} \begin{bmatrix} x \\ y \\ z \end{bmatrix}.$$

Thus the form has matrix

$$\mathbf{A} = \begin{bmatrix} 1 & -\frac{1}{2} & \frac{1}{2} \\ -\frac{1}{2} & 1 & -\frac{1}{2} \\ \frac{1}{2} & -\frac{1}{2} & 1 \end{bmatrix}$$

and so we are interested in the eigenvalues of $\mathbf{A}$. By solving the characteristic equation $\det(\mathbf{A} - \lambda\mathbf{I}) = 0$ we obtain $\lambda_1 = \frac{1}{2}$, $\lambda_2 = \frac{1}{2}$, $\lambda_3 = 2$. Thus, the eigenvalues are all positive, the matrix $\mathbf{A}$ is positive definite and the equation $\mathbf{x}^T\mathbf{A}\mathbf{x} = 1$ represents an ellipsoid.

It follows that there exists a transformation to coordinates $(X, Y, Z)$ such that the equation takes the form $\frac{1}{2}X^2 + \frac{1}{2}Y^2 + 2Z^2 = 1$. By comparison with the standard form of the equation of the ellipsoid the lengths of the semi-axes must be $\sqrt{2}, \sqrt{2}, 1/\sqrt{2}$.

4.4.2 Show that the quadratic forms

$$Q_1 = 2x^2 + 2y^2 + 5z^2 - 8xy - 6xz + 6yz$$
$$Q_2 = 2x^2 + 2y^2 + 2z^2 - 2xz + 2yz$$

can be simultaneously diagonalized, and express them in this form.

We have $Q_1 = \mathbf{x}^T\mathbf{A}\mathbf{x}$ and $Q_2 = \mathbf{x}^T\mathbf{B}\mathbf{x}$ where

$$\mathbf{x} = \begin{bmatrix} x \\ y \\ z \end{bmatrix}, \quad \mathbf{A} = \begin{bmatrix} 2 & -4 & -3 \\ -4 & 2 & 3 \\ -3 & 3 & 5 \end{bmatrix}, \quad \text{and } \mathbf{B} = \begin{bmatrix} 2 & 0 & -1 \\ 0 & 2 & 1 \\ -1 & 1 & 2 \end{bmatrix}.$$

We first require (Theorem 4.22) to show that at least one of **A** and **B** is positive definite The eigenvalues may be found from the characteristic equation in the usual way and we find that:

**A** has eigenvalues $-2, 1{\cdot}33, 9{\cdot}67$.

**B** has eigenvalues $0{\cdot}38, 2{\cdot}62, 3$.

It follows that since the eigenvalues of **B** are all positive, it is positive definite and the required diagonalization is possible.

We thus have

$$Q_1 = \mu_1 X^2 + \mu_2 Y^2 + \mu_3 Z^2$$
$$Q_2 = X^2 + Y^2 + Z^2$$

where $\mu_1, \mu_2, \mu_3$ are the roots of $\det(\mathbf{A} - \mu\mathbf{B}) = 0$. In this case

$$\mathbf{A} - \mu\mathbf{B} = \begin{bmatrix} 2 - 2\mu & -4 & -3 + \mu \\ -4 & 2 - 2\mu & 3 - \mu \\ -3 + \mu & 3 - \mu & 5 - 2\mu \end{bmatrix}$$

and $\det(\mathbf{A} - \mu\mathbf{B}) = 0$ gives $\mu = -1, 3, 0$. Hence the quadratic forms can be expressed as

$$Q_1 = -X^2 + Y^2$$
$$Q_2 = X^2 + Y^2 + Z^2.$$

4.4.3 If

$$\mathbf{A} = \begin{bmatrix} a_{11} & a_{12} & \cdots & a_{1n} \\ a_{21} & a_{22} & \cdots & a_{2n} \\ \cdot & & & \\ \cdot & & & \\ \cdot & & & \\ a_{n1} & a_{n2} & \cdots & a_{nn} \end{bmatrix}$$

and **D** is the diagonal matrix formed from the diagonal elements of **A**, that is

$$\mathbf{D} = \begin{bmatrix} a_{11} & 0 & \dots & 0 \\ 0 & a_{22} & & \\ \cdot & & \cdot & \\ \cdot & & & \cdot \\ \cdot & & & \cdot \\ 0 & & & a_{nn} \end{bmatrix},$$

show that if **A** is positive definite then so too is **D**.

We show that if **A** is positive definite then each $a_{ii} > 0$. We know that $\mathbf{x}^T\mathbf{A}\mathbf{x} > 0$ for all **x**; if we choose **x** so that its $i$th element is 1 and all other elements are zero then $\mathbf{x}^T\mathbf{A}\mathbf{x} = a_{ii}$, thus proving $a_{ii} > 0$ for all $i$.

Thus for arbitrary **x** we have

$$\mathbf{x}^T\mathbf{D}\mathbf{x} = \sum_{i=1}^{n} a_{ii}x_i^2$$

$$> 0 \qquad \text{for all non-zero } \mathbf{x} \text{ since all } a_{ii} > 0.$$

*Exercises*

4.4.1 Find an orthogonal transformation which transforms the quadratic form $2x^2 + y^2 + z^2 + 2x(y + z)$ to diagonal form.

4.4.2 Determine the type of surface represented by the equation

$$4x^2 + 4y^2 + 2z^2 - 4xy + 4xz - 4yz = 1.$$

4.4.3 Show that the following quadratic forms can be simultaneously diagonalized and obtain possible diagonal forms:

$$2x^2 + y^2 - z^2 + 4xy - 2yz$$

$$x^2 + 2y^2 + 2z^2 + 2xy + 2yz.$$

4.4.4 If $Q = \mathbf{x}^T\mathbf{A}\mathbf{x}$ where **A** is symmetric of order $n$ and

$$\mathbf{x} = \begin{bmatrix} x_1 \\ x_2 \\ \cdot \\ \cdot \\ \cdot \\ x_n \end{bmatrix}$$

show that

$$\begin{bmatrix} \dfrac{\partial Q}{\partial x_1} \\ \dfrac{\partial Q}{\partial x_2} \\ \cdot \\ \cdot \\ \cdot \\ \dfrac{\partial Q}{\partial x_n} \end{bmatrix} = 2\mathbf{Ax}.$$

## 4.5 Eigenvalue Bounds and Inequalities

So far, the only method we have for determining eigenvalues (and eigenvectors) is to solve the characteristic equation. We have already mentioned that this method cannot be recommended for matrices of order higher than about 3, and in Section 4.7 we shall be describing briefly some of the practical methods available for the determination of eigenvalues and eigenvectors. For some of these methods it is useful to have a starting approximation and in other contexts it can be useful to have some rough idea of the magnitudes of the eigenvalues obtained in some simple manner. For these reasons we devote this section to a study of results concerned with inequalities involving eigenvalues, most of which will give a bound to the maximum value an eigenvalue of a given matrix can have.

THEOREM 4.23

*If* $\lambda$ *is an eigenvalue of the square matrix* $\mathbf{A}$ *of order* $n$ *then*

$$|\lambda| \leqslant \max_i \sum_{j=1}^{n} |a_{ij}|$$

*and*

$$|\lambda| \leqslant \max_j \sum_{i=1}^{n} |a_{ij}|.$$

(These inequalities just state that the magnitude of every eigenvalue is less than the greatest row or column sum of magnitudes of elements.)

*Proof*

Suppose the eigenvector of $\mathbf{A}$ corresponding to the eigenvalue $\lambda$ is $\mathbf{x}$ and suppose that the component of $\mathbf{x}$ which is largest in magnitude is the $m$th so that $|x_i| \leqslant |x_m|$ for $1 \leqslant i \leqslant n$. But taking the $m$th component of $\mathbf{Ax} = \lambda\mathbf{x}$ gives

$$(\mathbf{Ax})_m = \lambda x_m$$

and thus

$$\sum_{j=1}^{n} a_{mj} x_j = \lambda x_m.$$

So

$$|\lambda x_m| = \left| \sum_{j=1}^{n} a_{mj}x_j \right|$$

giving

$$|\lambda|\,|x_m| \leqslant \sum_{j=1}^{n} |a_{mj}|\,|x_j|$$

since the modulus of a sum is less than or equal to the sum of the moduli. Thus

$$|\lambda| \leqslant \sum_{j=1}^{n} |a_{mj}| \left| \frac{x_j}{x_m} \right| .$$

But

$$\left| \frac{x_j}{x_m} \right| \leqslant 1 \qquad \text{and so} \qquad |\lambda| \leqslant \sum_{j=1}^{n} |a_{mj}|.$$

Since $m$ is not known (without $\mathbf{x}$ being known) this result is not particularly useful as it stands, but it implies that

$$|\lambda| \leqslant \max_{m} \sum_{j=1}^{n} |a_{mj}|$$

which proves the first part of the theorem.

The second part of the theorem may be proved in exactly the same way, using the fact that $\lambda$ is also an eigenvalue of the matrix $\mathbf{A}^{\mathrm{T}}$ and starting with the eigenvalue equation $\mathbf{A}^{\mathrm{T}}\mathbf{y} = \lambda\mathbf{y}$. ■

THEOREM 4.24 (Gerschgorin's theorem)

If $\lambda$ is an eigenvalue of the square matrix $\mathbf{A}$ of order $n$ then $\lambda$ must satisfy

$$|\lambda - a_{ii}| \leqslant \sum_{\substack{j=1 \\ j \neq i}}^{n} |a_{ij}| \qquad \text{for some } i$$

and

$$|\lambda - a_{ii}| \leqslant \sum_{\substack{j=1 \\ j \neq i}}^{n} |a_{ji}| \qquad \text{for some } i.$$

(This result implies that if we let

$$\sum_{\substack{j=1 \\ j \neq i}}^{n} |a_{ij}| = R_i \qquad \text{and} \qquad \sum_{\substack{j=1 \\ j \neq i}}^{n} |a_{ji}| = S_i$$

then every eigenvalue must lie inside at least one of the circles $|\lambda - a_{ii}| = R_i$ *and* one of the circles $|\lambda - a_{ii}| = S_i$ in the complex $\lambda$ plane.)

*Proof*

As in the previous proof let us suppose that $\mathbf{x}$ is the eigenvector of $\mathbf{A}$ corresponding to the eigenvalue $\lambda$ and that its component of greatest magnitude is $x_m$ so that $|x_i| \leqslant |x_m|$ for $1 \leqslant i \leqslant n$.

Then $\mathbf{Ax} = \lambda\mathbf{x}$ again gives, by taking the $m$th component of each side,

$$\sum_{j=1}^{n} a_{mj}x_j = \lambda x_m$$

which can be rewritten, by subtracting $a_{mm}x_m$ from both sides, in the form

$$(\lambda - a_{mm})x_m = \sum_{\substack{j=1 \\ j\neq m}}^{n} a_{mj}x_j.$$

Thus

$$|\lambda - a_{mm}|\,|x_m| \leqslant \sum_{\substack{j=1 \\ j\neq m}}^{n} |a_{mj}|\,|x_j|$$

giving, since $|x_j| \leqslant |x_m|$,

$$|\lambda - a_{mm}| \leqslant \sum_{\substack{j=1 \\ j\neq m}}^{n} |a_{mj}|.$$

And again, as in the previous theorem, this implies that for some $i$ we must have

$$|\lambda - a_{ii}| \leqslant \sum_{\substack{j=1 \\ j\neq i}}^{n} |a_{ij}|$$

which is the first of the results of the theorem.

The second result can be obtained in a similar manner by considering the equation $\mathbf{A}^{\mathrm{T}}\mathbf{y} = \lambda\mathbf{y}$. ■

*Example*

4.5.1 Apply Theorems 4.23 and 4.24 to obtain as much information as possible about the eigenvalues of

$$\text{(i)}\ \mathbf{A} = \begin{bmatrix} 3 & -4 & 2 \\ -4 & 6 & 8 \\ 2 & 8 & -1 \end{bmatrix} \quad \text{and} \quad \text{(ii)}\ \mathbf{A} = \begin{bmatrix} 1 & 6 & 2 \\ 2 & -3 & -1 \\ 0 & 4 & 1 \end{bmatrix}.$$

(i) We first note that $\mathbf{A}$ is real and symmetric so the eigenvalues are real. Theorem 4.23 says that every eigenvalue has magnitude less than the greatest row or column sum of magnitudes; in this case we only need consider row sums, since the symmetry implies that column sums are the same.

The row sums are:

Row 1 $\quad |3| + |-4| + |2| = 9$

Row 2 $\quad |-4| + |6| + |8| = 18$

Row 3 $\quad |2| + |8| + |-1| = 11.$

Thus if $\lambda$ is an eigenvalue we have that $|\lambda| \leqslant 18$.

Theorem 4.24 says that λ satisfies

$$|\lambda - a_{ii}| \leqslant \sum_{\substack{j=1\\ j\neq i}}^{n} |a_{ji}|$$

for some $i$ (again there is only one distinct result due to the symmetry of **A**). For $i = 1$ this gives

$$|\lambda - a_{11}| \leqslant |a_{21}| + |a_{31}|;$$

hence

$$|\lambda - 3| \leqslant 6. \qquad \text{(I)}$$

For $i = 2$ we obtain

$$|\lambda - a_{22}| \leqslant |a_{12}| + |a_{32}|$$

yielding

$$|\lambda - 6| \leqslant 12. \qquad \text{(II)}$$

For $i = 3$ we have

$$|\lambda - a_{33}| \leqslant |a_{13}| + |a_{23}|$$

so that

$$|\lambda + 1| \leqslant 10. \qquad \text{(III)}$$

The regions represented by the inequalities (I), (II) and (III) may, remembering that λ is real, be represented as shown in Figure 4.5.1.

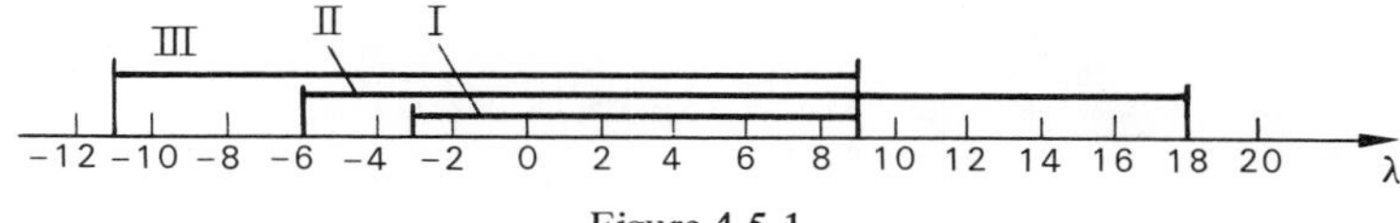


Figure 4.5.1

Since the theorem guarantees each eigenvalue to lie in some one of these three intervals, we see that we can deduce that $-11 \leqslant \lambda \leqslant 18$, which is a stronger result than that obtained from Theorem 4.23..

(ii) Here the row and column sums are given by:

| | |
|---|---|
| Row 1 | $\lvert 1 \rvert + \lvert 6 \rvert + \lvert 2 \rvert = 9$ |
| Row 2 | $\lvert 2 \rvert + \lvert -3 \rvert + \lvert -1 \rvert = 6$ |
| Row 3 | $\lvert 0 \rvert + \lvert 4 \rvert + \lvert 1 \rvert = 5$ |
| Column 1 | $\lvert 1 \rvert + \lvert 2 \rvert + \lvert 0 \rvert = 3$ |
| Column 2 | $\lvert 6 \rvert + \lvert -3 \rvert + \lvert 4 \rvert = 13$ |
| Column 3 | $\lvert 2 \rvert + \lvert -1 \rvert + \lvert 1 \rvert = 4.$ |

Thus the rows give the stronger result and we can deduce that any eigenvalue λ is such that $|\lambda| \leqslant 9$.

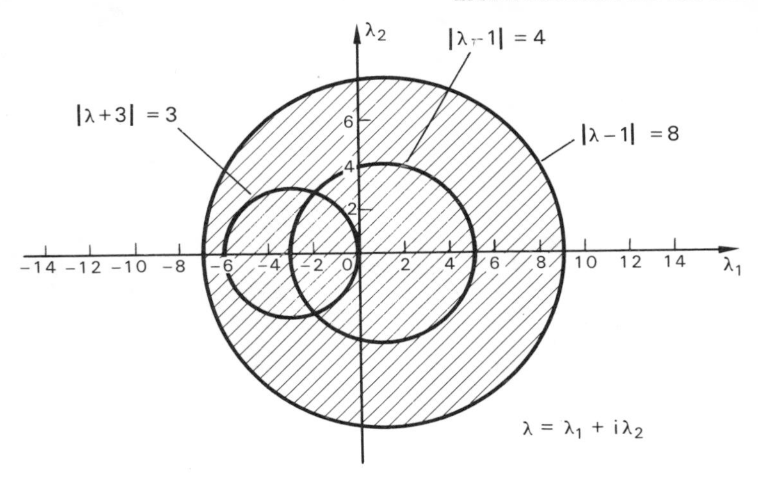


Figure 4.5.2

Theorem 4.24 now gives different domains for the rows and columns.

Rows lead to the inequalities

$$|\lambda - 1| \leqslant 8, \qquad |\lambda + 3| \leqslant 3, \qquad |\lambda - 1| \leqslant 4$$

while columns give

$$|\lambda - 1| \leqslant 2, \qquad |\lambda + 3| \leqslant 10, \qquad |\lambda - 1| \leqslant 3.$$

The first set of inequalities states that each eigenvalue must lie in the region shown shaded in Figure 4.5.2 while the second set give the region shown in Figure 4.5.3.

Thus each eigenvalue must lie in the overlap of these two regions shown in Figure 4.5.4.

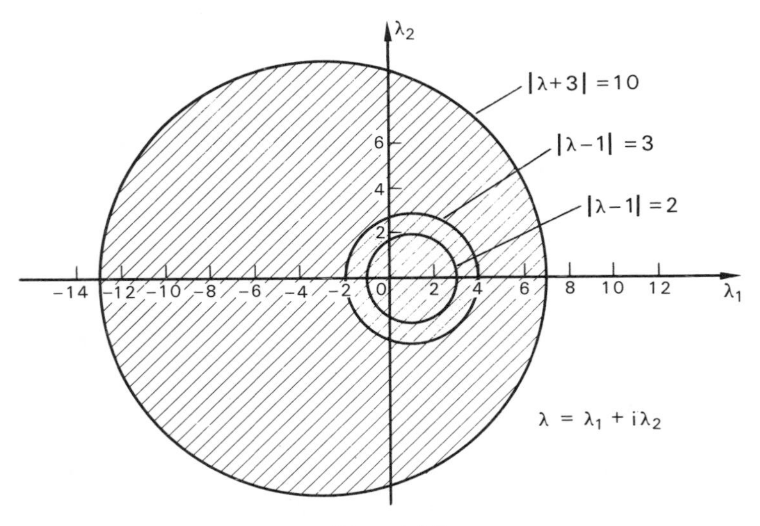


Figure 4.5.3

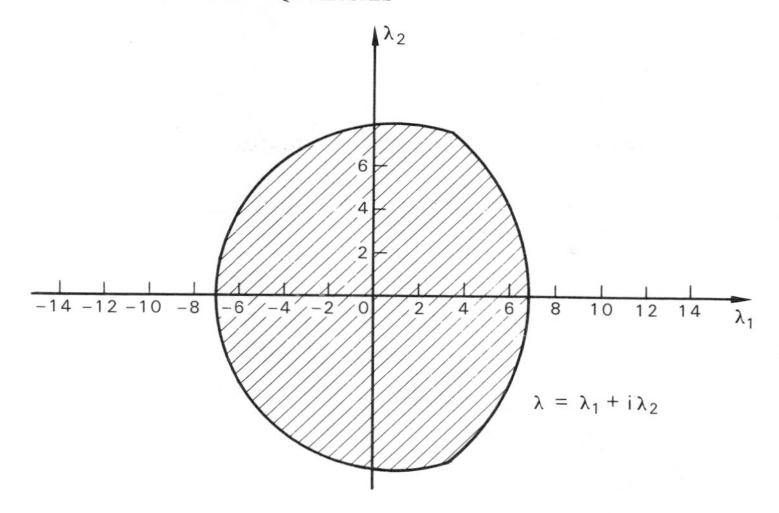


Figure 4.5.4

THEOREM 4.25

If **A** is a square matrix of order $n$ then any eigenvalue $\lambda$ satisfies the inequality

$$|\lambda|^2 \leqslant \sum_{i,j=1}^{n} |a_{ij}|^2.$$

*Proof*

We first prove the following lemma.

LEMMA

If **A** and **B** are matrices conformable for multiplication then

$$\mathrm{Tr}\,\{(\mathbf{AB})^+(\mathbf{AB})\} \leqslant \mathrm{Tr}\,(\mathbf{A}^+\mathbf{A})\,\mathrm{Tr}\,(\mathbf{B}^+\mathbf{B}).$$

*Proof of Lemma*

We first show that if $v_k, w_k$ $(1 \leqslant k \leqslant n)$ are complex numbers then

$$\left|\sum_{k=1}^{n} \bar{v}_k w_k\right|^2 \leqslant \left(\sum_{k=1}^{n} |v_k|^2\right)\left(\sum_{k=1}^{n} |w_k|^2\right).$$

This is proved by noting that

$$Q(t) = \sum_{k=1}^{n} (t|v_k| + |w_k|)^2$$

is positive for all real $t$. But

$$Q(t) = t^2 \sum_{k=1}^{n} |v_k|^2 + 2t \sum_{k=1}^{n} |v_k||w_k| + \sum_{k=1}^{n} |w_k|^2;$$

considering this as a quadratic in $t$, it will be positive for all $t$ if and only if the equation $Q(t) = 0$ has no real roots, which implies that the discriminant must be negative. Thus we have

$$\left(\sum_{k=1}^{n} |v_k||w_k|\right)^2 < \left(\sum_{k=1}^{n} |v_k|^2\right)\left(\sum_{k=1}^{n} |w_k|^2\right).$$

But

$$\left|\sum_{k=1}^{n} \bar{v}_k w_k\right|^2 \leqslant \left(\sum_{k=1}^{n} |\bar{v}_k w_k|\right)^2$$

by repeated use of the triangle inequality $|a + b| \leqslant |a| + |b|$

$$= \left(\sum_{k=1}^{n} |v_k||w_k|\right)^2$$

$$< \left(\sum_{k=1}^{n} |v_k|^2\right)\left(\sum_{k=1}^{n} |w_k|^2\right)$$

which is what we wished to prove†.

If we now choose $v_k = a_{ik}$ and $w_k = \bar{b}_{kj}$ we obtain the inequality

$$\left|\sum_{k=1}^{n} \bar{a}_{ik}\bar{b}_{kj}\right|^2 \leqslant \left(\sum_{k=1}^{n} |a_{ik}|^2\right)\left(\sum_{k=1}^{n} |b_{kj}|^2\right).$$

On the assumption that **A** is of order $p \times n$ and **B** of order $n \times q$ we now sum this inequality with respect to $i$ from 1 to $p$ and with respect to $j$ from 1 to $q$ to obtain

$$\sum_{i=1}^{p} \sum_{j=1}^{q} \left|\sum_{k=1}^{n} \bar{a}_{ik}\bar{b}_{kj}\right|^2 \leqslant \left(\sum_{i=1}^{p} \sum_{k=1}^{n} |a_{ik}|^2\right)\left(\sum_{j=1}^{q} \sum_{k=1}^{n} |b_{kj}|^2\right). \tag{4.5.1}$$

The proof of the lemma will be complete when we show that this is the same as

$$\text{Tr}\,\{(\mathbf{AB})^{+}(\mathbf{AB})\} \leqslant \text{Tr}\,(\mathbf{A}^{+}\mathbf{A})\,\text{Tr}\,(\mathbf{B}^{+}\mathbf{B}).$$

† We note that if we write

$$\mathbf{v} = \begin{bmatrix} v_1 \\ v_2 \\ \cdot \\ \cdot \\ \cdot \\ v_n \end{bmatrix}, \qquad \mathbf{w} = \begin{bmatrix} w_1 \\ w_2 \\ \cdot \\ \cdot \\ \cdot \\ w_n \end{bmatrix}$$

we obtan $(\mathbf{v}^{+}\mathbf{w})(\mathbf{w}^{+}\mathbf{v}) \leqslant (\mathbf{v}^{+}\mathbf{v})(\mathbf{w}^{+}\mathbf{w})$ which is one form of a well known inequality—the so-called Cauchy–Schwarz inequality.

But

$$\operatorname{Tr}(\mathbf{A}^+\mathbf{A}) = \sum_{k=1}^{n} (\mathbf{A}^+\mathbf{A})_{kk} \qquad \text{by definition}$$

$$= \sum_{k=1}^{n} \left( \sum_{i=1}^{p} A^+_{ki} A_{ik} \right)$$

$$= \sum_{k=1}^{n} \sum_{i=1}^{p} \bar{a}_{ik} a_{ik}$$

$$= \sum_{k=1}^{n} \sum_{i=1}^{p} |a_{ik}|^2.$$

Similarly

$$\operatorname{Tr}(\mathbf{B}^+\mathbf{B}) = \sum_{j=1}^{q} \sum_{k=1}^{n} |b_{kj}|^2,$$

and since

$$(\mathbf{AB})_{ij} = \sum_{k=1}^{n} a_{ik} b_{kj} = c_{ij},$$

say, it follows that

$$\operatorname{Tr}\{(\mathbf{AB})^+(\mathbf{AB})\} = \sum_{i=1}^{p} \sum_{j=1}^{q} |c_{ij}|^2$$

$$= \sum_{i=1}^{p} \sum_{j=1}^{q} \left| \sum_{k=1}^{n} a_{ik} b_{kj} \right|^2$$

$$= \sum_{i=1}^{p} \sum_{j=1}^{q} \left| \sum_{k=1}^{n} \bar{a}_{ik} \bar{b}_{kj} \right|^2$$

(since $|z| = |\bar{z}|$) and thus the inequality (4.5.1) becomes

$$\operatorname{Tr}\{(\mathbf{AB})^+(\mathbf{AB})\} \leqslant \operatorname{Tr}(\mathbf{A}^+\mathbf{A}) \operatorname{Tr}(\mathbf{B}^+\mathbf{B})$$

which completes the proof of the lemma.

*Proof of Theorem*

In the lemma let us choose $\mathbf{B} = \mathbf{x}$, where $\mathbf{x}$ is the eigenvector of $\mathbf{A}$ with eigenvalue $\lambda$. Then we have

$$\operatorname{Tr}\{(\mathbf{Ax})^+(\mathbf{Ax})\} \leqslant \operatorname{Tr}(\mathbf{A}^+\mathbf{A}) \operatorname{Tr}(\mathbf{x}^+\mathbf{x}).$$

But $\mathbf{Ax} = \lambda\mathbf{x}$ so $(\mathbf{Ax})^+ = \bar{\lambda}\mathbf{x}^+$ and hence we obtain, for the left-hand side of the above inequality,

$$\operatorname{Tr}\{(\mathbf{Ax})^+(\mathbf{Ax})\} = \operatorname{Tr}(\bar{\lambda}\lambda\mathbf{x}^+\mathbf{x})$$

$$= \operatorname{Tr}(|\lambda|^2)$$

since $\mathbf{x}^+\mathbf{x} = 1$, $\mathbf{x}$ being a normalized eigenvector

$$= |\lambda|^2.$$

As far as the right-hand side of the inequality is concerned, we have already shown in the lemma that

$$\text{Tr}\,(\mathbf{A}^+\mathbf{A}) = \sum_{i,j=1}^{n} |a_{ij}|^2$$

and we have noted above that $\mathbf{x}^+\mathbf{x} = 1$, implying that $\text{Tr}\,(\mathbf{x}^+\mathbf{x}) = 1$.

Thus we have

$$|\lambda|^2 \leqslant \sum_{i,j=1}^{n} |a_{ij}|^2$$

which is the required result. ■

The result of the above theorem may be strengthened considerably, as we show in the following theorem.

THEOREM 4.26 (Schur's inequality)

If **A** is a square matrix of order $n$ with eigenvalues $\lambda_1, \lambda_2, \ldots, \lambda_n$ then

$$\sum_{k=1}^{n} |\lambda_k|^2 \leqslant \sum_{i,j=1}^{n} |a_{ij}|^2.$$

*Proof*

We make use of Theorem 4.13 which states that a unitary matrix **T** can be found such that $\mathbf{T}^{-1}\mathbf{A}\mathbf{T} = \mathbf{U}$ where **U** is upper triangular; **U** thus has its eigenvalues (and therefore those of **A**) as its diagonal elements. Hence

$$\sum_{k=1}^{n} |\lambda_k|^2 = \sum_{i=1}^{n} |U_{ii}|^2$$

$$\leqslant \sum_{i,j=1}^{n} |U_{ij}|^2$$

this latter expression containing the off-diagonal terms of **U** as well

$$= \text{Tr}\,(\mathbf{U}^+\mathbf{U})$$

a result already proved in Theorem 4.25. But

$$\begin{aligned}\mathbf{U}^+\mathbf{U} &= (\mathbf{T}^{-1}\mathbf{A}\mathbf{T})^+(\mathbf{T}^{-1}\mathbf{A}\mathbf{T})\\ &= \mathbf{T}^+\mathbf{A}^+\mathbf{T}^{-1+}\mathbf{T}^{-1}\mathbf{A}\mathbf{T}\\ &= \mathbf{T}^{-1}\mathbf{A}^+\mathbf{T}\mathbf{T}^{-1}\mathbf{A}\mathbf{T} \qquad \text{since } \mathbf{T} \text{ is unitary } \mathbf{T}^+ = \mathbf{T}^{-1}\\ &= \mathbf{T}^{-1}\mathbf{A}^+\mathbf{A}\mathbf{T}.\end{aligned}$$

Thus $\mathbf{U}^+\mathbf{U}$ is similar to $\mathbf{A}^+\mathbf{A}$. Therefore it has the same eigenvalues (by Theorem 4.11) and hence the same trace (by Theorem 4.10).

So we have $\text{Tr}\,(\mathbf{U}^+\mathbf{U}) = \text{Tr}\,(\mathbf{A}^+\mathbf{A})$ and thus

$$\sum_{k=1}^{n} |\lambda_k|^2 \leqslant \text{Tr}\,(\mathbf{A}^+\mathbf{A}) = \sum_{i,j=1}^{n} |a_{ij}|^2$$

as already noted in the proof of the previous theorem. ∎

*Example*

4.5.2 Use Theorem 4.26 to obtain bounds on the eigenvalues of the matrices considered in Example 4.5.1:

$$\text{(i)}\ \mathbf{A} = \begin{bmatrix} 3 & -4 & 2 \\ -4 & 6 & 8 \\ 2 & 8 & 1 \end{bmatrix} \qquad \text{(ii)}\ \mathbf{A} = \begin{bmatrix} 1 & 6 & 2 \\ 2 & -3 & -1 \\ 0 & 4 & 1 \end{bmatrix}.$$

(i) In this case the eigenvalues are real and so

$$\sum_{i=1}^{n} |\lambda_i|^2 \leqslant \sum_{i,j=1}^{n} |a_{ij}|^2$$

gives

$$\lambda_1^2 + \lambda_2^2 + \lambda_3^2 \leqslant 9 + 16 + 4 + 16 + 36 + 64 + 4 + 64 + 1 = 224.$$

For the largest eigenvalue $\lambda_m$ we thus have

$$\lambda_m^2 \leqslant 224$$

and so $|\lambda_m| \leqslant 14{\cdot}97$ which is a stronger result than previously obtained.

(ii) Here we obtain

$$\lambda_1^2 + \lambda_2^2 + \lambda_3^2 \leqslant 1 + 36 + 4 + 4 + 9 + 1 + 16 + 1 = 72$$

giving for the eigenvalue of largest modulus

$$|\lambda_m|^2 \leqslant 72 \qquad \text{and hence} \qquad |\lambda_m| \leqslant 8{\cdot}49,$$

again a stronger result than previously obtained.

*Exercises*

4.5.1 Use the theorems of this section to obtain as much information as possible about the eigenvalues of the following matrices:

$$\text{(i)} \begin{bmatrix} 2 & -1 & 0 \\ -1 & 2 & 1 \\ 0 & 1 & 1 \end{bmatrix} \quad \text{(ii)} \begin{bmatrix} 3 & 2+\mathrm{i} & 2\mathrm{i} \\ 2-\mathrm{i} & 4 & 1-\mathrm{i} \\ -2\mathrm{i} & 1+\mathrm{i} & 2 \end{bmatrix} \quad \text{(iii)} \begin{bmatrix} 3 & -9 & 2 \\ 8 & 6 & 1 \\ 4 & 3 & 11 \end{bmatrix}.$$

4.5.2 Use Theorem 4.14 to show that if $\mathbf{A}$ is Hermitian then for any vector $\mathbf{x}$ we have

$$\lambda_{\min} \leqslant \frac{\mathbf{x}^+\mathbf{A}\mathbf{x}}{\mathbf{x}^+\mathbf{x}} \leqslant \lambda_{\max}$$

where $\lambda_{\min}, \lambda_{\max}$ are respectively the smallest and largest eigenvalues of $\mathbf{A}$.

## 4.6 Jacobi and Gauss–Seidel Convergence Criteria

In this section we use and develop some of the ideas of eigenvalue theory to discuss the problem arising in Section 3.3 of the convergence of the Jacobi and Gauss–Seidel iterative schemes for solving the systems of equations $\mathbf{Ax} = \mathbf{b}$. In that section we were led to consider the matrix $\mathbf{B}$ given by

$$\mathbf{B} = -\begin{bmatrix} 0 & a_{12}/a_{11} & a_{13}/a_{11} & \cdots & & a_{1n}/a_{11} \\ a_{21}/a_{22} & 0 & a_{23}/a_{22} & & & \\ a_{31}/a_{33} & a_{32}/a_{33} & 0 & & & \\ \vdots & & & & & \\ a_{n1}/a_{nn} & a_{n2}/a_{nn} & & \cdots & a_{n,n-1}/a_{nn} & 0 \end{bmatrix}.$$

We also introduced the upper and lower triangular matrices $\mathbf{B}_\text{U}$ and $\mathbf{B}_\text{L}$ such that $\mathbf{B} = \mathbf{B}_\text{U} + \mathbf{B}_\text{L}$ and we were then able to show that

(i) a necessary and sufficient condition for the convergence of the Jacobi iterative sequence is that $\mathbf{B}^r \to \mathbf{0}$ as $r \to \infty$, and

(ii) a necessary and sufficient condition for the convergence of the Gauss–Seidel iterative sequence is that $\{(\mathbf{I} - \mathbf{B}_\text{L})^{-1}\mathbf{B}_\text{U}\}^r \to \mathbf{0}$ as $r \to \infty$.

What we stated, but did not prove, was that a sufficient condition for either of the above conditions to be true was given by either

$$(a) \quad \sum_{i=1}^{n} |b_{ij}| < 1 \qquad \text{for} \quad 1 \leqslant j \leqslant n$$

or

$$(b) \quad \sum_{j=1}^{n} |b_{ij}| < 1 \qquad \text{for} \quad 1 \leqslant i \leqslant n$$

or

$$(c) \quad \sum_{i,j=1}^{n} |b_{ij}|^2 < 1.$$

It is the purpose of the present section to provide a proof of that statement. We first prove a theorem which relates the condition that $\mathbf{B}^r \to \mathbf{0}$ to a condition on its eigenvalues

THEOREM 4.27

If $\mathbf{A}$ is a square matrix then a necessary and sufficient condition for $\mathbf{A}^r \to \mathbf{0}$ as $r \to \infty$ is that all the eigenvalues of $\mathbf{A}$ should be less than one in magnitude.

*Proof*

We shall provide a proof for the situation in which all the eigenvalues of $\mathbf{A}$ are distinct, making use of Theorem 4.12 which says that $\mathbf{A}$ is then similar to a diagonal matrix. In the more general case a rather more complicated proof may be given, making use of Theorem 4.16 which states that in general $\mathbf{A}$ is similar to a block diagonal matrix.

We have, then, $\mathbf{T}^{-1}\mathbf{AT} = \mathbf{D}$ where $\mathbf{D}$ is diagonal with diagonal elements equal to the eigenvalues $\lambda_1, \lambda_2, \ldots, \lambda_n$ of $\mathbf{A}$. Then

$$\begin{aligned}\lim_{r\to\infty} \mathbf{A}^r &= \mathbf{T}\left(\lim_{r\to\infty} \mathbf{T}^{-1}\mathbf{A}^r\mathbf{T}\right)\mathbf{T}^{-1}\\ &= \mathbf{T}\left\{\lim_{r\to\infty} (\mathbf{T}^{-1}\mathbf{AT})^r\right\}\mathbf{T}^{-1}\\ &= \mathbf{T}\left(\lim_{r\to\infty} \mathbf{D}^r\right)\mathbf{T}^{-1}.\end{aligned}$$

But

$$\mathbf{D}^r = \begin{bmatrix} \lambda_1^r & 0 & 0 & \ldots & 0\\ 0 & \lambda_2^r & 0 & \ldots & 0\\ 0 & 0 & \lambda_3^r & \ldots & 0\\ \cdot & & & & \\ \cdot & & & & \\ \cdot & & & & \\ 0 & \ldots & \ldots & \ldots & \lambda_n^r \end{bmatrix}$$

and so $\mathbf{D}^r \to \mathbf{0}$ if and only if each $\lambda_i^r \to 0$, that is if each $|\lambda_i| < 1$.

In that case:

(i) if each $|\lambda_i| < 1$ then $\lim_{r\to\infty} \mathbf{A}^r = \mathbf{TOT}^{-1} = \mathbf{0}$.

(ii) if $\lim_{r\to\infty} \mathbf{A}^r = \mathbf{0}$ then $\mathbf{T}(\lim_{r\to\infty} \mathbf{D}^r)\mathbf{T}^{-1} = \mathbf{0}$ giving $\lim_{r\to\infty} \mathbf{D}^r = \mathbf{0}$ and thus each $|\lambda_i| < 1$.

(i) proves the sufficiency and (ii) the necessity of the condition. ■

We now note that Theorems 4.23 and 4.25 which say that if $\lambda$ is an eigenvalue of the matrix $\mathbf{A}$ then

(*a*) $|\lambda| \leqslant$ greatest row sum of magnitudes of elements of $\mathbf{A}$

(*b*) $|\lambda| \leqslant$ greatest column sum of magnitudes of elements of $\mathbf{A}$

(*c*) $|\lambda|^2 \leqslant$ sum of squared magnitudes of all elements of $\mathbf{A}$,

will now respectively guarantee the sufficiency of the three conditions (*a*), (*b*), (*c*) given previously for the convergence of the Jacobi method. For, taken together with Theorem 4.27, they guarantee that all eigenvalues of $\mathbf{B}$ are less than 1 in magnitude and hence $\mathbf{B}^r \to \mathbf{0}$ as $r \to \infty$ which is just what was required for the validity of the Jacobi method.

Since we have already noted that a necessary and sufficient condition for the convergence of the Gauss–Seidel method is that $\{(\mathbf{I} - \mathbf{B}_L)^{-1}\mathbf{B}_U\}^r \to \mathbf{0}$ as $r \to \infty$ it follows from Theorem 4.27 that a necessary and sufficient condition for this convergence is that the eigenvalues of $(\mathbf{I} - \mathbf{B}_L)^{-1}\mathbf{B}_U$ should be less than 1 in magnitude. We shall now relate the eigenvalues of this matrix to the eigenvalues of $\mathbf{B} = \mathbf{B}_L + \mathbf{B}_U$.

The eigenvalues of $(\mathbf{I} - \mathbf{B}_L)^{-1}\mathbf{B}_U$ are given by the roots $\mu$ of its characteristic equation

$$\det\{(\mathbf{I} - \mathbf{B}_L)^{-1}\mathbf{B}_U - \mu\mathbf{I}\} = 0$$

which is equivalent to

$$\det[(\mathbf{I} - \mathbf{B}_L)^{-1}\{\mathbf{B}_U - \mu(\mathbf{I} - \mathbf{B}_L)\}] = 0$$

or

$$\det(\mathbf{I} - \mathbf{B}_L)^{-1}\det\{\mathbf{B}_U - \mu(\mathbf{I} - \mathbf{B}_L)\} = 0$$

which reduces to

$$\det\{\mathbf{B}_U - \mu(\mathbf{I} - \mathbf{B}_L)\} = 0$$

(since $(\mathbf{I} - \mathbf{B}_L)$ is non-singular, so is its inverse) or

$$\det\{\mathbf{B}_U + \mu\mathbf{B}_L - \mu\mathbf{I}\} = 0.$$

But

$$\det(\mathbf{B}_U + \mu\mathbf{B}_L - \mu\mathbf{I}) = \det\{\sqrt{(\mu)}\mathbf{B}_U + \sqrt{(\mu)}\mathbf{B}_L - \mu\mathbf{I}\}$$

since by the expansion rules for the determinant the same number of elements from the upper and lower triangular parts ($\mathbf{B}_U$ and $\mathbf{B}_L$) respectively enter into an arbitrary term in the expansion; thus the factor $\mu$ appearing with $\mathbf{B}_L$ can be shared in any way between $\mathbf{B}_U$ and $\mathbf{B}_L$ provided the product gives $\mu$: $\det(\mu_1\mathbf{B}_U + \mu_2\mathbf{B}_L) = \det(\nu_1\mathbf{B}_U + \nu_2\mathbf{B}_L)$ provided $\mu_1\mu_2 = \nu_1\nu_2$. But

$$\det\{\sqrt{(\mu)}\mathbf{B}_U + \sqrt{(\mu)}\mathbf{B}_L - \mu\mathbf{I}\} = (\sqrt{\mu})^n \det\{\mathbf{B}_U + \mathbf{B}_L - \sqrt{(\mu)}\mathbf{1}\}.$$

Hence the roots of the characteristic equation are given by the roots of the equation $\det\{\mathbf{B} - \sqrt{(\mu)}\mathbf{I}\} = 0$, since $\mathbf{B} = \mathbf{B}_U + \mathbf{B}_L$. But the eigenvalues of $\mathbf{B}$ are given by $\det(\mathbf{B} - \lambda\mathbf{I}) = 0$. Hence $\lambda = \sqrt{\mu}$, $\mu = \lambda^2$ and $|\mu| = |\lambda|^2$.

Thus if $\mathbf{B}$ is such that all its eigenvalues are less than 1 in magnitude then so are all the eigenvalues of $(\mathbf{I} - \mathbf{B}_L)^{-1}\mathbf{B}_U$ and hence the sufficient conditions derived above for the Jacobi method are also sufficient conditions for the Gauss–Seidel method.

We conclude this section by proving another sufficient condition for the convergence of the Gauss–Seidel method: if $\mathbf{A}$ is a real symmetric positive definite matrix then the method converges.

To prove this we first of all note that $\mathbf{A}$ can be written in the form $\mathbf{A} = \mathbf{L} + \mathbf{D} + \mathbf{L}^T$ where $\mathbf{L}$ is the strictly lower triangular part of $\mathbf{A}$ and $\mathbf{D}$ is the diagonal part. Then from the definition of the associated matrix $\mathbf{B}$ we have that $\mathbf{B}_L = -\mathbf{D}^{-1}\mathbf{L}$ and $\mathbf{B}_U = -\mathbf{D}^{-1}\mathbf{L}^T$.

Thus

$$\begin{aligned}(\mathbf{I} - \mathbf{B}_L)^{-1}\mathbf{B}_U &= -(\mathbf{I} + \mathbf{D}^{-1}\mathbf{L})^{-1}\mathbf{D}^{-1}\mathbf{L}^T \\ &= -(\mathbf{D} + \mathbf{L})^{-1}\mathbf{L}^T\end{aligned}$$

(making use of the fact that $\mathbf{D}^{-1}$ is diagonal and so commutes with $\mathbf{L}$). The requirement that the eigenvalues of $(\mathbf{I} - \mathbf{B}_L)^{-1}\mathbf{B}_U$ are less than 1 in magnitude will thus be satisfied if those of $(\mathbf{D}+\mathbf{L})^{-1}\mathbf{L}^{\mathrm{T}}$ are, since they differ only in sign. Let $\lambda$ be an eigenvalue and $\mathbf{u}$ the corresponding eigenvector of this last matrix so that

$$(\mathbf{D}+\mathbf{L})^{-1}\mathbf{L}^{\mathrm{T}}\mathbf{u} = \lambda\mathbf{u}. \tag{4.6.1}$$

There is, of course, no guarantee that $\lambda$, $\mathbf{u}$ are real. But the fact that $\mathbf{A}$ is positive definite guarantees that

$$\mathbf{u}^{+}\mathbf{A}\mathbf{u} > 0$$

and thus

$$\mathbf{u}^{+}\mathbf{L}\mathbf{u} + \mathbf{u}^{+}\mathbf{D}\mathbf{u} + \mathbf{u}^{+}\mathbf{L}^{\mathrm{T}}\mathbf{u} > 0. \tag{4.6.2}$$

Using (4.6.1) gives $\mathbf{L}^{\mathrm{T}}\mathbf{u} = \lambda\mathbf{D}\mathbf{u} + \lambda\mathbf{L}\mathbf{u}$ and hence

$$\mathbf{u}^{+}\mathbf{L}^{\mathrm{T}}\mathbf{u} = \lambda\mathbf{u}^{+}\mathbf{D}\mathbf{u} + \lambda\mathbf{u}^{+}\mathbf{L}\mathbf{u}. \tag{4.6.3}$$

Taking the conjugate of both sides gives, since $\mathbf{L}$ and $\mathbf{D}$ are both real

$$\mathbf{u}^{+}\mathbf{L}\mathbf{u} = \bar{\lambda}\mathbf{u}^{+}\mathbf{D}\mathbf{u} + \bar{\lambda}\mathbf{u}^{+}\mathbf{L}^{\mathrm{T}}\mathbf{u}. \tag{4.6.4}$$

Hence, by considering (4.6.3) and (4.6.4) as simultaneous equations for $\mathbf{u}^{+}\mathbf{L}\mathbf{u}$ and $\mathbf{u}^{+}\mathbf{L}^{\mathrm{T}}\mathbf{u}$ we obtain

$$\mathbf{u}^{+}\mathbf{L}\mathbf{u} = \frac{\lambda(1+\lambda)}{1-\lambda\bar{\lambda}}\,\mathbf{u}^{+}\mathbf{D}\mathbf{u}$$

and

$$\mathbf{u}^{+}\mathbf{L}^{\mathrm{T}}\mathbf{u} = \frac{\bar{\lambda}(1+\lambda)}{1-\lambda\bar{\lambda}}\,\mathbf{u}^{+}\mathbf{D}\mathbf{u}.$$

Substituting these expressions into equation (4.6.2) we obtain

$$\frac{(1+\lambda)(1+\bar{\lambda})}{1-\lambda\bar{\lambda}}\,\mathbf{u}^{+}\mathbf{D}\mathbf{u} > 0.$$

But we have made the assumption that $\mathbf{A}$ is positive definite, and hence so is $\mathbf{D}$ (see Example 4.4.3). Thus

$$\frac{(1+\lambda)(1+\lambda)}{1-\lambda\bar{\lambda}} > 0.$$

But $(1+\lambda)(1+\bar{\lambda})$ is of the form $z\bar{z} = |z|^2$ and hence is positive; thus we have $1-\lambda\bar{\lambda} > 0$, $\lambda\bar{\lambda} < 1$ or $|\lambda|^2 < 1$, giving $|\lambda| < 1$ as required.

## 4.7 Numerical Methods for the Determination of Eigenvalues and Eigenvectors

In the previous sections of this chapter we have developed eigenvalue theory and used it in various contexts. It was implicitly assumed in some of these applications that given a matrix $\mathbf{A}$ the eigenvalues and eigenvectors would be known. In practice the only method we have as yet at our disposal is based on use of the characteristic equation, and we have already noted that this method cannot be recommended for matrices of greater order than 3. It is the purpose of the present section to remedy this defect and indicate what methods are available in practice for matrices of arbitrary order.

Proper study of this topic would require more space than is available here and the interested reader should refer to one of the specialized texts for a detailed treatment. What we shall do here is indicate some modern methods which might well be encountered in practice in computer implementations. No effort is directed towards a rigorous consideration of error analysis.

In common with most areas of numerical analysis there is no single best method to cover all situations. This is partially due to the fact that what is actually required may well vary from one problem to another. For example, sometimes we may wish to know only the eigenvalue of largest modulus, sometimes the eigenvalue of lowest modulus, sometimes all eigenvalues but no eigenvectors and sometimes all eigenvalues and eigenvectors. In addition some methods are specially suited for certain types of matrices such as real symmetric or triple diagonal. We shall attempt to indicate the areas of application of each of the methods we describe.

### METHOD 1: Power method

This is an iterative method, proceeding via a sequence of successive approximations. In its basic form it yields the eigenvalue of largest modulus together with the corresponding eigenvector, but in certain circumstances it can be extended to yield all eigenvalues and eigenvectors. The result on which the method is based is contained in the following theorem.

### THEOREM 4.28

Let $\mathbf{A}$ be a square matrix of order $n$ with eigenvalue of highest modulus $\lambda$ and corresponding eigenvector $\mathbf{x}$. Then, provided the eigenvectors form an independent set, if we construct the sequence of vectors $\mathbf{v}^{(0)}, \mathbf{v}^{(1)}, \ldots \mathbf{v}^{(k)}, \ldots$ according to the rule

$$\mathbf{v}^{(k)} = \mathbf{A}\mathbf{v}^{(k-1)}/\| \mathbf{A}\mathbf{v}^{(k-1)} \|$$

with $\mathbf{v}^{(0)}$ (almost) arbitrary, we have the results

$$\lim_{k\to\infty} \mathbf{v}^{(k)} = \mathbf{x}$$

$$\lim_{k\to\infty} \| \mathbf{A}\mathbf{v}^{(k)} \| = |\lambda|.$$

(We note that earlier results guarantee the eigenvectors to form an independent set if either

(i) $\mathbf{A}$ has distinct eigenvalues (Theorem 4.7)

or

(ii) $\mathbf{A}$ has linear elementary divisors (Theorem 4.16).)

*Proof*

Let the eigenvalues $\lambda_i$ $(1 \leqslant i \leqslant n)$ of $\mathbf{A}$ be labelled in such a way that

$$|\lambda_1| \geqslant |\lambda_2| \geqslant |\lambda_3| \geqslant \ldots \geqslant |\lambda_n|$$

and denote the corresponding eigenvectors by

$$\mathbf{x}^{(1)}, \mathbf{x}^{(2)}, \ldots \mathbf{x}^{(n)} \qquad \text{so that} \qquad \mathbf{A}\mathbf{x}^{(i)} = \lambda_i \mathbf{x}^{(i)}.$$

Since the vectors $\mathbf{x}^{(1)}, \mathbf{x}^{(2)}, \ldots \mathbf{x}^{(n)}$ are independent it follows that $\mathbf{v}^{(0)}$ can be written in the form

$$\mathbf{v}^{(0)} = C_1\mathbf{x}^{(1)} + C_2\mathbf{x}^{(2)} + \ldots + C_n\mathbf{x}^{(n)}.$$

Suppose first that $|\lambda_1| > |\lambda_2|$. Then we shall see that for the method to work we must have $C_1 \neq 0$; this is the restriction implied by 'almost' arbitrary in the statement of the theorem. Making these assumptions, then, we have $\mathbf{v}^{(0)} = C_1\mathbf{x}^{(1)} + C_2\mathbf{x}^{(2)} + \ldots + C_n\mathbf{x}^{(n)}$ where $C_1 \neq 0$ and $|\lambda_1| > |\lambda_2| \geqslant |\lambda_3| \geqslant \ldots \geqslant |\lambda_n|$.

We note now that our defining equation for $\mathbf{v}^{(k)}$, namely $\mathbf{v}^{(k)} = \mathbf{A}\mathbf{v}^{(k-1)}/\|\mathbf{A}\mathbf{v}^{(k-1)}\|$ may be written in the form $\mathbf{v}^{(k)} = \beta_k\mathbf{A}\mathbf{v}^{(k-1)}$ where $\beta_k$ is merely a normalizing factor to ensure that $\|\mathbf{v}^{(k)}\| = 1$.

Thus we have

$$\begin{aligned}\mathbf{v}^{(1)} &= \beta_1\mathbf{A}\mathbf{v}^{(0)}\\ &= \beta_1\mathbf{A}(C_1\mathbf{x}^{(1)} + C_2\mathbf{x}^{(2)} + \ldots + C_n\mathbf{x}^{(n)})\\ &= \beta_1(C_1\mathbf{A}\mathbf{x}^{(1)} + C_2\mathbf{A}\mathbf{x}^{(2)} + \ldots + C_n\mathbf{A}\mathbf{x}^{(n)})\\ &= \beta_1(C_1\lambda_1\mathbf{x}^{(1)} + C_2\lambda_2\mathbf{x}^{(2)} + \ldots + C_n\lambda_n\mathbf{x}^{(n)}).\end{aligned}$$

Then

$$\begin{aligned}\mathbf{v}^{(2)} &= \beta_2\mathbf{A}\mathbf{v}^{(1)}\\ &= \beta_2\beta_1\mathbf{A}(C_1\lambda_1\mathbf{x}^{(1)} + C_2\lambda_2\mathbf{x}^{(2)} + \ldots + C_n\lambda_n\mathbf{x}^{(n)})\\ &= \beta_2\beta_1(C_1\lambda_1^2\mathbf{x}^{(1)} + C_2\lambda_2^2\mathbf{x}^{(2)} + \ldots + C_n\lambda_n^2\mathbf{x}^{(n)}).\end{aligned}$$

Similarly

$$\mathbf{v}^{(3)} = \beta_3\beta_2\beta_1(C_1\lambda_1^3\mathbf{x}^{(1)} + C_2\lambda_2^3\mathbf{x}^{(2)} + \ldots + C_n\lambda_n^3\mathbf{x}^{(n)})$$

and in general

$$\begin{aligned}\mathbf{v}^{(k)} &= \beta_k\beta_{k-1}\ldots\beta_2\beta_1(C_1\lambda_1^k\mathbf{x}^{(1)} + C_2\lambda_2^k\mathbf{x}^{(2)} + \ldots + C_n\lambda_n^k\mathbf{x}^{(n)})\\ &= \alpha_k(C_1\lambda_1^k\mathbf{x}^{(1)} + C_2\lambda_2^k\mathbf{x}^{(2)} + \ldots + C_n\lambda_n^k\mathbf{x}^{(n)}),\end{aligned}$$

where $\alpha_k = \beta_k\beta_{k-1}\ldots\beta_2\beta_1$.

Now, for large values of $k$, $|\lambda_1|^k \gg |\lambda_2|^k$ and so we have approximately

$$\mathbf{v}^{(k)} \simeq \alpha_k C_1\lambda_1^k\mathbf{x}^{(1)}.$$

But $\alpha_k$ is just a constant chosen so as to ensure that $\|\mathbf{v}^{(k)}\| = 1$, and since $\|\mathbf{x}^{(k)}\| = 1$ we must therefore have $\mathbf{v}^{(k)} \simeq \mathbf{x}^{(1)}$, or, since we have used $\mathbf{x}$ for the eigenvector corresponding to the eigenvalue of largest modulus, $\mathbf{v}^{(k)} \simeq \mathbf{x}$. This result will be exactly true in the limit as $k \to \infty$, so $\lim_{k\to\infty}\mathbf{v}^{(k)} = \mathbf{x}$.

Also, since $\mathbf{v}^{(k)} \simeq \mathbf{x}$ then $\mathbf{A}\mathbf{v}^{(k)} \simeq \lambda\mathbf{x}$ and so

$$\|\mathbf{A}\mathbf{v}^{(k)}\| \simeq \|\lambda\mathbf{x}\| = |\lambda|\,\|\mathbf{x}\| = |\lambda|.$$

Again the result is exactly true in the limit as $k \to \infty$ giving

$$\lim_{k\to\infty}\|\mathbf{A}\mathbf{v}^{(k)}\| = |\lambda|.$$

If now the condition $|\lambda_1| > |\lambda_2|$ is not satisfied we may assume that

$$|\lambda_1| = |\lambda_2| = \ldots = |\lambda_r| > |\lambda_{r+1}| \geqslant |\lambda_{r+2}| \geqslant \ldots \geqslant |\lambda_n|.$$

If in fact $\lambda_1 = \lambda_2 = \ldots = \lambda_r$ then the above proof can be modified in a comparatively trivial manner; in this case the method will work provided the starting vector is of the form $\mathbf{v}^{(0)} = C_1\mathbf{x}^{(1)} + C_2\mathbf{x}^{(2)} + \ldots + C_r\mathbf{x}^{(r)} + C_{r+1}\mathbf{x}^{(r+1)} + \ldots + C_n\mathbf{x}^{(n)}$ with $C_1, C_2, \ldots C_r$ not all zero. Of course the eigenvector obtained is just one of the $r$ corresponding to the eigenvalue of multiplicity $r$. If, on the other hand, $\lambda_1, \lambda_2, \ldots \lambda_r$ are not all equal then no such proof is possible and the method is inapplicable. ∎

We now see how Theorem 4.28 may be used in practice:

(i) Start with some arbitrary vector $\mathbf{v}^{(0)}$, (hoping that it satisfies the criterion $C_1 \neq 0$).

(ii) Form the sequence of vectors $\mathbf{v}^{(1)}, \mathbf{v}^{(2)}, \ldots \mathbf{v}^{(k)}$ stopping when two successive vectors agree to within some predetermined tolerance. $\mathbf{v}^{(k)}$ will give an approximation to the desired eigenvector.

(iii) Form $\mathbf{A}\mathbf{v}^{(k)}$; then $\|\mathbf{A}\mathbf{v}^{(k)}\|$ gives an approximation to $|\lambda|$, where $\lambda$ is the desired eigenvalue. The sign of $\lambda$ is determined by noting the signs of the first (non-zero) elemen of $\mathbf{v}^{(k)}$ and $\mathbf{A}\mathbf{v}^{(k)}$; since $\mathbf{A}\mathbf{v}^{(k)} \simeq \mathbf{A}\mathbf{x} = \lambda\mathbf{x} \simeq \lambda\mathbf{v}^{(k)}$ these will be of the same sign if $\lambda < 0$. and of opposite sign if $\lambda < 0$.

We note the following points about the application of the method:

(*a*) We have assumed above that the eigenvalues are real.

(*b*) Some criterion has to be used for stopping the iteration, such as $\|\mathbf{v}^{(k)} - \mathbf{v}^{(k-1)}\| < \epsilon$ where $\epsilon$ is some given parameter. However there is no guarantee that if this test is used a small value for $\epsilon$ will lead to a good approximation for $\mathbf{x}$ and $\lambda$. Ideally we should carry out some error analysis to relate the accuracy of the approximation to $\epsilon$.

(*c*) Some initial vector $\mathbf{v}^{(0)}$ has to be chosen. In the absence of any information about the eigenvector we might as well choose

$$\mathbf{v}^{(0)} = \begin{bmatrix} 1 \\ 0 \\ 0 \\ \cdot \\ \cdot \\ \cdot \\ 0 \end{bmatrix}.$$

It would then make sense to carry out the method again with a different starting vector, checking that we obtain the same results.

(*d*) Strictly speaking, as we noted above, $\mathbf{v}^{(0)}$ has to contain a component of $\mathbf{x}^{(1)}$ for the method to work. However, in practice, rounding errors in the arithmetical processes involved will introduce a small component even if there were none present to begin with. In such a situation convergence will take place, albeit slowly. If convergence seems to be slow it might be worthwhile to try a different starting vector.

(*e*) Convergence will always be slow if the dominant eigenvalue is not much greater than the next largest eigenvalue. In such a case changing the starting vector will in general have little or no effect on the rate of convergence.

We shall now see how these ideas work out in practice by considering the following example.

*Example*

4.7.1 Apply the power method to the following matrices:

$$\text{(i)}\begin{bmatrix}1 & 0 & 0\\ 2 & 2 & 0\\ -2 & 2 & 3\end{bmatrix} \qquad \text{(ii)}\begin{bmatrix}2 & 0 & 0\\ 0 & -3 & 1\\ 0 & 1 & -3\end{bmatrix} \qquad \text{(iii)}\begin{bmatrix}2 & 1 & -1\\ 0 & -2 & -2\\ 1 & 1 & 0\end{bmatrix}.$$

(i) Let us choose as arbitrary starting vector

$$\mathbf{v}^{(0)} = \begin{bmatrix}1\\ 0\\ 0\end{bmatrix}.$$

Then $\mathbf{v}^{(1)} = \mathbf{A}\mathbf{v}^{(0)}/\|\mathbf{A}\mathbf{v}^{(0)}\|$. But

$$\mathbf{A}\mathbf{v}^{(0)} \begin{bmatrix}1 & 0 & 0\\ 2 & 2 & 0\\ -2 & 2 & 3\end{bmatrix}\begin{bmatrix}1\\ 0\\ 0\end{bmatrix} = \begin{bmatrix}1\\ 2\\ -2\end{bmatrix}.$$

So $\|\mathbf{A}\mathbf{v}^{(0)}\| = \sqrt{(1+4+4)} = 3$ and hence

$$\mathbf{v}^{(1)} = \tfrac{1}{3}\begin{bmatrix}1\\ 2\\ -2\end{bmatrix} = \begin{bmatrix}0{\cdot}33333\\ 0{\cdot}66667\\ -0{\cdot}66667\end{bmatrix}$$

correct to five decimal places. We now proceed to calculate $\mathbf{v}^{(2)} = \mathbf{A}\mathbf{v}^{(1)}/\|\mathbf{A}\mathbf{v}^{(1)}\|$.

$$\mathbf{A}\mathbf{v}^{(1)} = \begin{bmatrix}1 & 0 & 0\\ 2 & 2 & 0\\ -2 & 2 & 3\end{bmatrix}\begin{bmatrix}0{\cdot}33333\\ 0{\cdot}66667\\ -0{\cdot}66667\end{bmatrix}$$

$$= \begin{bmatrix}0{\cdot}33333\\ 2{\cdot}00000\\ -1{\cdot}33333\end{bmatrix},$$

and

$$\|\mathbf{A}\mathbf{v}^{(1)}\| = \sqrt{\{(0{\cdot}33333)^2 + (2{\cdot}00000)^2 + (-1{\cdot}33333)^2\}}$$
$$= 2{\cdot}42670,$$

giving

$$\mathbf{v}^{(2)} = \begin{bmatrix}0{\cdot}13736\\ 0{\cdot}82416\\ -0{\cdot}54944\end{bmatrix}.$$

Continuing in this way we obtain the following table; the calculations were in fact performed by computer working to 11 significant digits whereas the results are expressed correct to five decimal places.

| $k$ | $\mathbf{v}^{(k)}$ | | | $\|\mathbf{Av}^{(k-1)}\|$ |
|---|---|---|---|---|
| 1 | 0·33333 | 0·66667 | −0·66667 | 3·00000 |
| 2 | 0·13736 | 0·82416 | −0·54944 | 2·42670 |
| 3 | 0·07053 | 0·98748 | −0·14107 | 1·94742 |
| 4 | 0·02772 | 0·83173 | 0·55449 | 2·54414 |
| 5 | 0·00750 | 0·46511 | 0·88522 | 3·69567 |
| 6 | 0·00203 | 0·25589 | 0·96670 | 3·69388 |
| 7 | 0·00059 | 0·14967 | 0·98874 | 3·44665 |
| 8 | 0·00018 | 0·09167 | 0·99579 | 3·27817 |
| 9 | 0·00006 | 0·05785 | 0·99833 | 3·17567 |
| 10 | 0·00002 | 0·03720 | 0·99931 | 3·11271 |
| 11 | 0·00001 | 0·02422 | 0·99971 | 3·07320 |
| 12 | 0·00000 | 0·01590 | 0·99987 | 3·04794 |
| 13 | 0·00000 | 0·01049 | 0·99994 | 3·03158 |
| 14 | 0·00000 | 0·00695 | 0·99998 | 3·02089 |
| 15 | 0·00000 | 0·00461 | 0·99999 | 3·01385 |
| 16 | 0·00000 | 0·00306 | 1·00000 | 3·00920 |
| 17 | 0·00000 | 0·00204 | 1·00000 | 3·00612 |
| 18 | 0·00000 | 0·00136 | 1·00000 | 3·00407 |
| 19 | 0·00000 | 0·00090 | 1·00000 | 3·00271 |
| 20 | 0·00000 | 0·00060 | 1·00000 | 3·00181 |
| 21 | 0·00000 | 0·00040 | 1·00000 | 3·00120 |
| 22 | 0·00000 | 0·00027 | 1·00000 | 3·00080 |
| 23 | 0·00000 | 0·00018 | 1·00000 | 3·00053 |
| 24 | 0·00000 | 0·00012 | 1·00000 | 3·00036 |
| 25 | 0·00000 | 0·00008 | 1·00000 | 3·00024 |
| 26 | 0·00000 | 0·00005 | 1·00000 | 3·00016 |
| 27 | 0·00000 | 0·00004 | 1·00000 | 3·00011 |
| 28 | 0·00000 | 0·00002 | 1·00000 | 3·00007 |
| 29 | 0·00000 | 0·00002 | 1·00000 | 3·00005 |
| 30 | 0·00000 | 0·00001 | 1·00000 | 3·00003 |
| 31 | 0·00000 | 0·00001 | 1·00000 | 3·00002 |
| 32 | 0·00000 | 0·00000 | 1·00000 | 3·00001 |
| 33 | 0·00000 | 0·00000 | 1·00000 | 3·00001 |
| 34 | 0·00000 | 0·00000 | 1·00000 | 3·00001 |
| 35 | 0·00000 | 0·00000 | 1·00000 | 3·00000 |
| 36 | 0·00000 | 0·00000 | 1·00000 | 3·00000 |
| 37 | 0·00000 | 0·00000 | 1·00000 | 3·00000 |
| 38 | 0·00000 | 0·00000 | 1·00000 | 3·00000 |
| 39 | 0·00000 | 0·00000 | 1·00000 | 3·00000 |
| 40 | 0·00000 | 0·00000 | 1·00000 | 3·00000 |

These results show slow convergence to the eigenvector

$$\begin{bmatrix} 0{\cdot}00000 \\ 0{\cdot}00000 \\ 1{\cdot}00000 \end{bmatrix}$$

and the corresponding eigenvalue 3·00000. If we start with the initial vector

$$\begin{bmatrix} 2 \\ -1 \\ 0 \end{bmatrix}$$

we obtain the following results

| $k$ | $\mathbf{v}^{(k)}$ | | | $\|\mathbf{Av}^{(k-1)}\|$ |
|---|---|---|---|---|
| 1 | 0·30151 | 0·30151 | −0·90453 | 6·63325 |
| 2 | 0·10102 | 0·40406 | −0·90914 | 2·98481 |
| 3 | 0·04295 | 0·42954 | −0·90203 | 2·35173 |
| 4 | 0·01996 | 0·43912 | −0·89821 | 2·15197 |
| 5 | 0·00964 | 0·44333 | −0·89630 | 2·07105 |
| 6 | 0·00474 | 0·44531 | −0·89536 | 2·03440 |
| 7 | 0·00235 | 0·44627 | −0·89489 | 2·01693 |
| 8 | 0·00117 | 0·44675 | −0·89466 | 2·00840 |
| 9 | 0·00058 | 0·44698 | −0·89454 | 2·00418 |
| 10 | 0·00029 | 0·44710 | −0·89449 | 2·00209 |
| 11 | 0·00015 | 0·44716 | −0·89446 | 2·00104 |
| 12 | 0·00007 | 0·44718 | −0·89444 | 2·00052 |
| 13 | 0·00004 | 0·44720 | −0·89443 | 2·00026 |
| 14 | 0·00002 | 0·44721 | −0·89443 | 2·00013 |
| 15 | 0·00001 | 0·44721 | −0·89443 | 2·00007 |
| 16 | 0·00000 | 0·44721 | −0·89443 | 2·00003 |
| 17 | 0·00000 | 0·44721 | −0·89443 | 2·00002 |
| 18 | 0·00000 | 0·44721 | −0·89443 | 2·00001 |
| 19 | 0·00000 | 0·44721 | −0·89443 | 2·00000 |
| 20 | 0·00000 | 0·44721 | −0·89443 | 2·00000 |
| 21 | 0·00000 | 0·44721 | −0·89443 | 2·00000 |
| 22 | 0·00000 | 0·44721 | −0·89443 | 2·00000 |
| 23 | 0·00000 | 0·44721 | −0·89443 | 2·00000 |
| 24 | 0·00000 | 0·44721 | −0·89443 | 2·00000 |
| 25 | 0·00000 | 0·44721 | −0·89443 | 2·00000 |
| 26 | 0·00000 | 0·44721 | −0·89443 | 2·00000 |
| 27 | 0·00000 | 0·44721 | −0·89443 | 2·00000 |
| 28 | 0·00000 | 0·44721 | −0·89443 | 2·00000 |
| 29 | 0·00000 | 0·44721 | −0·89443 | 2·00000 |
| 30 | 0·00000 | 0·44721 | −0·89443 | 2·00000 |
| 31 | 0·00000 | 0·44721 | −0·89443 | 2·00000 |
| 32 | 0·00000 | 0·44721 | −0·89443 | 2·00000 |
| 33 | 0·00000 | 0·44721 | −0·89443 | 2·00000 |
| 34 | 0·00000 | 0·44721 | −0·89443 | 2·00001 |
| 35 | 0·00000 | 0·44721 | −0·89443 | 2·00001 |
| 36 | 0·00000 | 0·44721 | −0·89443 | 2·00001 |
| 37 | 0·00000 | 0·44720 | −0·89443 | 2·00002 |
| 38 | 0·00000 | 0·44719 | −0·89444 | 2·00003 |
| 39 | 0·00000 | 0·44719 | −0·89444 | 2·00004 |
| 40 | 0·00000 | 0·44717 | −0·89445 | 2·00006 |
| 41 | 0·00000 | 0·44715 | −0·89446 | 2·00009 |
| 42 | 0·00000 | 0·44712 | −0·89447 | 2·00014 |
| 43 | 0·00000 | 0·44707 | −0·89450 | 2·00021 |
| 44 | 0·00000 | 0·44700 | −0·89453 | 2·00032 |
| 45 | 0·00000 | 0·44689 | −0·89459 | 2·00048 |
| 46 | 0·00000 | 0·44673 | −0·89467 | 2·00072 |
| 47 | 0·00000 | 0·44649 | −0·89479 | 2·00108 |
| 48 | 0·00000 | 0·44613 | −0·89497 | 2·00162 |
| 49 | 0·00000 | 0·44559 | −0·89524 | 2·00242 |
| 50 | 0·00000 | 0·44478 | −0·89564 | 2·00363 |
| 51 | 0·00000 | 0·44358 | −0·89624 | 2·00544 |
| 52 | 0·00000 | 0·44178 | −0·89713 | 2·00815 |
| 53 | 0·00000 | 0·43910 | −0·89844 | 2·01218 |
| 54 | 0·00000 | 0·43514 | −0·90036 | 2·01820 |
| 55 | 0·00000 | 0·42932 | −0·90315 | 2·02712 |

| | | | | |
|---|---|---|---|---|
| 56 | 0·00000 | 0·42084 | −0·90713 | 2·04029 |
| 57 | 0·00000 | 0·40867 | −0·91268 | 2·05956 |
| 58 | 0·00000 | 0·39156 | −0·92015 | 2·08738 |
| 59 | 0·00000 | 0·36823 | −0·92974 | 2·12676 |
| 60 | 0·00000 | 0·33769 | −0·94126 | 2·18086 |
| 61 | 0·00000 | 0·29990 | −0·95397 | 2·25205 |
| 62 | 0·00000 | 0·25629 | −0·96660 | 2·34029 |
| 63 | 0·00000 | 0·20993 | −0·97772 | 2·44163 |
| 64 | 0·00000 | 0·16478 | −0·98633 | 2·54811 |
| 65 | 0·00000 | 0·12436 | −0·99224 | 2·65001 |
| 66 | 0·00000 | 0·09079 | −0·99587 | 2·73931 |
| 67 | 0·00000 | 0·06458 | −0·99791 | 2·81189 |
| 68 | 0·00000 | 0·04504 | −0·99899 | 2·86749 |
| 69 | 0·00000 | 0·03098 | −0·99952 | 2·90827 |
| 70 | 0·00000 | 0·02109 | −0·99978 | 2·93726 |
| 71 | 0·00000 | 0·01426 | −0·99990 | 2·95745 |
| 72 | 0·00000 | 0·00960 | −0·99995 | 2·97131 |
| 73 | 0·00000 | 0·00644 | −0·99998 | 2·98072 |
| 74 | 0·00000 | 0·00431 | −0·99999 | 2·98708 |
| 75 | 0·00000 | 0·00288 | −1·00000 | 2·99136 |
| 76 | 0·00000 | 0·00193 | −1·00000 | 2·99423 |
| 77 | 0·00000 | 0·00129 | −1·00000 | 2·99614 |
| 78 | 0·00000 | 0·00086 | −1·00000 | 2·99743 |
| 79 | 0·00000 | 0·00057 | −1·00000 | 2·99828 |
| 80 | 0·00000 | 0·00038 | −1·00000 | 2·99886 |
| 81 | 0·00000 | 0·00025 | −1·00000 | 2·99924 |
| 82 | 0·00000 | 0·00017 | −1·00000 | 2·99949 |
| 83 | 0·00000 | 0·00011 | −1·00000 | 2·99966 |
| 84 | 0·00000 | 0·00008 | −1·00000 | 2·99977 |
| 85 | 0·00000 | 0·00005 | −1·00000 | 2·99985 |
| 86 | 0·00000 | 0·00003 | −1·00000 | 2·99990 |
| 87 | 0·00000 | 0·00002 | −1·00000 | 2·99993 |
| 88 | 0·00000 | 0·00001 | −1·00000 | 2·99996 |
| 89 | 0·00000 | 0·00001 | −1·00000 | 2·99997 |
| 90 | 0·00000 | 0·00001 | −1·00000 | 2·99998 |
| 91 | 0·00000 | 0·00000 | −1·00000 | 2·99999 |
| 92 | 0·00000 | 0·00000 | −1·00000 | 2·99999 |
| 93 | 0·00000 | 0·00000 | −1·00000 | 2·99999 |
| 94 | 0·00000 | 0·00000 | −1·00000 | 3·00000 |
| 95 | 0·00000 | 0·00000 | −1·00000 | 3·00000 |
| 96 | 0·00000 | 0·00000 | −1·00000 | 3·00000 |
| 97 | 0·00000 | 0·00000 | −1·00000 | 3·00000 |
| 98 | 0·00000 | 0·00000 | −1·00000 | 3·00000 |
| 99 | 0·00000 | 0·00000 | −1·00000 | 3·00000 |
| 100 | 0·00000 | 0·00000 | −1·00000 | 3·00000 |

In this case convergence to the highest eigenvalue and eigenvector is much slower. Indeed at around $k = 20$ it looks as if convergence is to

$$\begin{bmatrix} 0{\cdot}00000 \\ 0{\cdot}44721 \\ -0{\cdot}89443 \end{bmatrix}$$

with eigenvalue 2·00000. In fact the given matrix has eigenvalues 1, 2, 3; and in this case the starting vector was a linear combination of eigenvectors belonging to the eigenvalues 1 and 2, containing no component of that for the highest eigenvalue. Thus it is only the effect of rounding error that introduces such a component and we have to wait for around 95 iterations before we get convergence to the true answer correct to five decimal places.

We note also that convergence would in any case be expected to be comparatively slow since the eigenvalue of highest modulus (3) is not much larger than the next highest (2).

(ii) With starting vector

$$\mathbf{v}^{(0)} = \begin{bmatrix} 1 \\ 0 \\ 0 \end{bmatrix}$$

we obtain the following

| $k$ | | $\mathbf{v}^{(k)}$ | | $\lVert \mathbf{A}\mathbf{v}^{(k-1)} \rVert$ |
|---|---|---|---|---|
| 1 | 1·00000 | 0·00000 | 0·00000 | 2·00000 |
| 2 | 1·00000 | 0·00000 | 0·00000 | 2·00000 |
| 3 | 1·00000 | 0·00000 | 0·00000 | 2·00000 |
| 4 | 1·00000 | 0·00000 | 0·00000 | 2·00000 |
| 5 | 1·00000 | 0·00000 | 0·00000 | 2·00000 |
| 6 | 1·00000 | 0·00000 | 0·00000 | 2·00000 |
| 7 | 1·00000 | 0·00000 | 0·00000 | 2·00000 |
| 8 | 1·00000 | 0·00000 | 0·00000 | 2·00000 |
| 9 | 1·00000 | 0·00000 | 0·00000 | 2·00000 |
| 10 | 1·00000 | 0·00000 | 0·00000 | 2·00000 |
| 11 | 1·00000 | 0·00000 | 0·00000 | 2·00000 |
| 12 | 1·00000 | 0·00000 | 0·00000 | 2·00000 |
| 13 | 1·00000 | 0·00000 | 0·00000 | 2·00000 |
| 14 | 1·00000 | 0·00000 | 0·00000 | 2·00000 |
| 15 | 1·00000 | 0·00000 | 0·00000 | 2·00000 |
| 16 | 1·00000 | 0·00000 | 0·00000 | 2·00000 |
| 17 | 1·00000 | 0·00000 | 0·00000 | 2·00000 |
| 18 | 1·00000 | 0·00000 | 0·00000 | 2·00000 |
| 19 | 1·00000 | 0·00000 | 0·00000 | 2·00000 |
| 20 | 1·00000 | 0·00000 | 0·00000 | 2·00000 |

With starting vector

$$\mathbf{v}^{(0)} = \begin{bmatrix} 1 \\ 1 \\ 1 \end{bmatrix}$$

on the other hand we obtain

| $k$ | | $\mathbf{v}^{(k)}$ | | $\lVert \mathbf{A}\mathbf{v}^{(k-1)} \rVert$ |
|---|---|---|---|---|
| 1 | 0·57735 | −0·57735 | −0·57735 | 3·46410 |
| 2 | 0·57735 | 0·57735 | 0·57735 | 2·00000 |
| 3 | 0·57735 | −0·57735 | −0·57735 | 2·00000 |
| 4 | 0·57735 | 0·57735 | 0·57735 | 2·00000 |
| 5 | 0·57735 | −0·57735 | −0·57735 | 2·00000 |
| 6 | 0·57735 | 0·57735 | 0·57735 | 2·00000 |
| 7 | 0·57735 | −0·57735 | −0·57735 | 2·00000 |

| | | | | |
|---|---|---|---|---|
| 8 | 0·57735 | 0·57735 | 0·57735 | 2·00000 |
| 9 | 0·57735 | −0·57735 | −0·57735 | 2·00000 |
| 10 | 0·57735 | 0·57735 | 0·57735 | 2·00000 |
| 11 | 0·57735 | −0·57735 | −0·57735 | 2·00000 |
| 12 | 0·57735 | 0·57735 | 0·57735 | 2·00000 |
| 13 | 0·57735 | −0·57735 | −0·57735 | 2·00000 |
| 14 | 0·57735 | 0·57735 | 0·57735 | 2·00000 |
| 15 | 0·57735 | −0·57735 | −0·57735 | 2·00000 |
| 16 | 0·57735 | 0·57735 | 0·57735 | 2·00000 |
| 17 | 0·57735 | −0·57735 | −0·57735 | 2·00000 |
| 18 | 0·57735 | 0·57735 | 0·57735 | 2·00000 |
| 19 | 0·57735 | −0·57735 | −0·57735 | 2·00000 |
| 20 | 0·57735 | 0·57735 | 0·57735 | 2·00000 |
| 21 | 0·57735 | −0·57735 | −0·57735 | 2·00000 |
| 22 | 0·57735 | 0·57735 | 0·57735 | 2·00000 |
| 23 | 0·57735 | −0·57735 | −0·57735 | 2·00000 |
| 24 | 0·57735 | 0·57735 | 0·57735 | 2·00000 |
| 25 | 0·57735 | −0·57734 | −0·57736 | 2·00000 |
| 26 | 0·57735 | 0·57733 | 0·57737 | 2·00000 |
| 27 | 0·57735 | −0·57732 | −0·57738 | 2·00000 |
| 28 | 0·57735 | 0·57728 | 0·57742 | 2·00000 |
| 29 | 0·57735 | −0·57721 | −0·57749 | 2·00000 |
| 30 | 0·57735 | 0·57708 | 0·57763 | 2·00000 |
| 31 | 0·57735 | −0·57680 | −0·57790 | 2·00000 |
| 32 | 0·57735 | 0·57625 | 0·57845 | 2·00000 |
| 33 | 0·57735 | −0·57515 | −0·57955 | 2·00001 |
| 34 | 0·57734 | 0·57294 | 0·58174 | 2·00003 |
| 35 | 0·57731 | −0·56850 | −0·58611 | 2·00012 |
| 36 | 0·57717 | 0·55957 | 0·59478 | 2·00046 |
| 37 | 0·57664 | −0·54146 | −0·61181 | 2·00186 |
| 38 | 0·57451 | 0·50442 | 0·64460 | 2·00741 |
| 39 | 0·56622 | −0·42806 | −0·70438 | 2·02926 |
| 40 | 0·53634 | 0·27461 | 0·79808 | 2·11142 |
| 41 | 0·45152 | −0·01084 | −0·89220 | 2·37574 |
| 42 | 0·30685 | −0·29212 | 0·90582 | 2·94292 |
| 43 | 0·17282 | 0·50187 | −0·84750 | 3·55111 |
| 44 | 0·08947 | −0·60910 | 0·78803 | 3·86326 |
| 45 | 0·04514 | 0·65980 | −0·75008 | 3·96382 |
| 46 | 0·02262 | −0·68394 | 0·72919 | 3·99082 |
| 47 | 0·01132 | 0·69565 | −0·71829 | 3·99770 |
| 48 | 0·00566 | −0·70141 | 0·71273 | 3·99942 |
| 49 | 0·00283 | 0·70427 | −0·70993 | 3·99986 |
| 50 | 0·00142 | −0·70569 | 0·70852 | 3·99996 |
| 51 | 0·00071 | 0·70640 | −0·70781 | 3·99999 |
| 52 | 0·00035 | −0·70675 | 0·70746 | 4·00000 |
| 53 | 0·00018 | 0·70693 | −0·70728 | 4·00000 |
| 54 | 0·00009 | −0·70702 | 0·70720 | 4·00000 |
| 55 | 0·00004 | 0·70706 | −0·70715 | 4·00000 |
| 56 | 0·00002 | −0·70708 | 0·70713 | 4·00000 |
| 57 | 0·00001 | 0·70710 | −0·70712 | 4·00000 |
| 58 | 0·00001 | −0·70710 | 0·70711 | 4·00000 |
| 59 | 0·00000 | 0·70710 | −0·70711 | 4·00000 |
| 60 | 0·00000 | −0·70711 | 0·70711 | 4·00000 |
| 61 | 0·00000 | 0·70711 | −0·70711 | 4·00000 |
| 62 | 0·00000 | −0·70711 | 0·70711 | 4·00000 |
| 63 | 0·00000 | 0·70711 | −0·70711 | 4·00000 |
| 64 | 0·00000 | −0·70711 | 0·70711 | 4·00000 |
| 65 | 0·00000 | 0·70711 | −0·70711 | 4·00000 |

And, finally, with starting vector

$$\mathbf{v}^{(0)} = \begin{bmatrix} 0 \\ 0 \\ 1 \end{bmatrix}$$

we obtain

| $k$ | $\mathbf{v}^{(k)}$ | | | $\lVert \mathbf{A}\mathbf{v}^{(k-1)} \rVert$ |
|---|---|---|---|---|
| 1 | 0·00000 | 0·31623 | −0·94868 | 3·16228 |
| 2 | 0·00000 | −0·51450 | 0·85749 | 3·68782 |
| 3 | 0·00000 | 0·61394 | −0·78935 | 3·91077 |
| 4 | 0·00000 | −0·66162 | 0·74984 | 3·97686 |
| 5 | 0·00000 | 0·68468 | −0·72885 | 3·99416 |
| 6 | 0·00000 | −0·69597 | 0·71807 | 3·99854 |
| 7 | 0·00000 | 0·70156 | −0·71261 | 3·99963 |
| 8 | 0·00000 | −0·70434 | 0·70986 | 3·99991 |
| 9 | 0·00000 | 0·70572 | −0·70849 | 3·99998 |
| 10 | 0·00000 | −0·70642 | 0·70780 | 3·99999 |
| 11 | 0·00000 | 0·70676 | −0·70745 | 4·00000 |
| 12 | 0·00000 | −0·70693 | 0·70728 | 4·00000 |
| 13 | 0·00000 | 0·70702 | −0·70719 | 4·00000 |
| 14 | 0·00000 | −0·70706 | 0·70715 | 4·00000 |
| 15 | 0·00000 | 0·70709 | −0·70713 | 4·00000 |
| 16 | 0·00000 | −0·70710 | 0·70712 | 4·00000 |
| 17 | 0·00000 | 0·70710 | −0·70711 | 4·00000 |
| 18 | 0·00000 | −0·70710 | 0·70711 | 4·00000 |
| 19 | 0·00000 | 0·70711 | −0·70711 | 4·00000 |
| 20 | 0·00000 | −0·70711 | 0·70711 | 4·00000 |

Again we see how an injudicious choice of starting vector can lead to apparent convergence to what in fact is a wrong answer. In the first case the starting vector was an exact eigenvector for one of the lower eigenvalues, and the second contained no component of the relevant eigenvector. The third case gave relatively fast convergence to the answer, correct to five decimal places:

$$\mathbf{x} = \begin{bmatrix} 0{\cdot}00000 \\ -0{\cdot}70711 \\ 0{\cdot}70711 \end{bmatrix} \qquad \lambda = -4{\cdot}00000.$$

Notice that $\mathbf{v}^{(k)}$ oscillates in sign because $\lambda$ is negative; it is irrelevant which sign we choose for the normalized eigenvector. Notice also that convergence of the eigenvalue in this case is faster than that of the eigenvector to the same number of decimal places.

(iii) With the starting vector

$$\mathbf{v}^{(0)} = \begin{bmatrix} 1 \\ 0 \\ 0 \end{bmatrix}$$

we obtain result as follows

| $k$ | $\mathbf{v}^{(k)}$ | | | $\|\mathbf{Av}^{(k-1)}\|$ |
|---|---|---|---|---|
| 1 | 0·89443 | 0·00000 | 0·44721 | 2·23607 |
| 2 | 0·72761 | −0·48507 | 0·48507 | 1·84391 |
| 3 | 0·89443 | 0·00000 | 0·44721 | 0·54233 |
| 4 | 0·72761 | −0·48507 | 0·48507 | 1·84391 |
| 5 | 0·89443 | 0·00000 | 0·44721 | 0·54233 |
| 6 | 0·72761 | −0·48507 | 0·48507 | 1·84391 |
| 7 | 0·89443 | 0·00000 | 0·44721 | 0·54233 |
| 8 | 0·72761 | −0·48507 | 0·48507 | 1·84391 |
| 9 | 0·89443 | 0·00000 | 0·44721 | 0·54233 |
| 10 | 0·72761 | −0·48507 | 0·48507 | 1·84391 |
| 11 | 0·89443 | 0·00000 | 0·44721 | 0·54233 |
| 12 | 0·72761 | −0·48507 | 0·48507 | 1·84391 |
| 13 | 0·89443 | 0·00000 | 0·44721 | 0·54233 |
| 14 | 0·72761 | −0·48507 | 0·48507 | 1·84391 |
| 15 | 0·89443 | 0·00000 | 0·44721 | 0·54233 |
| 16 | 0·72761 | −0·48507 | 0·48507 | 1·84391 |
| 17 | 0·89443 | 0·00000 | 0·44721 | 0·54233 |
| 18 | 0·72761 | −0·48507 | 0·48507 | 1·84391 |
| 19 | 0·89443 | 0·00000 | 0·44721 | 0·54233 |
| 20 | 0·72761 | −0·48507 | 0·48507 | 1·84391 |
| 21 | 0·89443 | 0·00000 | 0·44721 | 0·54233 |
| 22 | 0·72761 | −0·48507 | 0·48507 | 1·84391 |
| 23 | 0·89443 | 0·00000 | 0·44721 | 0·54233 |
| 24 | 0·72761 | −0·48507 | 0·48507 | 1·84391 |
| 25 | 0·89443 | 0·00000 | 0·44721 | 0·54233 |
| 26 | 0·72761 | −0·48507 | 0·48507 | 1·84391 |
| 27 | 0·89443 | 0·00000 | 0·44721 | 0·54233 |
| 28 | 0·72761 | −0·48507 | 0·48507 | 1·84391 |
| 29 | 0·89443 | 0·00000 | 0·44721 | 0·54233 |
| 30 | 0·72761 | −0·48507 | 0·48507 | 1·84391 |
| 31 | 0·89443 | 0·00000 | 0·44721 | 0·54233 |
| 32 | 0·72761 | −0·48507 | 0·48507 | 1·84391 |
| 33 | 0·89443 | 0·00000 | 0·44721 | 0·54233 |
| 34 | 0·72761 | −0·48507 | 0·48507 | 1·84391 |
| 35 | 0·89443 | 0·00000 | 0·44721 | 0·54233 |
| 36 | 0·72761 | −0·48507 | 0·48507 | 1·84391 |
| 37 | 0·89443 | 0·00000 | 0·44721 | 0·54233 |
| 38 | 0·72761 | −0·48507 | 0·48507 | 1·84391 |
| 39 | 0·89443 | 0·00000 | 0·44721 | 0·54233 |
| 40 | 0·72761 | −0·48507 | 0·48507 | 1·84391 |
| 41 | 0·89443 | 0·00000 | 0·44721 | 0·54233 |
| 42 | 0·72761 | −0·48507 | 0·48507 | 1·84391 |
| 43 | 0·89443 | 0·00000 | 0·44721 | 0·54233 |
| 44 | 0·72761 | −0·48507 | 0·48507 | 1·84391 |
| 45 | 0·89443 | 0·00000 | 0·44721 | 0·54233 |
| 46 | 0·72761 | −0·48507 | 0·48507 | 1·84391 |
| 47 | 0·89443 | 0·00000 | 0·44721 | 0·54233 |
| 48 | 0·72761 | −0·48507 | 0·48507 | 1·84391 |
| 49 | 0·89443 | 0·00000 | 0·44721 | 0·54233 |
| 50 | 0·72761 | −0·48507 | 0·48507 | 1·84391 |

We see that the results constantly oscillate; this is a consequence of the fact that the matrix **A** in this case does not have a unique eigenvalue of largest modulus—the eigenvalues being 0, 1, −1.

As presented above, the power method gives only the eigenvalue of highest modulus and a corresponding eigenvector. Modifications can be made to enable the power method to give more information.

Firstly, a simple technique can be used to obtain the eigenvalue of lowest modulus (provided this is unique) and a corresponding eigenvector. Since $\mathbf{Ax} = \lambda\mathbf{x}$ implies that $\mathbf{A}^{-1}\mathbf{x} = (1/\lambda)\mathbf{x}$ (provided $\mathbf{A}$ is non-singular) we see that if $\mathbf{x}$ is an eigenvector of $\mathbf{A}$ with eigenvalue $\lambda$ then $\mathbf{x}$ is also an eigenvector of $\mathbf{A}^{-1}$ with eigenvalue $1/\lambda$. It thus follows that the eigenvalue of largest modulus of $\mathbf{A}^{-1}$ is the reciprocal of the eigenvalue of lowest modulus of $\mathbf{A}$; thus finding the eigenvalue of largest modulus of $\mathbf{A}^{-1}$ (using the power method as described above) leads us to the eigenvalue of lowest modulus of $\mathbf{A}$ and the corresponding eigenvector.

Secondly, the method may be extended to yield all eigenvalues and eigenvectors provided the eigenvalues have distinct moduli; thus we assume that if $\mathbf{A}$ is of order $n$ with eigenvalues $\lambda_1, \lambda_2, \ldots \lambda_n$ then $|\lambda_1| > |\lambda_2| > \ldots > |\lambda_n|$.

We have already seen that the power method as described above works only if the initial starting vector $\mathbf{v}^{(0)}$ contains a component of $\mathbf{x}^{(1)}$; if there were no such component but there were a component of $\mathbf{x}^{(2)}$ then in principle the power method should converge to give $\lambda_2$ and $\mathbf{x}^{(2)}$. However, as we have seen in Example 4.7.1 (i) the presence of rounding error introduces the component of $\mathbf{x}^{(1)}$ and gives eventual convergence to $\lambda_1$ and $\mathbf{x}^{(1)}$. In order to ensure convergence to $\lambda_2$ and $\mathbf{x}^{(2)}$ it would be necessary to cut out any such component introduced through rounding error. Assuming that $\lambda_1$ is known this may be achieved by multiplying any vector $\mathbf{v}^{(k)}$ of the sequence by the matrix $\mathbf{B}_1 = (\mathbf{A} - \lambda_1\mathbf{I})$. For if

$$\mathbf{v}^{(k)} = \alpha_1\mathbf{x}^{(1)} + \alpha_2\mathbf{x}^{(2)} + \ldots + \alpha_n\mathbf{x}^{(n)}$$

then

$$\begin{aligned}\mathbf{B}_1\mathbf{v}^{(k)} &= (\mathbf{A} - \lambda_1\mathbf{I})(\alpha_1\mathbf{x}^{(1)} + \alpha_2\mathbf{x}^{(2)} + \ldots + \alpha_n\mathbf{x}^{(n)})\\ &= \alpha_1(\mathbf{A} - \lambda_1\mathbf{I})\mathbf{x}^{(1)} + \alpha_2(\mathbf{A} - \lambda_1)\mathbf{x}^{(2)} + \ldots + \alpha_n(\mathbf{A} - \lambda_1\mathbf{I})\mathbf{x}^{(n)}\\ &= \alpha_2(\lambda_2 - \lambda_1)\mathbf{x}^{(2)} + \alpha_3(\lambda_3 - \lambda_1)\mathbf{x}^{(3)} + \ldots + \alpha_n(\lambda_n - \lambda_1)\mathbf{x}^{(n)}.\end{aligned}$$

So $\mathbf{B}_1\mathbf{v}^{(k)}$ contains no component of $\mathbf{x}^{(1)}$. Similarly any other eigenvector $\mathbf{x}^{(r)}$ can be cut out by multiplying by the matrix $\mathbf{B}_r = (\mathbf{A} - \lambda_r\mathbf{I})$. Then the power method with the vectors $\mathbf{x}^{(1)}, \mathbf{x}^{(2)}, \ldots \mathbf{x}^{(r)}$ cut out will converge to $\lambda_{r+1}$ and $\mathbf{x}^{(r+1)}$.

Thus the following scheme will provide a method which, for suitable matrices, will yield all eigenvalues and eigenvectors:

(i) Choose an arbitrary vector $\mathbf{u}$.

(ii) Taking $\mathbf{v}^{(0)} = \mathbf{u}$ apply the power method to obtain $\lambda_1, \mathbf{x}^{(1)}$.

(iii) Form $\mathbf{B} = (\mathbf{A} - \lambda_1\mathbf{I})$.

(iv) Taking $\mathbf{v}^{(0)} = \mathbf{Bu}$ apply the power method to obtain $\lambda_2, \mathbf{x}^{(2)}$. It will be necessary to multiply the vector $\mathbf{v}^{(k)}$ by the matrix $\mathbf{B}$ at regular intervals to ensure that no component of the unwanted vector creeps in through rounding error.

(v) Form $\mathbf{B} = (\mathbf{A} - \lambda_2\mathbf{I})(\mathbf{A} - \lambda_1\mathbf{I})$.

(vi) Apply the power method with $\mathbf{v}^{(0)} = \mathbf{Bu}$ to obtain $\lambda_3, \mathbf{x}^{(3)}$, multiplying the vector $\mathbf{v}^{(k)}$ by $\mathbf{B}$ at regular intervals to remove unwanted components.

(vii) Repeat stages (v) and (vi) as often as necessary with $\mathbf{B} = (\mathbf{A} - \lambda_i\mathbf{I})(\mathbf{A} - \lambda_{i-1}\mathbf{I}) \ldots (\mathbf{A} - \lambda_1\mathbf{I})$ to obtain $\lambda_{i+1}, \mathbf{x}^{(i+1)}$. Again, at regular intervals the vector $\mathbf{v}^{(k)}$ must be multiplied by $\mathbf{B}$ to remove unwanted components.

METHOD 2: Jacobi's method

The power method described above was a theoretically simple method which under certain circumstances could yield all eigenvalues and eigenvectors. In practice, however, it had certain disadvantages; for example it depended on the eigenvalues having distinct moduli, it could be slow to converge and it could lead to wrong results if there were a bad choice of starting vector and/or convergence criterion. We now proceed to describe a method which is more complicated theoretically, but which gives more reliable results for a certain class of matrices. It is a transformation method in the sense that it involves the transformation of the original matrix to a new matrix with the same eigenvalues; however the new matrix is of such a form that the eigenvalues are now easy to find.

Jacobi's method is valid only for real symmetric matrices. For such matrices we know that the eigenvalues are all real (Theorem 4.3). We then utilize the result of Theorem 4.15 that there exists an orthogonal matrix $\mathbf{T}$ such that $\mathbf{T}^{-1}\mathbf{AT} = \mathbf{D}$ where $\mathbf{D}$ is diagonal, thus having its eigenvalues (and hence those of $\mathbf{A}$) as its diagonal elements. Also since the eigenvectors of $\mathbf{D}$ are just the columns $\mathbf{e}^{(i)}$ of the unit matrix $\mathbf{I}$ ($\mathbf{I} = [\mathbf{e}^{(1)}\ \mathbf{e}^{(2)} \ldots \mathbf{e}^{(n)}]$) it follows from Theorem 4.11 that the eigenvectors of $\mathbf{A}$ are just $\mathbf{Te}^{(i)}$ $(1 \leqslant i \leqslant n)$. Of course these vectors are just the columns of $\mathbf{T}$.

Jacobi's method provides a means of obtaining $\mathbf{T}$ as the limit of a sequence of matrices. In practice by proceeding sufficiently far along the sequence we obtain an approximation to $\mathbf{T}$ which transforms $\mathbf{A}$ to a diagonal matrix to the required accuracy.

We thus construct a sequence of matrices $\mathbf{A}_k$ such that

$$\mathbf{A}_0 = \mathbf{A}$$

and

$$\mathbf{A}_k = \mathbf{U}_k^{-1}\mathbf{A}_{k-1}\mathbf{U}_k \qquad (k \geqslant 1).$$

Thus

$$\begin{aligned}\mathbf{A}_k &= \mathbf{U}_k^{-1}\mathbf{U}_{k-1}^{-1} \ldots \mathbf{U}_1^{-1}\mathbf{A}\mathbf{U}_1\mathbf{U}_2 \ldots \mathbf{U}_{k-1}\mathbf{U}_k \\ &= \mathbf{T}_k^{-1}\mathbf{A}\mathbf{T}_k\end{aligned}$$

where $\mathbf{T}_k = \mathbf{U}_1\mathbf{U}_2 \ldots \mathbf{U}_k$. We shall construct the $\mathbf{U}$ matrices in such a way that, for sufficiently large $k$, $\mathbf{A}_k$ is very nearly diagonal; then its diagonal elements are approximations to the eigenvalues and the columns of $\mathbf{T}_k$ are approximations to the eigenvectors. Every $\mathbf{U}_k$ and hence every $\mathbf{T}_k$ will be seen to be orthogonal, so that $\mathbf{U}_k^{-1} = \mathbf{U}_k^{\mathrm{T}}$ and $\mathbf{T}_k^{-1} = \mathbf{T}_k^{\mathrm{T}}$.

At any stage (i.e. for any value of $k$) we aim to make $\mathbf{A}_k$ 'more diagonal' than $\mathbf{A}_{k-1}$ by destroying one of the off-diagonal elements, so that $\mathbf{A}$ will have a zero element in an off-diagonal position where there was a non-zero element in $\mathbf{A}_{k-1}$.

We shall prove first that the sum of the squares of all the elements of $\mathbf{A}_k$ is the same as the sum of the squares of all the elements of $\mathbf{A}_{k-1}$; this will help in showing that when an off-diagonal element is reduced to zero the matrix is in fact made more diagona
The appropriate result is contained in the following theorem.

THEOREM 4.29

If $\mathbf{A}$ and $\mathbf{B}$ are square matrices of order $n$ related by a similarity transformation then

$$\sum_{i,j=1}^{n} a_{ij}^2 = \sum_{i,j=1}^{n} b_{ij}^2.$$

*Proof*

For a square matrix $\mathbf{A}$ of order $n$, if we form the product $\mathbf{A}^2$ and sum the diagonal elements we find that we obtain the result

$$\operatorname{Tr}(\mathbf{A}^2) = \sum_{i,j=1}^{n} a_{ij}^2.$$

Similarly of course

$$\operatorname{Tr}(\mathbf{B}^2) = \sum_{i,j=1}^{n} b_{ij}^2.$$

But if $\lambda_i (1 \leqslant i \leqslant n)$ are the eigenvalues of $\mathbf{A}$ and $\mu_i (1 \leqslant i \leqslant n)$ are the eigenvalues of $\mathbf{B}$ then Theorem 4.10 tells us that

$$\operatorname{Tr}(\mathbf{A}^2) = \sum_{i=1}^{n} \lambda_i^2 \qquad \text{and} \qquad \operatorname{Tr}(\mathbf{B}^2) = \sum_{i=1}^{n} \mu_i^2.$$

And Theorem 4.11 tells us that since $\mathbf{A}$ and $\mathbf{B}$ are similar they have the same eigenvalues; thus $\operatorname{Tr}(\mathbf{A}^2) = \operatorname{Tr}(\mathbf{B}^2)$ and so

$$\sum_{i,j=1}^{n} a_{ij}^2 = \sum_{i,j=1}^{n} b_{ij}^2. \qquad \blacksquare$$

Let us now see the effect of choosing

$$\mathbf{U}_k = \begin{bmatrix} 1 &&&&&&&&&&& \\ & 1 &&&&&&&&&& \\ && \ddots &&&&&&&&& \\ &&& 1 &&&&&&&& \\ &&&& \cos\theta_k &&&& \sin\theta_k &&& \\ &&&&& 1 &&&&&& \\ &&&&&& \ddots &&&&& \\ &&&&&&& 1 &&&& \\ &&&& -\sin\theta_k &&&& \cos\theta_k &&& \\ &&&&&&&&& 1 && \\ &&&&&&&&&& \ddots & \\ &&&&&&&&&&& 1 \end{bmatrix} \begin{matrix} \\ \\ \\ \\ \leftarrow \text{row } p \\ \\ \\ \\ \leftarrow \text{row } q \\ \\ \\ \\ \end{matrix}$$

$$\qquad\qquad \uparrow \text{column } p \qquad\qquad \uparrow \text{column } q$$

all other elements being zero. This is a generalized rotation matrix—when applied to the vector

$$\begin{bmatrix} x_1 \\ x_2 \\ \cdot \\ \cdot \\ \cdot \\ x_n \end{bmatrix}$$

it gives $\mathbf{y} = \mathbf{U}_k\mathbf{x}$, where

$$y_i = x_i \qquad (i \neq p, q)$$

$$y_p = x_p \cos\theta_k + x_q \sin\theta_k$$

$$y_q = -x_p \sin\theta_k + x_q \cos\theta_k.$$

Comparing these equations with the result of Example 2.8.5 we see that we can talk about $\mathbf{U}_k$ representing a rotation of amount $\theta_k$ in the $pq$-plane. We also see immediately that $\mathbf{U}_k$ is orthogonal; for it is a trivial matter to verify that $\mathbf{U}_k^{\mathrm{T}}\mathbf{U}_k = \mathbf{U}_k\mathbf{U}_k^{\mathrm{T}} = \mathbf{I}$. Thus we have $\mathbf{U}_k^{-1} = \mathbf{U}_k^{\mathrm{T}}$.

If now we consider the equation $\mathbf{A}_k = \mathbf{U}_k^{\mathrm{T}}\mathbf{A}_{k-1}\mathbf{U}_k$ and denote the elements of $\mathbf{A}_k$ by $a_{ij}^{(k)}$ and those of $\mathbf{A}_{k-1}$ by $a_{ij}^{(k-1)}$ it follows in a simple but slightly tedious manner on writing out the matrices involved that we obtain the results

$$\left.\begin{aligned} a_{pj}^{(k)} &= a_{pj}^{(k-1)}\cos\theta_k - a_{qj}^{(k-1)}\sin\theta_k \\ a_{qj}^{(k)} &= a_{pj}^{(k-1)}\sin\theta_k + a_{qj}^{(k-1)}\cos\theta_k \end{aligned}\right\} \quad j \neq p \text{ or } q \tag{4.7.1}$$

$$\left.\begin{aligned} a_{ip}^{(k)} &= a_{ip}^{(k-1)}\cos\theta_k - a_{iq}^{(k-1)}\sin\theta_k \\ a_{iq}^{(k)} &= a_{ip}^{(k-1)}\sin\theta_k + a_{iq}^{(k-1)}\cos\theta_k \end{aligned}\right\} \quad i \neq p \text{ or } q \tag{4.7.2}$$

$$\begin{aligned} a_{pp}^{(k)} &= a_{pp}^{(k-1)}\cos^2\theta_k - 2a_{pq}^{(k-1)}\sin\theta_k\cos\theta_k + a_{qq}^{(k-1)}\sin^2\theta_k \\ a_{qq}^{(k)} &= a_{pp}^{(k-1)}\sin^2\theta_k + 2a_{pq}^{(k-1)}\sin\theta_k\cos\theta_k + a_{qq}^{(k-1)}\cos^2\theta_k \\ a_{pq}^{(k)} &= (a_{pp}^{(k-1)} - a_{qq}^{(k-1)})\sin\theta_k\cos\theta_k + a_{pq}^{(k-1)}(\cos^2\theta_k - \sin^2\theta_k) \end{aligned} \tag{4.7.3}$$

$$a_{ij}^{(k)} = a_{ij}^{(k-1)} \qquad (i \neq p \text{ and } j \neq q). \tag{4.7.4}$$

We can represent in a diagrammatic form the elements altered in $\mathbf{A}_k$ as compared with $\mathbf{A}_{k-1}$ (see Figure 4.7.1).

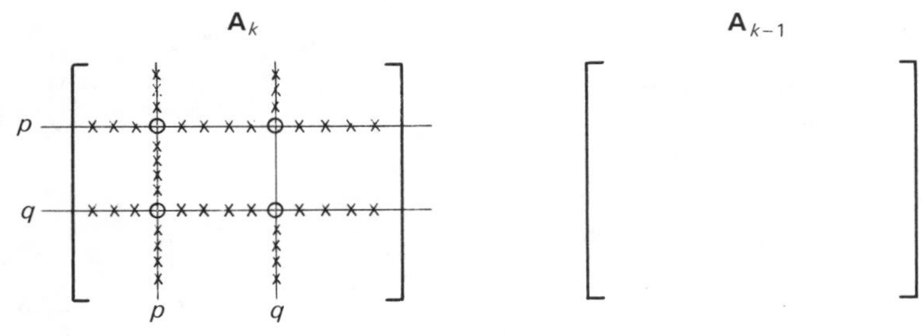


Figure 4.7.1

We can also show diagrammatically which elements in $\mathbf{A}_{k-1}$ contribute to the new elements in $\mathbf{A}_k$ (see Figure 4.7.2).

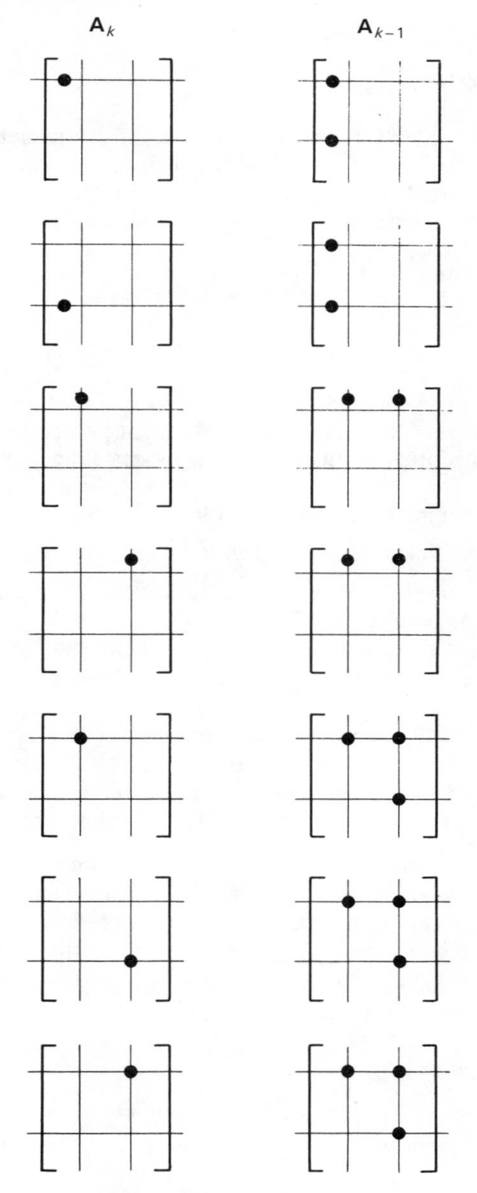


Figure 4.7.2

It is the last of the diagrams in Figure 4.7.2 which is the appropriate one for present considerations. By using the last of equations (4.7.3) we may reduce the off-diagonal element $a_{pq}^{(k)}$ to zero by choosing $\theta_k$ so that

$$(a_{pp}^{(k-1)} - a_{qq}^{(k-1)}) \sin \theta_k \cos \theta_k + a_{pq}^{(k-1)}(\cos^2 \theta_k - \sin^2 \theta_k) = 0.$$

Using the facts that $\sin 2\theta_k = 2 \sin \theta_k \cos \theta_k$ and $\cos 2\theta_k = \cos^2 \theta_k - \sin^2 \theta_k$ this reduces to

$$\tan 2\theta_k = \frac{2a_{pq}^{(k-1)}}{a_{qq}^{(k-1)} - a_{pp}^{(k-1)}}. \tag{4.7.5}$$

Substituting $\theta_k$ as determined by equation (4.7.5) into $\mathbf{U}_k$ thus reduces a pair of off-diagonal elements in $\mathbf{A}_k$ to zero (remember that $\mathbf{A}_k$ is symmetric). Unfortunately, the next step in the process while creating a new pair of zeros may well introduce non-zero contributions to formerly zero positions. However, from the equations (4.7.1) and (4.7.2) determining the transformed matrix we may deduce that

$$(a_{pj}^{(k)})^2 + (a_{qj}^{(k)})^2 = (a_{pj}^{(k-1)})^2 + (a_{qj}^{(k-1)})^2 \qquad (j \neq p \text{ or } q),$$

and

$$(a_{ip}^{(k)})^2 + (a_{iq}^{(k)})^2 = (a_{ip}^{(k-1)})^2 + (a_{iq}^{(k-1)})^2 \qquad (i \neq p \text{ or } q).$$

Thus the sum of squares of off-diagonal elements is reduced by $2(a_{pq}^{(k-1)})^2$ and since the total sum of squares remains the same the diagonal elements must be increased by the same amount. Hence the new matrix is in fact a better approximation to the desired diagonal matrix.

In practice the Jacobi method would be utilized as follows:

At every stage of the iteration, starting with $\mathbf{A}_0 = \mathbf{A}$, we should

(i) choose a non-zero off-diagonal element, thus defining $p$ and $q$;

(ii) calculate $\theta_k$ from equation (4.7.5) and hence obtain $\mathbf{U}_k$;

(iii) calculate those elements of $\mathbf{A}_k$ which differ from those of $\mathbf{A}_{k-1}$ using the basic transformation equations (4.7.1)–(4.7.3); also calculate $\mathbf{T}_k$ from $\mathbf{T}_k = \mathbf{T}_{k-1}\mathbf{U}_k$ ($\mathbf{T}_0 = \mathbf{I}$);

(iv) decide if the iteration has proceeded far enough (for example if the sum of the squares of the off-diagonal elements of $\mathbf{A}_k$ is near enough to zero); if it has, the diagonal elements of $\mathbf{A}_k$ give approximations to the eigenvalues and the columns of $\mathbf{T}_k$ give approximations to the eigenvectors, otherwise the iteration is repeated from step (i).

The only point that remains to be considered is how to choose $p$ and $q$ in step (i) above. Suitable strategies are either to choose $a_{pq}^{(k-1)}$ as the largest element or, to give greater efficiency in computer implementation, to choose $a_{pq}^{(k-1)}$ to be the first element found in a systematic search greater than some prescribed threshold value.

In a proper study of Jacobi's method, its convergence should be studied, but this topic is beyond the scope of the present work.

*Example*

4.7.2 Consider the Jacobi method applied to the matrix

$$\mathbf{A} = \begin{bmatrix} 11 & 2 & 8 \\ 2 & 2 & -10 \\ 9 & -10 & 5 \end{bmatrix}.$$

Even for as low order a matrix as this, the calculations involved are exceedingly tedious, and the computations described below were all performed by computer. The arithmetic was performed to eleven significant decimal digits; the results quoted are rounded to two decimal places.

The criterion used to select the element to be annihilated was to choose the element of largest modulus at that stage.

Thus the first step is to choose $p = 2$, $q = 3$ (so that $a_{pq} = a_{23} = -10$ is annihilated). From equation (4.7.5) we then have

$$\tan 2\theta_1 = \frac{2a_{23}^{(0)}}{a_{33}^{(0)} - a_{22}^{(0)}} = \frac{2 \times (-10)}{5 - 2} = -\frac{20}{3}.$$

To use this in the calculation of $\sin\theta_1$ and $\cos\theta_1$ which are required in the $\mathbf{U}$ matrix it is better from a computational point of view not to calculate $\theta_1$ but instead to make use of the results that if

$$\tan 2\alpha = \lambda/\mu \qquad (\mu > 0)$$

then

$$\cos\alpha = [\tfrac{1}{2}\{1 + \mu/(\lambda^2 + \mu^2)^{1/2}\}]^{1/2}$$

and

$$\sin\alpha = \frac{\lambda}{[2\{\lambda^2 + \mu^2 + \mu(\lambda^2 + \mu^2)^{1/2}\}]^{1/2}}.$$

In this case $\lambda = -20$, $\mu = 3$ giving $\cos\theta_1 = 0{\cdot}76$ and $\sin\theta_1 = -0{\cdot}65$; thus

$$\mathbf{U}_1 = \begin{bmatrix} 1 & 0 & 0 \\ 0 & 0{\cdot}76 & -0{\cdot}65 \\ 0 & 0{\cdot}65 & 0{\cdot}76 \end{bmatrix}$$

and so

$$\mathbf{A}_1 = \mathbf{U}_1^{\mathrm{T}}\mathbf{A}\mathbf{U}_1 = \begin{bmatrix} 1 & 0 & 0 \\ 0 & 0{\cdot}76 & 0{\cdot}65 \\ 0 & -0{\cdot}65 & 0{\cdot}76 \end{bmatrix} \begin{bmatrix} 11 & 2 & 8 \\ 2 & 2 & -10 \\ 8 & -10 & 5 \end{bmatrix} \begin{bmatrix} 1 & 0 & 0 \\ 0 & 0{\cdot}76 & -0{\cdot}65 \\ 0 & 0{\cdot}65 & 0{\cdot}76 \end{bmatrix}$$

$$= \begin{bmatrix} 11 & 6{\cdot}74 & -4{\cdot}76 \\ 6{\cdot}74 & -6{\cdot}61 & 0 \\ -4{\cdot}76 & 0 & 13{\cdot}61 \end{bmatrix}.$$

The next stage is to eliminate what is now the off-diagonal element of largest modulus $>-6{\cdot}74$. So we take $p = 1$, $q = 2$ and proceed as before, calculating

$$\mathbf{U}_2 = \begin{bmatrix} 0{\cdot}95 & -0{\cdot}32 & 0 \\ 0{\cdot}32 & 0{\cdot}95 & 0 \\ 0 & 0 & 1 \end{bmatrix}$$

and

$$\mathbf{A}_2 = \mathbf{U}_2^{\mathrm{T}}\mathbf{A}_1\mathbf{U}_2 = \begin{bmatrix} 13{\cdot}28 & 0 & -4{\cdot}50 \\ 0 & -8{\cdot}89 & -1{\cdot}53 \\ -4{\cdot}50 & -1{\cdot}53 & 13{\cdot}61 \end{bmatrix}.$$

Proceeding in this manner we obtain the following results:

| $k$ | $\mathbf{A}_k$ | Sum of squares of off-diagonal elements |
|---|---|---|
| 0 | $\begin{bmatrix} 11 & 2 & 8 \\ 2 & 2 & -10 \\ 8 & -10 & 5 \end{bmatrix}$ | 336·00 |
| 1 | $\begin{bmatrix} 11 & 6{\cdot}74 & -4{\cdot}76 \\ 6{\cdot}74 & -6{\cdot}61 & 0 \\ -4{\cdot}76 & 0 & 13{\cdot}61 \end{bmatrix}$ | 136·00 |
| 2 | $\begin{bmatrix} 13{\cdot}28 & 0 & -4{\cdot}50 \\ 0 & -3{\cdot}89 & -1{\cdot}53 \\ -4{\cdot}50 & -1{\cdot}53 & 13{\cdot}61 \end{bmatrix}$ | 45·25 |
| 3 | $\begin{bmatrix} 8{\cdot}94 & -1{\cdot}06 & 0 \\ -1{\cdot}06 & -8{\cdot}89 & 1{\cdot}10 \\ 0 & 1{\cdot}10 & 17{\cdot}95 \end{bmatrix}$ | 4·66 |
| 4 | $\begin{bmatrix} 8{\cdot}94 & -1{\cdot}06 & 0{\cdot}04 \\ -1{\cdot}06 & -8{\cdot}94 & 0 \\ 0{\cdot}04 & 0 & 18{\cdot}00 \end{bmatrix}$ | 2·24 |
| 5 | $\begin{bmatrix} 9{\cdot}00 & 0 & 0{\cdot}04 \\ 0 & -9{\cdot}00 & 0{\cdot}00 \\ 0{\cdot}04 & 0{\cdot}00 & 18{\cdot}00 \end{bmatrix}$ | 0·00 |
| 6 | $\begin{bmatrix} 9{\cdot}00 & 0{\cdot}00 & 0 \\ 0{\cdot}00 & -9{\cdot}00 & 0{\cdot}00 \\ 0 & 0{\cdot}00 & 18{\cdot}00 \end{bmatrix}$ | 0·00 |

Thus we see that we obtain convergence to the eigenvalues 9·00, –9·00, 18·00. Computing the sequence of $\mathbf{U}_k$ would give the matrix $\mathbf{T}_6$ containing the corresponding eigenvectors as its columns:

$$\mathbf{T}_6 = \begin{bmatrix} 0{\cdot}67 & -0{\cdot}33 & 0{\cdot}67 \\ 0{\cdot}67 & 0{\cdot}67 & -0{\cdot}33 \\ -0{\cdot}33 & 0{\cdot}67 & 0{\cdot}67 \end{bmatrix}.$$

METHOD 3: QR method

Jacobi's method as described above is a reasonably reliable method, but can take a comparatively long computation time and, more importantly, is valid for real symmetric matrices only. We now describe a modern (1961) method which is both extremely stable numerically and is applicable to a general matrix. Although the reader should be aware of the fact that for special types of matrices such as tridiagonal or sparse there may well be preferable methods, it is still true that the QR method provides the basis for a good general prupose procedure for finding eigenvalues and eigenvectors.

The fundamental property which the method utilizes is that any real matrix **A** can be written in the form **A** = **QU** where **Q** is orthogonal and **U** is upper triangular.† (This is in contrast to the situation holding in Theorems 2.14 and 2.15 where the decomposition into the form **LU** was dependent on certain submatrices of **A** having non-zero determinants.) This result is proved in the following theorem.

THEOREM 4.30

For an arbitrary real square matrix **A** there exists an orthogonal matrix **Q** and an upper triangular matrix **U** such that **A** = **QU**.

*Proof*

We first re-introduce the 'rotation' matrix as used in Jacobi's method, but denote it by $\mathbf{R}(p, q, \theta)$ to indicate which rows contain the non-zero off-diagonal elements:

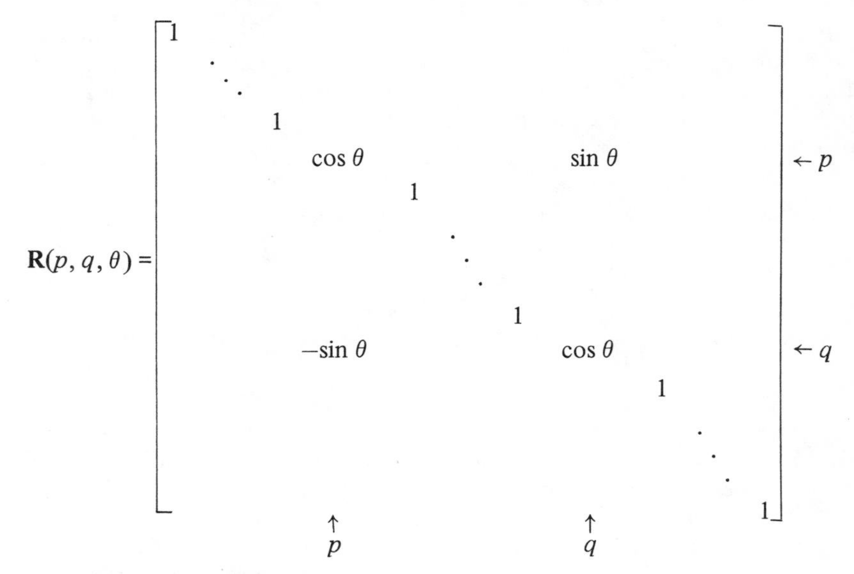

(We recall also that **R** is orthogonal.)

The effect of multiplying a vector **x** by $\mathbf{R}(p, q, \theta)$ is that if $\mathbf{y} = \mathbf{Rx}$ then

$$
\begin{aligned}
y_p &= r_{pp}x_p + r_{pq}x_q \\
&= \cos\theta\, x_p + \sin\theta\, x_q \\
y_q &= r_{qp}x_p + r_{qq}x_q \\
&= -\sin\theta\, x_p + \cos\theta\, x_q \\
y_i &= x_i \qquad (i \neq p \text{ or } q).
\end{aligned}
$$

We may thus choose $\theta$ in such a way as to make either $y_p$ or $y_q$ equal to zero. In addition we see that if $x_p = x_q = 0$ then $y_p = y_q = 0$ also.

† The name of the method derives from the fact that the notation of **R** for upper (right) triangular matrix was used in the original presentation by Francis (J. G. F. Francis, "The QR Transformation, A Unitary Analogue to the LR Transformation" Computer Journal Vol. 4 (1961).)

Thus, if we write

$$\mathbf{A} = [\mathbf{x}^{(1)} \quad \mathbf{x}^{(2)} \quad \ldots \quad \mathbf{x}^{(n)}]$$

then

$$\mathbf{R}(p, q, \theta)\mathbf{A} = [\mathbf{y}^{(1)} \quad \mathbf{y}^{(2)} \quad \ldots \quad \mathbf{y}^{(n)}]$$

and by a suitable choice of $\theta$ we can reduce to zero an element in either row $p$ or row $q$. It thus follows that with suitable $\theta$'s we can reduce to zero all the elements in the first column after the first element (since $\mathbf{R}(1, 1, \theta)$ is not defined) by forming the product

$$\mathbf{R}(n, 1, \theta_{n,1})\mathbf{R}(n-1, 1, \theta_{n-1,1}) \ldots \mathbf{R}(3, 1, \theta_{31})\mathbf{R}(2, 1, \theta_{21})\mathbf{A}.$$

If we now multiply this by $\mathbf{R}(3, 2, \theta_{32})$ we will be able, by a suitable choice of $\theta_{32}$, to reduce the element in the third row and second column to zero; also since the element in these positions in the first column are now both zero, they will remain zero after the multiplication by $\mathbf{R}$. Proceeding in this way we see that we obtain an upper triangular matrix by forming the product

$$\mathbf{U} = \mathbf{R}(n, n-1, \theta_{n,n-1})\mathbf{R}(n, n-2, \theta_{n,n-2})\mathbf{R}(n-1, n-2, \theta_{n-1,n-2}) \ldots$$

$$\mathbf{R}(3, 1, \theta_{31})\mathbf{R}(2, 1, \theta_{21})\mathbf{A}$$

$$= \left( \prod_{j=1}^{n-1} \prod_{i=j+1}^{n} \mathbf{R}(i, j, \theta_{ij}) \right) \mathbf{A}.$$

However, since each $\mathbf{R}$ is an orthogonal matrix it follows that the product of $\mathbf{R}$'s is also orthogonal and thus we have the result that $\mathbf{U} = \mathbf{SA}$ where $\mathbf{S}$ is orthogonal and $\mathbf{U}$ is upper triangular.

Thus, since $\mathbf{S}$ is orthogonal we have $\mathbf{A} = \mathbf{S}^T\mathbf{U}$ and so $\mathbf{A} = \mathbf{QU}$ letting $\mathbf{Q} = \mathbf{S}^T$ This proves the required result, since the transpose of an orthogonal matrix is also orthogonal. ■

If we now make use of the above theorem we may construct a sequence of matrices $\mathbf{A}_0, \mathbf{A}_1, \mathbf{A}_2, \ldots$ defined by

$$\mathbf{A}_0 = \mathbf{A} = \mathbf{Q}_0\mathbf{U}_0$$

and

$$\mathbf{A}_k = \mathbf{Q}_k\mathbf{U}_k = \mathbf{U}_{k-1}\mathbf{Q}_{k-1} \qquad (k \geqslant 1).$$

This just means that we start with $\mathbf{A}_0 = \mathbf{A}$, write it in the form $\mathbf{Q}_0\mathbf{U}_0$ to obtain $\mathbf{Q}_0$ and $\mathbf{U}_0$ and then obtain $\mathbf{A}_1$ by taking the product $\mathbf{U}_0\mathbf{Q}_0$ in the reverse order. And we proceed in this manner to construct as many terms in the sequence as we please:

$$
\begin{aligned}
\mathbf{A} = \mathbf{A}_0 &= \mathbf{Q}_0\mathbf{U}_0 && \text{defines } \mathbf{Q}_0, \mathbf{U}_0 \\
\mathbf{A}_1 &= \mathbf{U}_0\mathbf{Q}_0 && \text{defines } \mathbf{A}_1 \\
&= \mathbf{Q}_1\mathbf{U}_1 && \text{defines } \mathbf{Q}_1, \mathbf{U}_1 \\
\mathbf{A}_2 &= \mathbf{U}_1\mathbf{Q}_1 && \text{defines } \mathbf{A}_2 \\
&= \mathbf{Q}_2\mathbf{U}_2 && \text{defines } \mathbf{Q}_2, \mathbf{U}_2 \\
&\vdots \\
\mathbf{A}_k &= \mathbf{U}_{k-1}\mathbf{Q}_{k-1} && \text{defines } \mathbf{A}_k \\
&= \mathbf{Q}_k\mathbf{U}_k && \text{defines } \mathbf{Q}_k, \mathbf{U}_k \\
&\vdots
\end{aligned}
$$

The usefulness of this sequence for the determination of eigenvalues is shown by the following theorem.

THEOREM 4.31

If $\mathbf{A}_0, \mathbf{A}_1, \ldots$ is a sequence of matrices such that $\mathbf{A}_0 = \mathbf{A}$ and $\mathbf{A}_k = \mathbf{Q}_k\mathbf{U}_k = \mathbf{U}_{k-1}\mathbf{Q}_{k-1}$ $(k = 1, 2 \ldots)$ and the product $\mathbf{Q}_0\mathbf{Q}_1\mathbf{Q}_2 \ldots \mathbf{Q}_k$ converges as $k \to \infty$ then $\mathbf{A}_k$ converges to an upper triangular matrix with the eigenvalues of $\mathbf{A}$ as its diagonal elements.

*Proof*

Since

$$
\begin{aligned}
\mathbf{A}_k &= \mathbf{Q}_k\mathbf{U}_k = \mathbf{U}_{k-1}\mathbf{Q}_{k-1} \\
\mathbf{A}_{k-1} &= \mathbf{Q}_{k-1}\mathbf{U}_{k-1}
\end{aligned}
$$

we have

$$\mathbf{U}_{k-1} = \mathbf{Q}_{k-1}^{-1}\mathbf{A}_{k-1}$$

(valid since $\mathbf{Q}_{k-1}$ is orthogonal and so possesses an inverse) and thus

$$\mathbf{A}_k = \mathbf{Q}_{k-1}^{-1}\mathbf{A}_{k-1}\mathbf{Q}_{k-1}.$$

Hence $\mathbf{A}_k$ is similar to $\mathbf{A}_{k-1}$, implying that all the $\mathbf{A}_k$ are similar to $\mathbf{A}_0 = \mathbf{A}$ and thus have the same eigenvalues as $\mathbf{A}$.

Also

$$\mathbf{A}_{k-1} = \mathbf{Q}_{k-2}^{-1}\mathbf{A}_{k-2}\mathbf{Q}_{k-2}$$

and thus we obtain

$$
\begin{aligned}
\mathbf{A}_{k+1} &= \mathbf{Q}_k^{-1}\mathbf{Q}_{k-1}^{-1} \ldots \mathbf{Q}_0^{-1}\mathbf{A}\mathbf{Q}_0\mathbf{Q}_1 \ldots \mathbf{Q}_k \\
&= \mathbf{P}_k^{-1}\mathbf{A}\mathbf{P}_k \qquad \text{writing } \mathbf{P}_k = \mathbf{Q}_0\mathbf{Q}_1 \ldots \mathbf{Q}_k
\end{aligned}
$$

Also, if the condition of the theorem is satisfied $\lim_{k\to\infty} \mathbf{P}_k$ exists and we may denote it by $\mathbf{P}$.

Then

$$\lim_{k\to\infty} \mathbf{Q}_k = \lim_{k\to\infty} \left(\mathbf{P}_{k-1}^{-1}\mathbf{P}_k\right)$$
$$= \left(\lim_{k\to\infty} \mathbf{P}_{k-1}^{-1}\right)\left(\lim_{k\to\infty} \mathbf{P}_k\right)$$
$$= \mathbf{P}^{-1}\mathbf{P}$$
$$= \mathbf{I}.$$

Now we may obtain two expressions for $\lim_{k\to\infty} \mathbf{A}_k$. Firstly, since $\mathbf{A}_k = \mathbf{P}_{k-1}^{-1}\mathbf{A}\mathbf{P}_{k-1}$ we have

$$\lim_{k\to\infty} \mathbf{A}_k = \left(\lim_{k\to\infty} \mathbf{P}_{k-1}^{-1}\right)\mathbf{A}\left(\lim_{k\to\infty} \mathbf{P}_{k-1}\right)$$
$$= \mathbf{P}^{-1}\mathbf{A}\mathbf{P}.$$

So in the limit $\mathbf{A}_k$ is similar to $\mathbf{A}$ and thus has the same eigenvalues as $\mathbf{A}$.

Secondly, since $\mathbf{A}_k = \mathbf{Q}_k\mathbf{U}_k$ we have

$$\lim_{k\to\infty} \mathbf{A}_k = \left(\lim_{k\to\infty} \mathbf{A}_k\right)\left(\lim_{k\to\infty} \mathbf{U}_k\right).$$

But

$$\lim_{k\to\infty} \mathbf{Q}_k = \mathbf{I}$$

giving

$$\lim_{k\to\infty} \mathbf{A}_k = \lim_{k\to\infty} \mathbf{U}_k$$

showing that, since every $\mathbf{U}_k$ is an upper triangular matrix the limit of $\mathbf{A}_k$ must also be an upper triangular matrix.

These two facts concerning $\lim_{k\to\infty} \mathbf{A}_k$ complete the proof of the theorem. ■

There are two points which remain to be discussed before the method can be used in practice:

(i) No mention has been made of the conditions under which $\mathbf{Q}_0\mathbf{Q}_1 \ldots \mathbf{Q}_k$ possesses a limit as $k \to \infty$. The investigation of the existence of this limit is beyond the scope of present considerations, and we content ourselves with noting that it does in fact exist for a large class of matrices.

(ii) The method depends crucially on finding a good algorithm for the decomposition of $\mathbf{A}_k$ into the form $\mathbf{Q}_k\mathbf{U}_k$. Again this point is beyond the scope of the present work.

For further details of both the above matters, the interested reader should consult one of the texts specializing in numerical methods for eigenvalue problems.

# Bibliography


1. Bellman, R. *Introduction to Matrix Analysis.* McGraw-Hill (1960).
2. Bronson, R. *Matrix Methods.* Academic Press (1969).
3. Cullen, G. C. *Matrices and Linear Transformations.* Addison-Wesley (1966).
4. Faddeev, D. K. and Faddeeva, V. N. *Computational Methods of Linear Algebra.* Freeman (1963).
5. Ferrar, W. L. *Algebra.* Oxford University Press (1941).
6. Ferrar, W. L. *Finite Matrices.* Oxford University Press (1951).
7. Finkbeiner, D. T. *Introduction to Matrices and Linear Transformations.* Freeman (1966).
8. Fox, L. *An Introduction to Numerical Linear Algebra.* Oxford University Press (1964).
9. Gantmacher, F. R. *The Theory of Matrices.* Chelsea (1959).
10. Gere, J. M. and Weaver, W. *Matrix Algebra for Engineers.* Van Nostrand (1965).
11. Goult, R. J., Hoskins, R. F., Milner, J. A. and Pratt, M. J. *Computational Methods in Linear Algebra.* Stanley Thornes (1974).
12. Gourlay, A. R. and Watson, G. A. *Computational Methods for Matrix Eigenproblems.* Wiley (1973).
13. Hadley, G. *Linear Algebra.* Addison-Wesley (1969).
14. Hohn, F. E. *Elementary Matrix Algebra.* Macmillan (N.Y.) (1958).
15. Householder, A. S. *The Theory of Matrices in Numerical Analysis.* Blaisdell (1964).
16. MacDuffee, C. C. *The Theory of Matrices.* Chelsea (1956).
17. Mirsky, L. *An Introduction to Linear Algebra.* Oxford University Press (1963).
18. Pipes, L. A. *Matrix Methods for Engineering.* Prentice-Hall (1963).
19. Ralston, A. *A First Course in Numerical Analysis.* McGraw-Hill (1965).
20. Wendroff, B. *Theoretical Numerical Analysis.* Academic Press (1966).
21. Wilkinson, J. H. *The Algebraic Eigenvalue Problem.* Oxford University Press (1965).
22. Wilkinson, J. H. and Reinsch, C. *Handbook for Automatic Computation.* Vol. II—Linear Algebra. Springer-Verlag (1971).

# Solutions to Exercises

*Chapter 1*

1.1.1 (i) $-192$ (ii) $0$ (iii) $-(a+b+c)(a^2+b^2+c^2-bc-ca-ab)$

1.1.2 (i) $x=-19/3$ $y=6$ $z=16/3$
(ii) $x=3$ $y=0$ $z=0$

1.2.1 $-156$

1.2.2 $-1$

1.2.3 $x=1$ $y=-1$ $z=2$ $u=0$

1.2.4 $-180$

*Chapter 2*

2.1.1
$$\mathbf{A}=\begin{bmatrix}0 & 1 & 2 & 3\\ 1 & 0 & 1 & 2\\ 2 & 1 & 0 & 1\end{bmatrix}$$

2.1.2 $A_{13}=5$ $A_{22}=9$ $A_{31}=2$ $B_{12}=2xy$
$B_{13}$ is undefined $B_{32}=x^2-y^2$ $C_{11}=\mathrm{e}^{\theta}$
$C_{31}=\sinh\theta$ $u_2=\beta$ $u_3=\gamma$

2.1.3 (i) Impossible (ii) $x=2$ $y=1$

2.2.1 (i) $\begin{bmatrix}0 & 7 & 5\\ 5 & -2 & 7\\ 9 & 7 & 0\end{bmatrix}$ (ii) $\begin{bmatrix}21 & 3 & 4\\ 3 & 3 & 5\\ 12 & -6 & 13\end{bmatrix}$

(iii) $\begin{bmatrix}4 & 1 & -3\\ -1 & 14 & 9\\ 10 & 9 & 17\end{bmatrix}$

2.2.2 (i) $\begin{bmatrix}6 & 15 & 36\\ -9 & 18 & 57\end{bmatrix}$ (ii) $[28]$

(iii) $\begin{bmatrix}12 & 20 & 24\\ 6 & 10 & 12\\ 3 & 5 & 6\end{bmatrix}$

2.2.3 $-2 \pm j\sqrt{6} \quad (j = \sqrt{-1})$

2.2.4 $a = 2 \quad b = 6 \quad c = 0 \quad d = -3$

2.2.5 $$\{\mathbf{R}(\theta)\}^n = \begin{bmatrix} \cos n\theta & \sin n\theta \\ -\sin n\theta & \cos n\theta \end{bmatrix}$$

2.3.2 $(\mathbf{A} + \mathbf{I})^3 = \mathbf{A}^3 + 3\mathbf{A}^2 + 3\mathbf{A} + \mathbf{I}$

$$(\mathbf{A} + \mathbf{I})^n = \sum_{r=0}^{n} {}^nC_r\mathbf{A}^r \quad \left({}^nC_r = \frac{n!}{(n-r)!r!}\right)$$

Analogue exists only if **A** and **B** commute.

2.5.1 $$\begin{bmatrix} -57 & 57 & 1 & 8 \\ 5 & 39 & 3 & 2 \\ 7 & 9 & 35 & 74 \\ -8 & 6 & 10 & -51 \end{bmatrix}$$

2.6.2 $\mathbf{L}_1\mathbf{L}_2$ is lower triangular

2.6.3 Neither **LU** nor **UL** has any special form.

2.6.4 (i) Not possible

(ii) $$\begin{bmatrix} 1 & 0 \\ 4 & 1 \end{bmatrix} \begin{bmatrix} 1 & 0 \\ 0 & 2 \end{bmatrix} \begin{bmatrix} 1 & 0 \\ 0 & 1 \end{bmatrix}$$

(iii) $$\begin{bmatrix} 1 & 0 & 0 \\ -\frac{1}{2} & 1 & 0 \\ 1 & -\frac{7}{5} & 1 \end{bmatrix} \begin{bmatrix} 2 & 0 & 0 \\ 0 & 5 & 0 \\ 0 & 0 & \frac{19}{10} \end{bmatrix} \begin{bmatrix} 1 & 2 & \frac{1}{2} \\ 0 & 1 & -\frac{3}{10} \\ 0 & 0 & 1 \end{bmatrix}$$

2.7.1 1

2.8.1 $$a^2 - bc \neq 0 \quad \mathbf{A}^{-1} = \frac{1}{a^2 - bc}\begin{bmatrix} a & -b \\ -c & a \end{bmatrix}$$

2.8.2 $$\tfrac{1}{15}\begin{bmatrix} 216 & 228 \\ -54 & -57 \end{bmatrix}$$

2.8.3 $$-\tfrac{1}{3}\begin{bmatrix} 15 & 3 & -3 \\ -16 & -4 & 3 \\ 13 & 4 & -3 \end{bmatrix}$$

2.8.9 $$\begin{bmatrix} 1 & 0 & -17 & 10 \\ 0 & 1 & -12 & 7 \\ 0 & 0 & 3 & -2 \\ 0 & 0 & 2 & -1 \end{bmatrix}$$

2.10.1 (i) $$\begin{bmatrix} 1 & 0 & 0 & 0 \\ 0 & 0 & 0 & 1 \\ 0 & 0 & 1 & 0 \\ 0 & 1 & 0 & 0 \end{bmatrix}$$ (ii) $$\begin{bmatrix} 1 & 0 & 0 & 0 \\ 0 & 1 & 0 & 0 \\ 0 & 0 & 99 & 0 \\ 0 & 0 & 0 & 1 \end{bmatrix}$$

(iii) $$\begin{bmatrix} 1 & 0 & -1 & 0 \\ 0 & 1 & 0 & 0 \\ 0 & 0 & 1 & 0 \\ 0 & 0 & 0 & 1 \end{bmatrix}$$

2.10.2 (i) $\begin{bmatrix} 1 & 0 & 0 \\ 0 & 0 & 1 \\ 0 & 1 & 0 \end{bmatrix}$ (ii) $\begin{bmatrix} 1 & 0 & 0 \\ 0 & -3 & 0 \\ 0 & 0 & 1 \end{bmatrix}$ (iii) $\begin{bmatrix} 1 & 2 & 0 \\ 0 & 1 & 0 \\ 0 & 0 & 1 \end{bmatrix}$

2.11.4 Not positive definite

*Chapter 3*

3.1.1 (i) $r(\mathbf{A}) = 3$ $r(\mathbf{A_b}) = 3$ Unique solution
(ii) $r(\mathbf{A}) = 2$ $r(\mathbf{A_b}) = 2$ Non-unique solution
(iii) $r(\mathbf{A}) = 2$ $r(\mathbf{A_b}) = 3$ No solution
(iv) $r(\mathbf{A}) = 3$ $r(\mathbf{A_b}) = 3$ Unique solution
(v) $r(\mathbf{A}) = 3$ $r(\mathbf{A_b}) = 4$ No solution

3.2.1 $x_1 = -1$ $x_2 = 1$ $x_3 = 3$ $x_4 = 5$

$$\mathbf{A}^{-1} = \frac{1}{8}\begin{bmatrix} -12 & 10 & -13 & 10 \\ 4 & -2 & 5 & -2 \\ 24 & -24 & 32 & -16 \\ 4 & -2 & 1 & -2 \end{bmatrix}$$

$\det(\mathbf{A}) = -8$

3.2.2 $x_1 = 1$ $x_2 = 0$ $x_3 = -1$ $x_4 = 2$
$\det(\mathbf{A}) = -32$

3.2.3

$$\mathbf{A}^{-1} = \begin{bmatrix} \frac{14}{85} & \frac{3}{85} & \frac{2}{85} \\ \frac{1}{34} & -\frac{1}{34} & \frac{5}{34} \\ -\frac{2}{85} & \frac{23}{255} & -\frac{13}{255} \end{bmatrix}$$

3.3.2 $w = 1{\cdot}20$ $x = 1{\cdot}37$ $y = 0{\cdot}0427$ $z = 0{\cdot}128$

*Chapter 4*

4.1.1 $\lambda^3 - \lambda^2 - 2\lambda = 0$

$$\lambda_1 = 0 \quad \mathbf{x}^{(1)} = \frac{1}{\sqrt{26}}\begin{bmatrix} 1 \\ 5 \\ 0 \end{bmatrix}$$

$$\lambda_2 = -1 \quad \mathbf{x}^{(2)} = \frac{1}{\sqrt{5}}\begin{bmatrix} 0 \\ 2 \\ 1 \end{bmatrix}$$

$$\lambda_3 = 2 \quad \mathbf{x}^{(3)} = \frac{1}{\sqrt{5}}\begin{bmatrix} 1 \\ 0 \\ -2 \end{bmatrix}$$

4.1.2 $\mathbf{A} : \lambda = 3, 3, -2$

$$\lambda = 3 \quad \mathbf{x}^{(1)} = \frac{1}{\sqrt{26}}\begin{bmatrix} 1 \\ 0 \\ -5 \end{bmatrix} \quad \mathbf{x}^{(2)} = \frac{1}{\sqrt{17}}\begin{bmatrix} 0 \\ 1 \\ 4 \end{bmatrix}$$

$\mathbf{B} : \lambda = 1, 1, 0$

$$\lambda = 1 \quad \mathbf{x} = \frac{1}{\sqrt{2}}\begin{bmatrix} 1 \\ 1 \\ 0 \end{bmatrix}$$

4.2.3
$$\mathbf{SAS}^{-1} = \begin{bmatrix} 2a & 0 & 0 \\ 0 & 2b & 0 \\ 0 & 0 & 2c \end{bmatrix}$$

4.3.1
$$\mathbf{A}^{-1} = \tfrac{1}{6}\begin{bmatrix} 26 & 44 & 64 \\ -22 & -37 & -56 \\ 8 & 14 & 22 \end{bmatrix}$$

4.3.2
$$\sin(\mathbf{A}) = \mathbf{A} - \frac{1}{3!}\mathbf{A}^3 + \frac{1}{5!}\mathbf{A}^5 - \frac{1}{7!}\mathbf{A}^7 + \ldots$$

$$\cos(\mathbf{A}) = \mathbf{I} - \frac{1}{2!}\mathbf{A}^2 + \frac{1}{4!}\mathbf{A}^4 - \frac{1}{6!}\mathbf{A}^6 + \ldots$$

4.3.3
$$\text{(i)} \begin{bmatrix} 1 & 126 & 210 \\ 0 & 64 & 105 \\ 0 & 0 & 1 \end{bmatrix} \qquad \text{(ii)} \begin{bmatrix} 9 & 6126 & 10210 \\ 0 & 3072 & 5105 \\ 0 & 0 & 9 \end{bmatrix}$$

$$\text{(iii)} \begin{bmatrix} 2{\cdot}718 & 9{\cdot}340 & 16{\cdot}350 \\ 0 & 7{\cdot}388 & 11{\cdot}700 \\ 0 & 0 & -0{\cdot}010 \end{bmatrix}$$

$$\text{(iv)} \begin{bmatrix} 0{\cdot}842 & 0{\cdot}136 & 0{\cdot}787 \\ 0 & 0{\cdot}909 & 2{\cdot}918 \\ 0 & 0 & -0{\cdot}842 \end{bmatrix}$$

4.3.4
$$\text{(i)} \begin{bmatrix} -100 & 101 & -201 \\ -99 & 100 & -199 \\ 1 & -1 & 2 \end{bmatrix}$$

$$\text{(ii)} \begin{bmatrix} 1{\cdot}368 & -1{\cdot}000 & 1{\cdot}368 \\ -0{\cdot}264 & 0{\cdot}632 & 0{\cdot}104 \\ -0{\cdot}632 & 0{\cdot}632 & -0{\cdot}264 \end{bmatrix}$$

4.3.5 $\exp(\mathbf{A})\exp(\mathbf{B}) = \exp(\mathbf{B})\exp(\mathbf{A}) = \exp(\mathbf{A} + \mathbf{B})$

$$= \begin{bmatrix} 74{\cdot}39 & 74{\cdot}02 \\ 74{\cdot}02 & 74{\cdot}39 \end{bmatrix}$$

Equality expected when **A** and **B** commute.

4.3.6
$$\exp(\mathbf{A}) = \begin{bmatrix} 1{\cdot}482 & 2{\cdot}833 \\ 0 & 1{\cdot}649 \end{bmatrix}$$

$$\ln(\mathbf{A}) = \begin{bmatrix} 0{\cdot}406 & 1{\cdot}099 \\ 0 & -0{\cdot}693 \end{bmatrix}$$

$$\sqrt{\mathbf{A}} = \begin{bmatrix} 1{\cdot}225 & 0{\cdot}518 \\ 0 & 0{\cdot}707 \end{bmatrix}$$

4.3.8 $\frac{1}{5}\begin{bmatrix} -3 & 3 \\ 2 & -2 \end{bmatrix}$

4.4.1 $\begin{bmatrix} \frac{1}{\sqrt{3}} & 0 & \frac{2}{\sqrt{6}} \\ -\frac{1}{\sqrt{3}} & \frac{1}{\sqrt{2}} & \frac{1}{\sqrt{6}} \\ \frac{1}{\sqrt{3}} & -\frac{1}{\sqrt{2}} & \frac{1}{\sqrt{6}} \end{bmatrix}$

4.4.2 Ellipsoid

4.4.3 $0{\cdot}5X^2 + 2Y^2$
$X^2 + Y^2 + Z^2$

# Index